Nineteenth-Century Geographies

Nineteenth-Century Geographies

THE TRANSFORMATION OF SPACE FROM THE VICTORIAN AGE TO THE AMERICAN CENTURY

EDITED BY

HELENA MICHIE

RONALD R. THOMAS

Rutgers University Press

New Brunswick, New Jersey, and London

Library of Congress Cataloging-in-Publication Data

Nineteenth-century geographies : the transformation of space from the Victorian Age to the American Century / edited by Helena Michie and Ronald R. Thomas.
   p. cm.
  Includes bibliographical references and index.
  ISBN 0-8135-3143-8 (cloth : alk. paper) — ISBN 0-8135-3144-6 (pbk : alk. paper)
  1. Human geography—History—19th century. 2. Geography—History—19th century. 3. Geographical perception—England—History—19th century. 4. Geographical perception—United States—History—19th century. 5. Geography in literature—History—19th century. I. Michie, Helena. II. Thomas, Ronald, 1940–

GF13.3.G7 N55 2002
304.2'3'09034—dc21

2002017880

British Cataloging-in-Publication information is available from the British Library.

Manufactured in the United States of America

To Theresa Munisteri, with thanks

CONTENTS

Domestic Fronts

Orientations

ACKNOWLEDGMENTS

This book grew out of an interdisciplinary conference, Nineteenth-Century Geographies, held at Rice University in 1998 and cosponsored by the Rice Center for the Study of Cultures and Trinity College of Hartford, Connecticut. Our first debt is to the conference speakers, whose intelligence, energy, and enthusiasm for the topic inspired us to undertake this collection of essays from a variety of disciplines. We would also like to thank the scholars who have for the last ten years been bringing the idea of geography to life in a new way that made our thinking about the conference and the collection possible: Derek Gregory, Anne McClintock, David Harvey, and, especially, Joseph Roach.

Scott Derrick and Mary Thomas patiently endured the literal geographic negotiations of this project, making into holidays our journeys to Santa Cruz, Dijon, New York, Connecticut, and Texas, where the ideas for this project were fleshed out. Ross and (later) Paul Michie-Derrick increased tenfold the holiday spirit of these visits. Thanks will always be due to Paul for holding off his arrival on the scene until the conference was over and the volume just begun. We are also grateful to our contributors for their excellent essays and enduring patience, as well as to Leslie Mitchner for her commitment to the success of this project from the beginning.

Our greatest debt is, without question, to Theresa Munisteri, our wise, tolerant, and skilled manuscript editor, to whom, with warmest appreciation, we dedicate this volume in the hope that she can finally begin the travel forestalled by her work with us.

Nineteenth-Century Geographies

Introduction

HELENA MICHIE AND RONALD R. THOMAS

$\mathcal{T}$he nineteenth century was a time of unprecedented discovery and exploration throughout the globe, a period when the "blank spaces of the earth" were systematically investigated, occupied, and exploited by the major imperial powers of Western Europe and the United States. Expeditions were launched in search of the Northwest Passage, treks to the North and South Poles were dispatched, the Indian subcontinent was surveyed, the sources of the great rivers of Africa were tracked down and mapped, the Far East was opened to Western markets, and Mount Everest was discovered by Western explorers. During this period, England expanded its influence and presence throughout the globe, settled Australia, placed India under crown rule, acquired colonies from Hong Kong to Cape Town, and controlled territory from Singapore to the Suez Canal. In the New World, this age when Britannia is said to have ruled the waves also witnessed the spectacular western expansion and settlement of the American frontier. Once the massive southwest territory was won from Mexico and a divisive civil war was resolved in favor of union, the United States went on to annex real estate from the Caribbean to the Philippine Islands. By century's end American territories and possessions stretched from the Atlantic Ocean to the westernmost regions of the Pacific Ocean and from Alaska in the north to Latin America in the south. This expansion consolidated the young nation's commercial and political might in anticipation of a contest with European powers for global domination. Fittingly, the end of this era so often identified with the triumphs of a British Empire upon which the sun never set saw the dawning of an age that would come to be known as the American Century.[1]

Together with and contributing to these fundamental advances in geographical knowledge and shifts in the geopolitical configuration of the globe, the experience of space was changing in dramatic ways as a result of new developments in technology, communication, and transportation. Inventions like the telegraph and the telephone, the railroad and the steamship, and the phonograph and the cinema would fundamentally alter the way space and time structured human life. In the years following the Napoleonic Wars, for example, some of these developments made possible British middle-class commercialization of pleasure travel at home and abroad, superseding the fashionable Grand Tour that had been reserved for the elite of the eighteenth century. While Italy remained a popular tourist destination, the sublime views of the alpine landscape and the attractions of mountaineering and natural spas established Switzerland as "the playground of Europe" by midcentury. In the latter part of the century, group excursions for middle-class Victorians were instituted on a large scale by Thomas Cook, with the popular Cook's tours running not only to the great sites of the Continent and the Orient but also to domestic seashore resorts, which (along with the Lake Country) took on the status of national treasures and restful retreats from modern life. The more remote parts of the American continent also became favored destinations for British and American travelers alike, in an early expression of modern "ecotourism." Especially sought after by these travelers were such natural sites as Niagara Falls, the Ohio and Mississippi Rivers, and the Grand Canyon or dramatic trips through the rugged terrain traversed by the transcontinental railway all the way to California. Settlement of the West and advances into the great natural resources of the frontier toward the end of the century provoked Congress to pass the Forest Reserve Act of 1891, commencing the modern conservation movement, which was founded upon a conception of natural space as a scarce commodity in need of preservation and protection. This movement led to the creation of the National Park Service to set aside and maintain millions of acres of territory as national parks, and established a new kind of park tourism indulged in by Americans and Europeans alike.

As a result of such developments, the nineteenth century was characterized by an overdetermined interest in place, both local and global. Geography, the science dedicated to studying the features of a place or region, was also undergoing a period of significant redefinition and expansion. Indeed, it may be argued that the modern professional discipline of geography was an invention of the nineteenth century, and so was the figure of the geographer. For centuries, individuals had been doing various kinds of geographical work: navigators, cartographers, explorers, surveyors, and a range of different describers of the earth and its characteristics. However, with the establishment of the first national geographical societies in the capitals of many European countries in the first half of the nineteenth century, and in the United States during the second half, and with the creation of the first academic chairs in the principal universities of those countries toward the end the century, the professional geographer and the modern academic discipline of geography were born.

Our title, *Nineteenth-Century Geographies: The Transformation of Space from the Victorian Age to the American Century,* combines the interrelated categories of time and space to identify the concerns of the essays in this book. From various avenues of approach, the essays explore developments in the complex relationship between these two categories during this particular period. Each section engages current critical debates about the embattled history of the discipline of geography as well as contemporary inquiries into understanding the geography of history. Each grouping of essays is concerned with the way geographical discourse expanded its mission over the course of the nineteenth century from the fundamental "cartographic project" of earth description to the task of incorporating other fields of inquiry into the representation of space—biology, ethnography, anthropology, politics, spatial science, aesthetics, and even literary criticism. We are equally concerned, however, with the ways that the methods and discursive practices of those other disciplines have in turn redrawn the boundaries of the field of geography and its approach to spatial representation and analysis.

"The nineteenth century" is no less problematic a temporal term than geography is a spatial category. The borders between one hundred-year span and another carry no real historical or conceptual explanatory value, and yet we continue to refer to such periods as if they had some coherent character or genuine classificatory status. The current popularity of the designation "the long nineteenth century" (like "the long twentieth") underscores the inadequacy and arbitrariness of the century as a significant temporal marker, expanding the era backwards as it does to 1789 and forward to 1914 to establish a period defined (at least in the West) by the beginning of the epoch of great nationalist revolutions, on the one hand, and the dawn of a new imperial age of worldwide conflict and the balance of powers, on the other. Even our subtitle, *The Transformation of Space from the Victorian Age to the American Century,* refers not to a specific place or even to a set of places but to a set of dynamic relationships between two nations going through conditions of dramatic transformation during the long nineteenth century, nations whose geographies, even in the most narrow sense, were experiencing radical revision over this time. As Britain established itself as the dominant commercial and political power throughout the world in the nineteenth century, America moved from being a collection of fledgling British colonies in a remote wilderness to becoming Britain's principal rival for global domination, expanding its own presence on the North American continent and exerting its influence well beyond that precinct. As the United States separated itself from Britain politically, it went on to emulate the mother country in the development of territorial, commercial, and imperial ambitions throughout the globe at the turn of the century.

With an eye toward understanding the geographical implications of the transition from the putative Victorian Age to the American Century during such far-reaching global transformations, our focus in this volume will be on the British and American roles in these developments, roles that were clearly related and yet took significantly divergent paths. The circumstances surrounding the establishment of the Royal Geographical Society in London in 1830 and the American Geographical and Statistical Society in New York more than two decades later in 1851 suggest a great deal about the changes taking place in the field of geography at the time and about the complex forces at work both inside and outside the field. Traces of these events can be detected in the distinctive images with which the two organizations identified themselves and with which they emblazoned the extensive body of publications—journals, proceedings, and reports—they produced.

The Royal Geographical Society (RGS) of London was founded in 1830—directly on the heels of similar organizations in Paris (1821) and Berlin (1828)—the year before Darwin set off on his five-year, history-making journey on the *Beagle.* The RGS was the direct offspring of the Raleigh Travellers' Club, an es-

FIGURE 1. Seal of the Royal Geographical Society of London, 1861.

tablished and very exclusive dining club made up of forty distinguished travelers, each one of whom would host a fortnightly dinner to celebrate and report upon a recent journey. In May of 1830, the group agreed to form a new society to promote "that most important and entertaining branch of knowledge"—geography.[2] The centrality of this quest for geographical knowledge as the driving force of the society is signified in the medallion-shaped seal that appears on the title page of the organization's proceedings (Figure 1). The oval seal is dominated by a full-length representation of the figure of Athena, the Greek goddess of wisdom. She appears in profile with a warrior's helmet on her head, clothed in flowing robes. Visible behind her and partially obscured by her tunic is a portion of the earth in the form of a globe, inscribed only with the geographer's lines of longitude and latitude. At her feet and just behind her is a sextant, the principal geographical instrument used by astronomers to measure angles and distances between objects and by navigators to establish location and calculate latitude at sea. In her right hand, just above the sextant, Athena holds a laurel wreath, the ancient symbol of victory and merit, which she appears to be bestowing upon the sextant. In her left hand, hanging at her side, she clutches a scrolled map, with the tracings of land masses evident upon the map as it unfurls before the globe just behind it.

The text surrounding these images is equally revealing. Written on the pedestal upon which Athena stands are the words "Royal Geographical Society of London" in small, fine letters. In much larger type, arching over her head is the Latin inscription *Ob Terras Reclusas.* The phrase may be translated to read either "to" or "from" these *terras reclusas,* these "hidden lands," "unknown territories," "secret places." Significantly, however, *reclusum* can also mean the opposite of hidden. Indeed, the older of the two Latin meanings of the word referred to that which has been opened, revealed, uncovered. Only later did that which is uncovered also come to mean that which is hidden, a meaning that suggests for the motto a particularly complex relationship between the familiar and the unknown. The image on the RGS seal seems to show Athena, armed with a map, stepping forward from the earth toward the new territory of geography and the instruments it deploys, emerging from one *terra reclusa* and entering into another. It suggests a time of progress and achievement for geography, a sense of expanding activity and meritorious discovery, crowned with the wreath of victory.

This dynamic representation of Athena not only invokes a classical precedent and imprimatur but also echoes popular contemporary images of the helmeted figure of Britannia, who appeared on coins during the period in much the same posture in which Athena is represented here. The invocation of the wise warrior Athena (or her Roman counterpart Minerva) on the seal and the combination of Latin and English text suggest Britain's rightful inheritance of the legacy of European domination from Greece and Rome. Taken together, the details of the seal implicitly connect the quest for and acquisition of secret knowledge about exotic lands with military conquest and power. In this, the seal reflects the early membership and orientation of the society, which was made up of a large contingent of military men and politicians, combined with a group of amateur explorers, armchair geographers, and a handful of academics. The foregrounding of the tools of geographic knowledge and representation, the sextant and the map, and the apparent honoring of those tools with the wreath of merit offer a powerful image of the importance given to the role of geography in the establishment and maintenance of Britain's still expanding empire.

As Britain's political capital provided the site where the RGS originated, the American Geographical and Statistical Society (AGS) found its home some twenty years later in New York, the capital of commerce and trade for the powerful economic force that the United States was in the process of becoming. Unlike the early membership of the RGS, many of the most influential founders of the AGS were industrialists and businessmen, who were joined by the collection of philanthropists, amateur explorers, military men, and scientists that made up the society.[3] Whereas the RGS had a political orientation in its collection and dissemination of geographical knowledge, the interests of the AGS moved more in the direction of identifying the connections between exploration and commerce.

The seal that appeared on the journal published by the AGS in 1859 is strikingly different from that of the RGS, though it contains some of the same elements

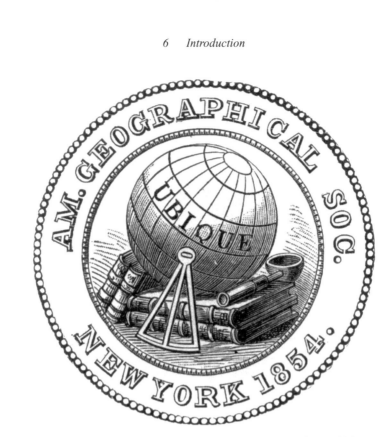

FIGURE 2. Seal of the American Geographical Society, 1898.

(Figure 2). The American version is shaped like a circle, the inside of which is filled almost entirely by a large globe. Like the smaller, partially visible globe on the RGS seal, this one is inscribed only with the lines of longitude and latitude. The seal's inclusion of this complete image of the globe suggests the scope of this society's interests. By omitting any naturalistic representations of the earth's details, like land masses or bodies of water, in favor of the rational grid of geometric lines superimposed upon the globe, the seal emphasizes a confidence in geography's power to order, contain, and even dominate space by imposing a systematic explanatory grid upon it. One hundred years later, significantly, in a version of the seal introduced in the 1940s, the globe would incorporate an outline of the Western Hemisphere on its surface, with only North America in plain view.

There are no ancient gods or human figures on the more direct and practical image of the AGS seal. The globe pictured here rests not behind the tunic of a god but upon two massive bound volumes, with two more books leaning against it on the left, accompanied by giant representations of a sextant, a telescope, and a mortar—a more complete inventory of the geographer's instruments. In another version of the seal (Figure 3), used in the proceedings and in the journal, the book leaning against the globe has "Census" written on its cover, underscoring, perhaps, the emerging importance of the subfield of "human geography" for the discipline and pointing to the expanding scope of the field of geography—from a mission

FIGURE 3. Seal of the American Geographical and Statistical Society. This design was used in the *Proceedings* of the AGS in the 1860s.

of physical description to one of more comprehensive explorations of the social dimensions and practical applications of spatial analysis.

Finally, written boldly and in large type across the globe in all the versions of the AGS seal is the single Latin word *Ubique,* "everywhere." The one-word motto suggests more aggressively, perhaps, than the *terras reclusas* of the RGS the uncircumscribed ambition of the AGS and its unconditional conviction about the primacy of place as the foundation for everything else. The phrase makes no distinction between hidden and familiar lands, between the exotic and the known. It affirms the ubiquity of the power of place. This American confidence in the nation's own manifest destiny as everywhere present is also expressed in the scale of the images that appear on the seal: the books and instruments of geographical knowledge and observation are nearly equal in size to the globe itself, seeming to support and, at the same time, to contain and control it, not so much through acts of physical conquest or reshaping but through the power of discourse itself.

Together, the seals of the two societies manifest a quality of geographic aspiration and purpose on the part of these two nations: one confident in its legacy, appealing to classical precedents and meritorious achievements to conquer brave new worlds; the other confident in its future, taking assurance in the possibilities the tools of geography hold for the eventual pervasiveness of its influence throughout the globe. With the hindsight offered by the long twentieth century, we can see the increasing dominance of the United States as a world power whose geographical and technological prominence was to shape future historical events. In that context, the choice of Washington, D.C., for the International Meridian Conference of 1884 to establish the zero point of the earth suggested a shift not only in geopolitical but also in geographical dominance. Although both prominent American geographical societies (the American Geographical Society and the National Geographic Society) saw themselves, even as late as 1899, as distinctly less important and less venerable than their British counterpart, by the 1890s geography in the United States was taking on a distinct local flavor and increasing significance.

The births of these sibling national geographical organizations were not unattended by conflict, competition, and theoretical difference from both within and without. Indeed, we can locate different definitional and ideological strands within the theory and practice of geography as early as the first half of the nineteenth century. Clearly the dominant motifs in Britain, and later in America, were exploration and expansion and the practices and attitudes—surveying, commerce, military expedition, imperialism—associated with these. If we look at the *Journal of the Royal Geographic Society,* for example, we will see that until the 1890s the vast majority of articles consisted of accounts of exploration, some self-consciously allied to the professional practices of geography, some more like travelogues. Until the late 1860s the medals and other prizes awarded by the society were almost exclusively for exploration or surveying projects or both; in the late 1860s prizes began more routinely to be granted for academic and lay publications. One of the first of these was also the first medal to be awarded by the RGS to a woman: in 1869 the Patron's Victoria Medal went to Mary Somerville in recognition of "her proficiency in those branches of science which form the basis of Physical Geography," as manifested in "her having published a most able work on that science."[4]

Exploration and the ideology of imperialism were of course closely linked, as W. R. Hamilton's 1842 presidential address to the RGS made abundantly clear:

> The study of geography is the most natural as it is the most useful, of all human pursuits. . . . I have observed before that, though geography may be devoid of the charms of other systematic sciences, though it does not lend itself to brilliant theories, though it scarcely admits of the most innocent speculations, though it treats mainly dry matters of fact, yet geography has other, and perhaps superior, claims. . . . It looks alone to truth as its object . . . the absolute and relative position of definite and fixed places. . . . It embraces the whole globe on which we live, and have our being; all the interests, all the occupations of men are, more or less, dependent upon it. It is the mainspring of all the operations of war, and of all negotiations of a state of peace; and in proportion as any one nation is the foremost to extend her acquaintance with the physical conformation of the earth, and the water which surrounds it, will ever be the opportunities she will possess, and the responsibilities she will incur, for extending her commerce, for enlarging her powers of civilizing the yet benighted

portions of the globe, and for bearing her part in forwarding and direct-
ing the destinies of mankind.[5]

Moving as it does from nature to conquest to civilization, this passage rehearses
the familiar epistemological moves of imperialism, in this case of both a national
and a disciplinary kind. The boundlessness of the ambitions of geography's gaze
echoes and is linked to the expansionist ambitions of empire as it colonizes physical
space. In their parallel ambitions to "embrace" "the whole globe," discipline and
nation become entangled with each other, as knowledge (the "acquaintance with
the physical conformation of the earth") becomes power.

The very vastness of the geographical enterprise, as it is outlined in
Hamilton's address, should alert us to a tension between "dry matters of fact" and
a more human, or perhaps in this case, inhumane, component of geographical study.
What Hamilton describes are, in fact, not mere facts, not mere mapping, but rather
the control, through knowledge, of entire populations. This control is the more sinister side of "human geography," which, according to most narratives of the history of the discipline, emerged as a subfield in the first half of the twentieth century.
Other Victorian geographers offer a more benign sense of the need to include the
study of specifically human relations to the environment in the discipline: a year
earlier, G. B. Greenough, Hamilton's predecessor as RGS president, made a distinction between "simple" and "compound" geography, a distinction in which
simple geography was understood as a branch of physical science, dependent on
classification and nomenclature, whereas compound geography could justifiably
"claim to investigate every subject which is connected directly or indirectly with
the earth," including the "study of man" in a range of contexts including "political, civil, statistical, ethnographical, philosophical, classical, and scripture geography."[6] Indeed, even Victorian definitions of geography often take the form of
lists like the one enumerated by Greenough. Geography becomes in this idiom the
basis for other disciplines, a point of departure for projects both vast and various,
just as the description of physical space becomes merely a starting point for the
endeavors of geography as a discipline.

These lists of the provenances and tools of geography, like many Victorian
lists, bear an uneasy and contradictory relationship to the supposed nineteenth-
century confidence in empiricism. One can read them simply for their capacious-
ness and take their ambition as evidence of empiricist hubris; one can also read
them—as, for example, others have read the details of realist fiction—as attempts
at command that reveal by their hyperbolic reliance on detail the impossibility of
any sense of mastery. A few geographers, such as Colonel Julian Jackson, were
explicit about the limits of even the most descriptive forms of geography: anticipating the postmodern deconstruction of the map, he noted in 1835 that maps themselves were not transcriptions of landscape but highly conventional ways of
representing it, and that human perception of physical reality was inevitably partial,
relational, and subjective.[7]

If doubts about the exploratory project were already part of the discourse of
the RGS at midcentury, by the end of the century they had taken a distinct shape.

There seemed to be a pervasive sense throughout the society's publications that in some ways this aspect of geography had been too successful, that the world as a place to explore had been used up, just as maps had become filled up with detail. This late crisis of Victorian geography led to a rearticulation, perhaps even more insistent and more nervous, of the ambitions of geography beyond exploration once the great roll call of discoveries had come to an end.

Geography in Britain and America might have taken a turn that made it less dependent on exploration by integrating it more fully into higher education, as it traditionally had been in Germany. In Britain, however, attempts to establish chairs in geography at Oxford and Cambridge were unsuccessful until the 1880s. Prior to that time, the primary educational accomplishments of the RGS had been the workshops it ran for explorers and travelers; by 1879 (just prior to the establishment of the first university chairs in geography), the RGS had even cut back severely on its program of public lectures in favor of giving "a more scientific direction to geography."[8] In America, the priority was given to the fostering of primary and secondary geographical education, with an emphasis on the physical geography of the United States as a form of civic education and as preparation for an informed (and patriotic) national citizenry.[9]

For the American geographers, inside and outside of the American Geographical Society and, later, the National Geographic Society, the geography of the home country was of central importance. Partly because the United States came relatively late to the discipline of geography, and partly because of the nation's own ever expanding borders and the sheer size of its territories, geography as it was undertaken and taught during the last decades of the century was first and foremost "home geography." The United States also had specific geographical resources with which to work: for example, census surveys from the 1860s and 1870s as well as information networks dating from Civil War campaigns. The United States, with its vast land mass and different climates, was also well suited to become a leader in weather mapping and in the formation of more-general meteorological enterprises, all of which came to rely on technologies mainly developed within the United States. Although the United States became increasingly interested in its international stature in the fields encompassed by and relating to geography, in the waning years of exploration the nation became more and more inwardly focused in terms of its resources.[10]

Through much of the eighteenth century, geography was historically understood to be a discipline of earth description, as its name implies. The task of geography was not primarily to explain or analyze or theorize over the physical features of the earth but merely to describe and locate them. Indeed, for geography, to describe the earth's land masses and rivers and their topography and location was to provide all the explanation necessary. In the nineteenth century, however, such a narrow understanding of the mission of geography was called into question. Anne-Marie Godlewska has described the era as "a period of uncertainty and extended dormancy" for geography, a time characterized by a "loss of direction . . . manifested in a loss of purpose, uncertainty about what to teach . . . discomfort with the role and nature of theory, loss of status, a jettisoning of both the most technical and most theoretical aspects of the field, a proliferation of exclusivist definitions of the discursive formation, and general uncertainty about how to assess research quality."[11] Godlewska attributes this turn of events to the rise in status of biology and geology over geography in the late eighteenth and early nineteenth centuries, a development that privileged analytical disciplines over descriptive ones, "a shift in focus from the mostly static depiction of form and location to an increasing appreciation of the mostly dynamic behaviors, interactions, and movements of living and nonliving phenomena" (5).

During the nineteenth century, the field of geography was clearly going through fundamental rethinking and revision, and at the same time it was also experiencing extraordinary activity and expansion, a fact that challenges Godlewska's characterization of geography's "dormancy" during this period. Shifts in disciplinary terrain and methodology presented challenges but did not render geography static or powerless. On the contrary, the nineteenth century was an age of tremendous renewal and activity in the field. In the plethora of geographic expeditions and explorations launched during the nineteenth century for a wide variety of scientific, political, commercial, and military purposes, the tools and discourse of geography were being appropriated by a host of other disciplines—for example, biology, anthropology, ethnology, physics, and literary and travel writing—and the influence of geography expanded well beyond the confines of the profession of geography itself.[12]

Tensions within the evolving discipline of geography are only one way of exploring profound changes, over the span of the nineteenth century, in more general attitudes toward space. While some of the essays in this collection take up the cartographic projects of the nineteenth century, they do so in relation to other cultural endeavors at describing, circumscribing, and living with and within physical place. Other essays focus on more private and indeed more intimate relations to place and placedness: from quotidian negotiations of urban and suburban landscapes on foot, by train, and by subway to what we might think of as the consumption of place in the form of tourism, spectacle, and exotic food. The essays remind us of specific changes in the landscape and in relations to it during the long nineteenth century, changes brought about by and reflected in an almost endless series of inventions for traveling through and chronicling space. The essays also deal more broadly with ideological shifts in ideas of place: the privatization of domestic space, the gendering and regendering of rooms, buildings, and the activities imagined to take place inside them. While attentive to these more local concerns of everyday life, this collection attempts its own placement of them within a culture of official maps and conquests and within a discourse that is in the broadest sense geographical.

In the course of looking back upon and interpreting a particular period of time from diverse disciplinary perspectives, these essays reflect and draw upon the new "place-centeredness" of other fields in the current moment, taking advantage of the increased emphasis so many of them have begun to put upon the significance of location for critical analysis. Social scientists and anthropologists like Erving Goffman, Anthony Giddens, and Clifford Geertz, for example, have emphasized for their disciplines the crucial significance of the relation between ritualized spaces and human bodies in the creation of social systems and social integration. Henri Lefebvre and Edward Soja demonstrate how problematic a concept physical space is for these and all disciplines that would critically examine the significance of place, since physical space presents itself to us as a material and natural constant but is in fact deeply implicated in human constructions and practices.[13] Like David Harvey and others, they argue that space is fundamentally a human construction, both constitutive and reflective of the complex relations of power in society, a claim that Michel Foucault has advanced repeatedly in his comprehensive "archaeology" of institutional spaces like the hospital, the school, and the prison. "The production of space (and the making of history) can thus be described as both the *medium* and the *outcome* of social action and relationship," Soja says in summarizing Lefebvre's *Production of Space.* "Spatiality and temporality, human geography and human history, intersect in a complex social process which creates a constantly evolving historical sequence of spatialities."[14]

The essays in this volume explore the discursive strategies in this sequence of spatialities, and in doing so they demonstrate how the Victorians' emerging new attitudes toward place anticipated and helped to shape the vocabulary of modern geography. Hints of these connections can be seen in cultural criticism like Richard Sennett's *The Fall of Public Man,* which explores how the "dead public spaces" of modernity contributed to the rise of the culture of intimacy.[15] Edward Said's analysis of Orientalism and the work of scholars in the fields of colonial and postcolonial studies have focused upon the ways the imagination of "exotic" spaces functions for the observer-occupier as well as for the gazed-upon Other in the culture of imperialism.[16] Historians of nationalism and the modern state draw our attention to the ways the map, the museum, and the census function as "institutions of power" in the grammar of colonialism and modern nation-building.[17] Franco Moretti's *Atlas of the European Novel* (1998) exemplifies one version of literary criticism's fascination with the significance of space, literally "mapping" the settings and distributions of novels over national boundaries to show how the novel functioned in the formation of modern nation-states by creating a dynamic story about the tension between local loyalties and transcending national interests.[18] Feminists like Susan Morgan, Doreen Massey, and Donna Haraway have critiqued the "positioned rationality" of dominant modes of thought to demonstrate how objectivity is always a view from somewhere, rooted in some position in social space.[19] A broadly influential field of interdisciplinary studies called "ecocriticism" has developed that combines scholarship from environmental science, the philosophy of nature, urban studies, landscape geography, and literary criticism.[20] A host of subfields of geography itself has arisen as well, including feminist geography, radical-Marxist geography, cultural geography, regional science, quantitative-spatial geography, humanistic geography, postmodern geography, and so on.

Although the essays in this volume make use of developments in nineteenth-century geographical thought, they are not themselves primarily about nineteenth-century geography. Rather, the essays are concerned both with the institutionalization of space and with everyday experiences of it. In this respect, the essays trace out not a history of geography but a genealogy of the sense of place in the nineteenth century and the resultant discourses that came to make up the discipline of modern geography. The "situated knowledges" these fields generate are understood to be dynamically contingent and multivalent, spatially as well as historically. They are functions of the places and institutional spaces the disciplines investigate as well as the particular sites from which those places and spaces are investigated.

One way of understanding the two kinds of analyses that together structure the argument of this collection of essays is embodied in the distinction Michel de Certeau draws between the terms "place" (*lieu*) and "space" (*espace*) as categories for geographical understanding. According to Certeau, "place" is governed by the law of the "proper," where things have their correct and distinct location, their appropriate adjacencies, their stable positions. "Space" is a more mobile and active concept, composed of forces and movements within a place, an effect produced by the operations that orient and make it function as a polyvalent unity of conflictual programs and proximities. "In short, space is a practiced place," he says, in the way that the place of the street geometrically defined by urban planners is transformed into a space by walkers.[21] Space, in this context, expresses the experiences and practices of individuals or groups in a particular place that may conform to or resist or redefine the officially sanctioned and structured purpose of that place.

Some of the essays here take as their point of departure the imposition of a state-sponsored geometry or governance of territory, the mapping of places to advance national power or economic control, and the implications for those who inhabit or will come to occupy those places. Others begin by looking at more-personal spatial experiences and practices within certain places and the ways they are structured by or secretly subvert the official intentions and ideologies of those places. One essay recounts the way the official place for mapping the globe in Britain is transformed into an urban development site by the staging of a world's fair at the location. Another shows how a world's fair in an American city is deployed by ingenious entrepreneurs as the rationale for the creation of the American suburb. As Certeau argues, the three or four centuries that preceded the nineteenth were characterized by the growing hegemony of the map, by the dominance of the principle of "place." As the essays in this volume collectively show, however, the nineteenth century offered a particularly potent array of opportunities for inventive spatial operations by individuals and groups to pose microgeographical phenomenologies of space against the macrogeographical strategies of nations and institutions.

In moving toward each other, geography and the humanities have begun the work called for by many, including Michel Foucault in *Power/Knowledge*. There Foucault offers a challenge that is taken up by the essays in this book: "A whole history remains to be written of spaces—which would at the same time be a history of powers . . . of the habitat."[22] Geographers like Derek Gregory and Trevor Barnes have taken up the project of that history of space and power not only by investigating the way spaces have been mapped, fought over, conquered, and imagined but also by examining the power of the spatial idiom itself as it is expressed and maintained by the more traditional objects of geographical inquiry, including, of course, the map and the survey.[23] Despite these trends toward the historical and the material in humanistic studies, however, histories of space and spaces remain to be written.

This is the critical space *Nineteenth-Century Geographies* attempts to occupy through a set of four groupings of essays, written by scholars from Britain, the United States, and Canada who represent the fields of geography, history, art history, anthropology, theater and performance, history of science, and literary studies. Of course, no one could hope to write a definitive history of place, but nineteenth-century Britain and the United States might be a good place[24] to begin such a project. As we have suggested here, nineteenth-century Anglo-American culture is situated at a turning point, a crossroads, if you like, in the imagination and deployment of space. *Nineteenth-Century Geographies* proposes that the transformations of nineteenth-century culture necessitated a radical reimagining of space and of human relations to it, a reimagining signaled in the shift over the course of the century from the dominance of European imperialism to the emergence of global economic domination by the United States.

Space was also reapportioned along what this book thinks of as an ever shifting "domestic front." In Britain, the doctrine of separate gendered spheres, however imperfect and uneven its application, created a spatial grid for the understanding of gender, for the deployment of time, and for the organization of towns, cities, and frontiers. In the United States, the relation between frontier and civilization or settlement, differently but spectacularly gendered, also provided a spatial idiom for differences between men and women. Although much has been written about the history of nineteenth-century cities—particularly London and New York—very little attention has been paid to the discourses made possible by the invocation of the urban and by a cultural mapping that made the city so crucial a term of contrast with the rural, with the natural, and, toward the end of the century, in a further refinement of the idiom, with the suburban.

These are among the changing terrains that *Nineteenth-Century Geographies* explores. Individually and collectively, the essays move the discussion of "geography" beyond the official and national spaces created by maps and natural boundaries to explore other cultural uses and metaphors of space—from domestic architecture to the regulation of city streets, to the association of certain spaces or regions with different epochs, to attitudes about migration and nativism, and to the mapping of the notion of space upon the human body.[25] The essays in the volume are grouped into sections in accordance with these concerns: Time Zones, Commodities and Exchanges, Domestic Fronts, and Orientations. Each section engages a critical issue about nineteenth-century space from several different perspectives.

The opening section of the collection, titled Time Zones, traces the ways place gets expressed as time, how geography becomes history. It opens with an essay by Ronald Thomas that explores the circumstances surrounding the official establishment of time zones as we now know them and the recognition of Greenwich as the zero point of longitude at the 1884 international conference on the prime meridian in Washington, D.C. Thomas interprets those events as a worldwide acknowledgment of the inherent politics (and economics) of the map and the dawn of a postnational (and postcolonial) organization of space. He reads the standardization of time zones in light of the infamous anarchist bombing of the Royal Greenwich Observatory, Conrad's fictional account of that event in his novel *The Secret Agent,* and Tony Blair's recent designation of Greenwich and the Millennium Dome Project sited there as "the home of time." Historian Ussama Makdisi's essay then examines how a particular geographical location—the Ottoman Orient—was defined during the nineteenth century by the competing historiographies of timelessness (the eternal Orient) and earliness (the premodern and developing Orient) through relentless Western encroachment on the Ottoman Empire and the Ottoman attempts to resist that encroachment. Makdisi shows how Ottoman officials made the case for independent reformation of the Ottoman Empire in reaction to nineteenth-century European travelers, politicians, and missionaries who configured the Orient as premodern, only to elicit in return works like Richard Burton's *Personal Narrative of a Pilgrimage to Al-Madinah and Meccah,* which responded to the advent of non-Western modernization with a romantic quest for and rediscovery of an uncorrupted and timeless Orient. Literary critic David Lipscomb follows with an analysis of the early-nineteenth-century historical novels of James Fenimore Cooper to show how time can be transformed into particular places rather than how places are made into expressions of time. He reads examples of this new nineteenth-century narrative form together with contemporary geopolitical maps of the United States to show how they translated fundamental historical transformations into the visual idiom of landscape and how Cooper did so by departing in fundamental ways from his British precursor, Walter Scott.

In establishing links between geography and temporality, these essays identify three national or regional traditions, each with a specific relation to geopolitical dominance: the "Oriental," the British, and the American. In establishing the Orient as both timeless and premodern, British explorers and travel writers wrote themselves into the center of history, a national position later given a sense of place in the geographical centrality of the prime meridian. Lipscomb's essay links British narratives of center and periphery to what he argues is a nascent American form of expressing dominance through time and space: the layering of present over past as a form of historical and geographic erasure.

A second grouping of essays, Commodities and Exchanges, explores the role of geographic origin as it was embodied in particular objects. The section includes an analysis by art historian Diane Dillon of the phenomenon of the fair map as souvenir at the 1893 World's Columbian Exposition in Chicago and the transformation of geography into spectacle; an investigation by critic Julie Fromer of the persistent effacement in literature and advertising of the "impure" Chinese origins of tea that would lead to the beverage's general designation (and acceptance) as "a typically English brew"; and a comparison by Jules Law of competing strategies of "mapping" the dynamics of African trade as they appear in Victorian travel writing and ethnography. Joseph Litvak's essay on ethnicity, assimilation, and the street in Victorian London concludes this section. Litvak begins with a discursive paradox, defined in this case by two nineteenth-century anti-Semitic fears: that the Jew would inevitable invade gentile space by disguising his or her Jewishness, and that the Jew, once assimilated, would become an object of spatial desire. Litvak's close reading of Anthony Trollope's *The Prime Minister* reveals a profound (and to Litvak's mind) representative anxiety that the Jew *as object* could be incorporated or "assimilated" by being turned into a "consumable" that threatened to penetrate and inhabit the gentile Victorian body politic, creating a "blockage" reminiscent of the Jewish street.

The set of essays titled Domestic Fronts moves the locus of the discussion from the public to the always porous but culturally resonant "private" sphere, identifying a series of encounters in which the domestic comes up against, absorbs, and is modified by a force troped as foreign. Helena Michie's article on Victorian honeymoons examines how the psychic, bodily, and legal displacements that accompanied marriage in the nineteenth century were negotiated through foreign travel and, in particular, through the incorporation of a highly canonical and culturally readable landscape comprising specific "sights" of European capitals. Historian Philippa Levine's "Erotic Geographies" extends the range of inquiry to colonial settings and looks specifically at how terms from the lexicon of the domestic—particularly house and home—function in British discussions of brothels in India and Egypt. Feminist critic Betty Joseph takes the discussion to the colonial interior by focusing on the narratives of development that became coded (at once) to gender and to nationality in the British occupation of the subcontinent during the nineteenth century. Martin Brückner's essay looks at how the discipline of geography itself was absorbed into domestic life. Through readings of specific nineteenth-century geography textbooks and maps, Brückner's article looks at how, in the antebellum United States, literacy began to be defined in terms of the map as well as of the alphabet. Robert Patten then offers an overarching literary and historical narrative of how the domestic is imagined and experienced from the beginning to the end of the nineteenth century. In his account, the landmarks or coordinates of identification shift from the house to the square to the street with the urbanization of the British landscape: although at the beginning of the century people thought of themselves as attached to and identifiable by a particular house, by the century's end the primary topos of location was the street address. Together, all five essays in this section identify critical moments of personal and national self-fashioning or refashioning as the foreign is imported home in the form of stories, souvenirs, textbooks, and legal discourse.

The final section, Orientations, takes as its explicit topic the relations between body, identity, and space. Geographer Mona Domosh begins by examining a set of very public spaces on the avenues of New York City to complicate their identification in much recent work on the city as sites of democracy and pleasure. Domosh argues that these public spaces were often controlled and maintained by private interests and that behavior on the street—particularly for women and racial minorities—was closely monitored. Domosh reads a variety of popular images of street life as well as Edith Wharton's *The Age of Innocence* in an attempt to map a gendered and racially inflected notion of politeness onto "public" urban space. Focusing upon accounts of famous nineteenth-century railroad accidents and emerging theories of psychic trauma in the period, Jill Matus goes on to show how the railroad and the tracks upon which it ran became the dominant structuring metaphor for explaining the relationship between the conscious and the unconscious, between psychic and somatic symptoms. This internalizing of the changing geographic conditions of the modern world and their meaning for personal identification is also the topic of Ana Vadillo's essay, which examines a group of women poets connected physically and in other ways by the expanding London Underground railway system. Vadillo looks at the role of the underground in maintaining poetic and personal relationships and at the means by which this feature of late-nineteenth-century London made its way into poetry as a trope for sexual independence and communication. Mary Pat Brady moves the focus from London urban life back to the American West at midcentury to explore early efforts to represent the southwestern United States in the work of John Bartlett and William Emory, members of the United States and Mexican Boundary Commission. Brady reveals how the tools of the imperialist geographic repertoire—mapmaking, ethnography, surveys—were deployed to erase the traces of three hundred years of Spanish and Mexican colonialism of native peoples, only to build a new colonial structure on the buried rubble of previously colonized bodies and discourses, as if upon virgin soil.

This final section concludes with a retrospection and a comparison, as Jon Hegglund's essay evaluates the British and American colonial projects by analyzing the spectacle of 1930s Hollywood films that look back at the British Empire. He argues that the visual idiom of the map is subsumed by the mise-en-scène of cinema in these films, establishing the immediacy of the cinematic image (as the medium of United States imperialism) at the expense of the specialized geographical and historical knowledge that characterized the British imperial cartographic project in the nineteenth century. Especially in the context of the essays that precede it, this essay encapsulates our interest in the transformation of the Royal Geographical Society's *terras reclusas* to the American Geographical Society's *ubique,* the move from a nineteenth-century imperial tactic of spatial discovery by occupation to one of territorial ubiquity through technology and representation.

It would be easy to see the nineteenth century in England and America as a period of rigid geographic demarcations, where home became divided spatially from work, the public from the private, the foreign from the domestic, the slum from the middle-class dwelling. *Nineteenth-Century Geographies* argues, however, for a Victorian crisis of spatial division, produced in part—perhaps ironically—by the appearance of new spaces and new ways to negotiate and categorize them. Thus, we see the Victorians—British and American—not only as dividing and categorizing space, as agents acting upon it, but also as human subjects changed in the very act of negotiating their relations to place. Many of the essays in this collection, for example, focus on the porousness of the human subject as spatial technologies (like the railway in Matus's essay) transform not only ways of thinking about the exterior landscape but ways of experiencing and imagining the process of consciousness itself. The collection is full of examples of the literal incorporation of space and the spatial: Julie Fromer's essay on the consumption of tea with its politically charged national history, Diane Dillon's account of how souvenirs transform past and future selves under the sign of memory, Helena Michie's essay on the transformation of the sights of Europe into new sexual identities. To be sure, the essays also focus on the refusal of incorporation: in Joseph Litvak's essay, the resistance to Jewish assimilation rehearses the defensive reiteration of bodily boundaries when Trollope characters, for example, are faced with the threat of penetration from outside, while in Mona Domosh's essay the threat of the breakdown between public and private produces the social policing of New York streets.

Together the essays in this volume argue that relations to place in the nineteenth century involved repeated renegotiations of the very terms with which Victorians seemed to keep the physical world epistemologically coherent. The abiding facts of Victorian imperialism, anti-Semitism, and misogyny, each of which produced forms of spatial division or conquest, existed, as it were, side by side with what we can only think of as a cultural disorientation. Robert Patten's essay can in this respect be read as a metonymy for the volume as a whole: we think of nineteenth-century culture as asking for directions in a world where landmarks and their meaning are in constant flux. It is, we think, instructive, that perhaps the most culturally dominant narrative of exploration and the topic of John Hegglund's essay—the journey of Henry Stanley to find David Livingstone—is, on one level, a story about getting lost or, more accurately, a story whose central mystery is whether the protagonist was lost or not. Philippa Levine's essay reminds us of the permeable, ephemeral boundaries between brothel and middle-class home, suggesting that official response to the terror of sexual misrecognition and disorientation took the form of an insistence on boundaries subject to different linguistic rules for different needs and occasions.

Victorian geography and the various spatial practices we might include under that term might well be read as a defense against getting lost, a "reorientation," as Michie puts it. One thinks of the almost hysterical production of maps Diane Dillon chronicles in her article on the World's Columbian Exposition: as plans for the future, guides to the bewildered (conveniently pocket-sized), and advertisements for real estate, these maps held out the ultimately illusory promise that the journey to and through the world's fair would be a sort of coming home. The attempt, chronicled by Ronald Thomas, to find a single, unitary "home of time" by establishing the prime meridian suggests an anxiety about being lost *in* time, an anxiety defended against, as Ussama Makdisi discusses, by the imperial project of identifying the "modern" West against "timeless" cultures in the Orient. The process of getting lost destabilizes the careful and culturally powerful division between public and private: as Betty Joseph suggests, the mapping of women onto the private realm can allow them to get lost, and this in turn can simultaneously bolster and undermine the visual economy of patriarchy. Together, the essays present the discourses of space operating in the nineteenth century as rather richly complicating the traditional notion of "separate spheres" often associated with the period. Rather than identifying clearly demarcated territories, the essays attempt to illuminate a host of uncanny habitats and productive border wars that contested and reconfigured the settled boundaries between inside and outside, the physical and the psychological, the familiar and the foreign.

In our initial account of the cross-pollination between geography and the humanities, we spoke as if the humanities themselves represented an agreed-upon set of values and styles. "Geography," in this narrative, enters to disrupt humanistic complacency by turning the gaze of humanists to physical space. The essays in this collection, however, do not represent a united front. Like all interdisciplinary efforts—even within a single academic category like the "humanities" or the "social sciences"—these essays reaffirm disciplinary boundaries at the moment of their crossing. The essays vary in terms of their relation to empiricism, to epistemology, and even to style. Although all the essays follow the contours of a single, if multistranded, argument, the experience of reading, say, Joseph Litvak's speculative argumentation might be very different from the more accumulative and incremental experience of moving from object to object in Diane Dillon's piece on the world's fair. Both argumentative structures, both dominant idioms of speculation and accumulation, lead appropriately enough to a similar place, as both articles end with and in real estate. With the experience of readership firmly in mind, we have also chosen to present essays that are slightly shorter than the academic norm and to include more of them: these linked choices reflect a desire to accommodate the reading of several essays in one sitting and to maximize the sense of conversation among them.

Notes

1. We have deliberately chosen, when discussing certain nineteenth-century texts and ideas, to use the controversial and multiply inflected term "America" rather than the more accurate "United States." As several of these essays will demonstrate, the insistence on the Americanness of the United States and on the metonymy that allowed the United States to embody and to stand (in) for "America" is an important feature of nineteenth-century geography and the many geographically inflected discourses we explore in this volume.

2. Ian Cameron, *To the Farthest Ends of the Earth: 150 Years of World Exploration by the Royal Geographical Society* (New York: E. P. Dutton, 1980), 16.

3. A significant number of the founders and early leaders of the organization were shipping magnates, merchants, and prominent figures in American publishing. See John Kirtland Wright, *Geography in the Making: The American Geographical Society, 1851–1951* (New York: American Geographical Society, 1952), 14–34.

4. *Journal of the Royal Geographic Society of London* 39 (1869): cxxxiii–cxxxiv.

5. W. R. Hamilton, "Address at the Anniversary Meeting, 23 May 1842," *Journal of the Royal Geographic Society of London* 12 (1842): lxxxviii–lxxxix.

6. Greenough's remarks appear in the *Journal of the Royal Geographic Society* 10 (1840): lxix–lxxi. See T. W. Freeman, "The Royal Geographical Society and the Development of Geography," in *Geography Yesterday and Tomorrow,* ed. E. H. Brown (Oxford: Oxford University Press, 1980).

7. Freeman, "Royal Geographical Society," 4.

8. *Proceedings of the Royal Geographical Society* 1 (1879): 453.

9. Tim Unwin, *The Place of Geography* (New York: Longman Scientific and Technical, 1992), 5–11.

10. See Wright, *Geography in the Making,* 112–133.

11. Anne-Marie Claire Godlewska, *Geography Unbound: French Geographic Science from Cassini to Humboldt* (Chicago: University of Chicago Press, 1999), 4.

12. See David N. Livingstone and Charles W. J. Withers, introduction to *Geography and Enlightenment,* ed. David N. Livingstone and Charles W. J. Withers (Chicago: University of Chicago Press, 1999), 1–32.

13. See, for example, Henri Lefebvre, *The Production of Space,* trans. Donald Nicholson-Smith (Oxford: Blackwell, 1991); Lefebvre, *Writings on Cities,* trans. Eleonore Kofman and Elizabeth Lebas (Cambridge, Mass.: Blackwell, 1996); Edward Soja, *Postmodern Geographies: The Reassertion of Space in Critical Social Theory* (New York: Verso, 1989); and Soja, *Postmetropolis: Critical Studies of Cities and Regions* (Oxford: Blackwell, 2000).

14. E. W. Soja, "The Spatiality of Social Life: Towards a Transformative Retheorization," in *Social Relations and Spatial Structures,* ed. Derek Gregory and John Urry (London: Macmillan, 1985), 90–127.

15. Richard Sennett, *The Fall of Public Man* (New York: Knopf, 1977).

16. Edward Said's work on the cultural significance of place has been considerable since the publication of his landmark *Orientalism* (New York: Pantheon, 1978).

17. Benedict Anderson, *Imagined Communities: Reflections on the Origin and Spread of Nationalism* (New York: Verso, 1991), chap. 10.

18. Franco Moretti, *Atlas of the European Novel, 1800–1900* (New York: Verso, 1998).

19. For the most comprehensive treatments of this perspective, see Doreen Massey, *Space, Place, and Gender* (Minneapolis: University of Minnesota Press, 1994); and Susan Morgan, *Place Matters: Gendered Geography in Victorian Women's Travel Books about Southeast Asia* (New Brunswick, N.J.: Rutgers University Press, 1996).

20. See, for example, Lawrence Buell, *The Environmental Imagination: Thoreau, Nature Writing, and the Formation of American Culture* (Cambridge, Mass.: Belknap Press of Harvard University Press, 1995); John Brinckerhoff Jackson, *Landscape in Sight: Looking at America,* ed. Helen Lefkowitz Horowitz (New Haven, Conn.: Yale University Press, 1997); Robert David Sack, *Homo Geographicus: A Framework for Action, Awareness, and Moral Concern* (Baltimore: Johns Hopkins University Press, 1997); and Yi-fu Tuan, *Space and Place: The Perspective of Experience* (Minneapolis: University of Minnesota Press, 1977).

21. Michel de Certeau, *The Practice of Everyday Life,* trans. Steven F. Rendall (Berkeley: University of California Press, 1984), 117.

22. Michel Foucault, *Power/Knowledge: Selected Interviews and Other Writings, 1972–1977,* ed. and trans. Colin Gordon (Brighton, Sussex: Harvester Press, 1980), 64.

23. See Derek Gregory, *Geographical Imaginations* (Oxford: Blackwell, 1994); and Trevor Barnes and Derek Gregory, eds., *Reading Human Geography: The Poetics and Politics of Inquiry* (London: Arnold, 1997).

24. We have tried throughout this introduction to avoid obvious geographical punning. The difficulty of the task and the fact that it has broken down here are instructive about the ubiquity of spatial discourses.

25. Again our emphasis is simultaneously on lived experience of space and on spatialized discourses, with the understanding that in daily life people draw from a lexicon of imaginatively and culturally available terms.

✳

Time Zones

The Home of Time

THE PRIME MERIDIAN, THE DOME OF THE MILLENNIUM, AND POSTNATIONAL SPACE

RONALD R. THOMAS

Telling Time at the End of the Map

On 19 June 1997, Prime Minister Tony Blair, wearing a hard hat and knee-high rubber boots, delivered an address at the polluted industrial site and deserted gasworks occupying the narrow peninsula between the northern and southern banks of the Thames River. In that address, he identified this derelict parcel of British soil as "the home of time."[1] Signs posted around the area reading "This Is Where Time Began" echoed the proclamation. Like Blair's speech, those signs referred to the immense construction zone set aside for the Millennium Dome, the ambitious building project that would produce the largest enclosed structure in the world (Figure 1). Twice the size of its closest competitor, more than a mile in circumference, and covering more than twenty acres of land, the Teflon-covered structure boasts an exhibition space large enough (according to the promotional materials describing it) to contain thirteen Royal Albert Halls and eighteen thousand double-decker London buses. As Blair's remarks make clear, however, the significance of the structure lies not in its command over space but in its claim upon time. The Dome was designed to mark Britain's spectacular celebration of the new millennium, which epoch, we were reminded in those same marketing materials, would begin in this place. "We may no longer have an empire," a *New York Times* article quoted one Londoner as saying with reference to the Dome, "but we still own time."[2]

Such assertions find their justification in the fact that the Millennium Dome was situated just down river from the Royal Observatory in Greenwich, the site of the prime meridian of longitude and the place where Greenwich Mean Time was established as an international standard in 1884. From that point onward, this spot would mark the official ground zero from which the earth would be mapped into longitudinal coordinates and divided into time zones. The Millennium Dome, proclaimed as a harbinger of the future and of the "New Britain," also looked

FIGURE 1. The Dome of the Millennium under construction, spring 1998. This photograph shows the polluted industrial site and the deserted gasworks occupying the site between the northern and southern banks of the Thames River. *(Photo: Brian Harris. © The Independent Picture Syndication. Used with permission.)*

backwards to that moment in British imperial history when London was officially acknowledged as the center of the civilized world. As such, the structure loomed as a collective memory of other great monuments of British global domination in the nineteenth century—not just the Royal Observatory at Greenwich, but the midcentury Great Exhibition at the Crystal Palace and the celebrated clock tower of Big Ben, completed with such fanfare in 1859. Part urban renewal project, part commercial venture, part patriotic shrine, part Anglo-Disney, the Dome presented itself as a monument to the national memory—to the time when Britain owned an expansive empire of territory and trade routes upon which the sun never set and, therefore, owned time as well.

The ambitious claims spoken in anticipation of the Dome's completion rang rather hollow after its notoriously disappointing debut on New Year's Eve of 2000 and in the context of Tony Blair's farewell visit to the site at the end of the year. In that final visit, Blair admitted that the original visions for the Millennium Dome as conceived in 1997 were, in fact, "too ambitious." He proceeded to thank the staff for its extraordinary achievements and dedication to the project amid all the controversy it generated, and he announced plans to close the attraction in light of an international high-tech consortium's plans to purchase it.[3] Those plans soon fell through, as did several other deals involving buyers ranging from the Tussaud's Group to the BBC to a Japanese bank to an international leisure-property development firm to a sports-and-entertainment management company. At the time of this writing, the fate of the now empty Dome was still undecided. Variously denounced in the press as a "great white elephant," "the butt of all jokes," "the amazing Blunderdome," and (by Prince Charles himself) as "a monstrous *blancmange*," the Dome of the Millennium has run out the clock of its short life under a cloud on the prime meridian, where, one newspaper headline proclaimed, it succeeded only in bringing "Mean Times in Greenwich."[4] In fact, although attendance at the exhibition over the year 2000 fell far short of inflated expectations, the Dome was immensely successful by any other standard. Like its legendary predecessor, the Great Exhibition of 1851, it was always intended to close after the first year of operation. During that year, the Dome succeeded in attracting nearly six million paying customers, ranking it the most popular paid attraction in Britain and one of the five or six highest in the world.[5]

At issue here, however, is not the relative success or failure of the enterprise but the terms in which its ambitious visions were originally conceived. In this essay, I treat this effort at reconstructing the story of time by situating its origins in this London site as a culminating episode in the twentieth-century critique of the nineteenth-century science of modern geography. The task of a truly critical geography, David Harvey argues, calls for joining aesthetic with theoretic perspectives in the discipline, for properly positioning geography at the point of intersection between space and time. A rigorously historical geography, Harvey adds, must bring together those perspectives "that give space priority over time with those that give time priority over space."[6] Accordingly, I begin this analysis of the Dome—an architectural structure explicitly designed to prioritize time by locating time's origins in a particular space—with a review of the back-story events that justify the claim that this site is the *place* where time begins: the circumstances leading up to the 1884 international conference on the prime meridian in Washington, D.C., when delegates from twenty-six nations met and agreed that the Greenwich observatory would mark the globe's zero-degree line of longitude. I then return to the prime meridian a decade later, when the observatory became the target of an anarchist bombing by a French radical who was killed in the course of carrying out the plot. I view that event through the lens of Joseph Conrad's 1906 treatment of it in *The Secret Agent,* that "simple tale of the nineteenth century" (as Conrad subtitled it) in which he presented the terrorist attack on the Greenwich observatory as an attack on the fictions underlying the modern science of geopolitical mapping. Finally, I glimpse briefly into the third millennium, in which the Dome functions, as designed, at once as a symbol for new geographies and for the "New Britain" at a time when the map of Europe is being redrawn in the context of the newly empowered transnational economic "state" called the European Economic Community.

This survey of three moments in the history of Britain's relationship to its geographical and political place in the world demonstrates that the designation of a universal prime meridian in 1884 not only brought together nineteenth-century scientific debates with nationalist controversies but also marked the end of an age of romantic nationalism and modern "cartographic culture" as the demands of new internationalist economies reconfigured national boundaries into convenient market designations. The history of the quest for longitude—for the longitudinal center of the globe—offers perhaps the most literal demonstration of J. Bryan Harley's claim that the power of the map is that it operates behind a mask of neutral science, especially in light of the geographical fact that was generally recognized but systematically obscured in the nineteenth-century debate about the prime meridian:

that longitude is fundamentally different from latitude because longitude cannot be calculated from the physical configuration of the globe.[7] Whereas the equator is the prime latitudinal line, formed by the series of points midway between the North and South Poles, longitude has no such equivalent fixed points from which a center line can be charted. The prime longitudinal meridian, therefore, might theoretically be situated anywhere on the globe, on any line running from pole to pole. The decision to locate the first meridian, consequently, is not a scientific procedure (as it is for the equator) but a political act. Even after most of the world had acknowledged that fact, many geographers continued to conceive of the enterprise in more objective scientific terms throughout the nineteenth century.[8] Indeed, this controversy became so highly charged that the official siting of the prime meridian in London in 1884 would not be recognized by some governments—most notably the French—for another twenty-seven years after the Washington conference sanctioned it. By its very nature, the plotting of the prime meridian is a contested story about the history of nineteenth-century politics as much as it is about geography, about time as much as it is about space.

Fixing Time and Identifying Nations

The 1884 congress on the prime meridian in Washington represents a culminating episode in the long story of longitude, a story that can be traced to the period when the globe began to be explored by European imperialist powers for purposes of profit and influence. The impetus to solve the problem of longitude was always primarily political and commercial, provoked by the effort to control trade routes and deliver goods in a safe and timely way. When Charles II took the throne in England in 1660, he also took command of the largest maritime fleet in the world and was determined to resolve the longitude problem for the safe and swift travel of his ships of war and trade. To that end, he inaugurated the Royal Society of London for Improving Natural Knowledge in 1662, established the Royal Observatory at Greenwich to address the longitude question, and appointed a twenty-seven-year-old astronomer—John Flamsteed—as the first royal "astronomical observator" to apply "the most exact Care and Diligence to rectifying the Tables of the Motions of the Heavens, and the Places of the fixed Stars, so as to find out the so-much desired Longitude at Sea, for perfecting the art of Navigation."[9] The founding of the Royal Observatory at Greenwich, like its predecessor (and principal competitor), the observatory in Paris, was a scientific project conceived quite explicitly to address a political and commercial problem, a heaven-ordained effort to map the globe and chart the seas from their origins in British soil.

Whether through reading elaborate maps of the stars or developing new time machines, the project of discovering longitude was from the beginning understood as the scientific problem of translating space into time, a project consistent with the Enlightenment conception of the earth as a great timepiece moving with predictable precision through a clockwork universe. It was not until 1828, more than a century after its founding, that the British Longitude Board established by Queen Anne was finally disbanded by the repeal of the Longitude Act, with the bulk of the prize money offered for the most effective system going to John "Longitude" Harrison, the eighteenth-century inventor of the remarkably accurate marine time-keeper called "Harrison's clock." By the early decades of the nineteenth century, the timepiece method for calculating relative longitude had proven its practical superiority over astronomical measurement so conclusively that the principal responsibility of the Longitude Board had become the testing and assigning of accurate chronometers to ships of the Royal Navy. When the *Beagle* set out to fix the longitudes of foreign lands in 1831, for example, twenty-two chronometers were assigned to the ship so that the crew might successfully accomplish this important geographic mission. By 1860 there were some two hundred ships in the Royal Navy armed with the more than eight hundred chronometers in the navy's possession.[10] The Royal Observatory at Greenwich, established to assemble the most dependable and widely used astronomical tables and publish them in the *Nautical Almanac,* became the nation's timekeeper in 1833, when it began providing the world's first visual time signal. Each day, at precisely 1 P.M., a large red ball would descend from atop a time turret constructed on the eastern end of the observatory, signaling the hour to all the navigators on ships harbored in the Thames so that they could accurately set their chronometers. The Royal Observatory, founded as Britain's official surveyor of space, had become the nation's official keeper of time.

The daily signal announced at 1 P.M. from the observatory was Greenwich time, however, which differed from London time by some twenty-three seconds, and from Plymouth time by another sixteen minutes. These differences were vigilantly noted and preserved in each locality. The very technological advances that had brought these places closer together in the nineteenth century—like the railroad and the telegraph—had also created a new problem that separated them. As the midcentury expansion of the railroad system combined with an increasingly efficient postal system and developments in rapid communication, local time differences had become increasingly significant and troublesome. While local times were staunchly and even patriotically adhered to throughout Britain (and across Europe), railroad companies voiced increasingly urgent demands to establish a more generally accepted "British time" to avoid the tremendous confusion created by recognizing so many distinct local times on increasingly complex train schedules. Consequently, in 1840 the Great Western Railway issued a controversial decree that London time would be kept exclusively in all its stations throughout the country and would be published in all its timetables. Five years later, the Liverpool and Manchester Railway Company petitioned Parliament to establish "uniformity of time for all ordinary and commercial purposes throughout the land," a proposal that was met with considerable opposition in local municipalities.[11]

By 1855, owing largely to the profound impact of train travel throughout the nation, 95 percent of the clocks in Britain were set by the agreed standard of Greenwich Mean Time. Another thirty years would pass before Parliament officially established a uniform time for the nation, however, the delay reflecting the customary response to what Harvey has called the phenomenon of "time-space compression" in modern society. This implosion of social space advanced by new technologies of travel and communication commonly engenders what Harvey refers to as a collective "crisis of identity," a phenomenon that reacts against the shrinking of the world and the erasure of differences with an enhanced level of devotion to localized memories of place, nation, region, or ethnic grouping.[12] In this instance of the phenomenon, an increasingly extended and complex map of railroad lines gradually imposed a standardized experience of space (calculated station by station) as an increasingly frequent and precise schedule of operation regimented one's sense of time (according to departure and arrival timetables). By the time the prime meridian conference convened in Washington in 1884, space had quite literally become a matter of time both in Britain and throughout the modern industrialized world. Once again, the demands of modern transportation and trade, not a consensus of scientific belief, forced the issue of establishing an international standard for a universal mapping of time and space over against more local points of orientation.

Prior to 1884, London was only one of more than a dozen recognized prime meridians by which navigators charted their location at sea. The zero point of the earth was, for various reasons of commerce and politics, sometimes calculated from Rome, Paris, Madrid, Lisbon, Washington, Jerusalem, Cape Verde, the Canary Islands, the Bering Strait, or Copenhagen, to name a few of the other frequently used sites. Because the Royal Observatory at Greenwich had historically produced and published the most accurate astronomical tables in its *Nautical Almanac,* however, Greenwich had, by custom, become one of the most frequently used first meridians for maritime purposes. But there remained these several other legitimate claimants to the home of time for other purposes. At the time of the Washington conference, for example, the United States Congress had for more than thirty years officially maintained two prime meridians: the first located at the Naval Observatory in Washington, used for all astronomical and land survey purposes, and the second at Greenwich, used for nautical calculations only.[13] Other nations, such as France and Italy, looked to their own capitals as the location for the prime meridian in much the same way. At the same time, many geographers advocated a more "neutral" scientific approach, one that would transcend these political loyalties, urging that the earth be treated as "a rotary globular clock" with a universal meridian. This scheme usually situated the first meridian at a line running through Palma in the Canary Islands, the defining point of the Renaissance prime meridian because it formed the line separating the Eastern and Western Hemispheres. Since this line would split the continents of Europe and Africa from North and South America and since it passed almost entirely through ocean rather than land, it seemed to present itself as a more universal and "scientific" first meridian than those several "domestic meridians" that ran through national capitals and invoked a kind of cultural patriotism as part of the argument for advocating them.[14]

At stake in these protracted debates about establishing a universal first line of longitude over against the numerous domestic meridians were not only issues of national sovereignty and competitiveness but also concerns about the cultural potency enjoyed by the map. David N. Livingstone has shown how the scientific project of European cartographers had historically sprung from what Bruno Latour called the imperialist impulse to establish national "centres of calculation" that "facilitate the mobilisation of information, and thereby domination, at a distance."[15] The Washington conference exposed this political fiction of the map, bowing to the authority of Britain's naval and commercial superiority as the primary basis of its decision. This action made explicit the conventional character of the natural geometry of longitude, revealing it and other natural boundaries to be that much more arbitrary and transient. Ironically, then, the triumph of British nationalism in the controversy over longitude in 1884 succeeded only in unmasking national distinctions as more contingent, and, therefore, that much more vulnerable to attack.

Amid these shifting boundary lines between natural and national categories of spatial ordering in the late nineteenth century, a number of unsuccessful international gatherings had been held prior to the Washington conference for the purpose of establishing a universal first meridian. The first, the International Geographical Congress, held in Antwerp in 1871, concluded by proposing that Greenwich be officially adopted as the first meridian because it was already widely used by merchants and traders owing to Britain's naval superiority. The same recommendation came out of subsequent international congresses: the one in Rome in 1875, a second in Venice in 1881, the 1883 International Geodesic Conference in Rome, and, finally, the Washington conference of 1884.[16] The year of the first of these gatherings of geographers and cartographers corresponded to the outbreak of the Franco-Prussian War; the second, with Britain's acquisition of major holdings in the Suez Canal and Victoria's being proclaimed empress of India; and the third, with the establishment of a threatening German presence in South Africa to rival that of Great Britain. In light of the geopolitical circumstances of these events and the international realignments (and threats to nationalism) they signaled, the earlier scientific congresses made it quite clear that the question of how to map the globe would not be resolved by scientists. The prime meridian issue had become a political problem of such magnitude that it required political action rather than another academic or scientific conference.

For this reason, when the U.S. government convened its international conference in 1884 for the purpose of "fixing a prime meridian and a universal day," the delegates consisted primarily of diplomats and politicians rather than geographers and cartographers.[17] This fundamental difference from earlier congresses was emphasized in the first day of debate, when the French delegation argued vehemently against inviting scientific specialists to take part in deliberations, since the questions to be taken up were "exclusively governmental"; although the French delegation "would be happy to extend any courtesy to men distinguished in science," it was "constrained to oppose the proposition" that scientists take part in the discussion on this principle.[18] The political and practical aspects of establishing a prime meridian, *as opposed to* the scientific aspects of the project, structured the entire debate of the Washington congress, a debate that sometimes veiled (and sometimes did not) the manifest competition between the French and British delegations on the matter. In opposing the motion to name Greenwich as the prime meridian, the French delegation first argued that the congress should be limited to making only the political decision of establishing *some* universal meridian, leaving the identification of a proper location for it to the scientists. Failing to win this point, the French then substituted their own motion, declaring that Greenwich was an inappropriate choice because it would be "at variance with the exclusively scientific principles which we are instructed to maintain," principles that called for a site with "a character of absolute neutrality," one that would "cut no great continent—neither Europe nor America."[19] Failing to carry that battle, the French then pressed the case that both historically *and* scientifically, Paris had a stronger claim than Greenwich for the honor.

But the dictates of politics and commerce ultimately prevailed. From the outset, the United States justified locating the conference in Washington, as the invitation explained, on the basis that America possessed "the greatest longitudinal extension of any country traversed by railway and telegraph lines."[20] Notably, all North American railroads had adopted a standard time based on Greenwich time only eighteen days before the invitations to the conference were sent out, a fact that underscores how deeply the issue of establishing a universal ordering of space and time was driven by the need to foster the movement of goods and the enhancement of trade.[21] Then, by referencing its own territorial expansion as well as its technological and economic advancement as the rationale for hosting the 1884 conference, the United States implied in starkly pragmatic terms its own emerging significance as the world's "centre of calculation" for negotiating the map of the globe. In his welcoming address, accordingly, the president of the conference, Admiral C.R.P. Rodgers (U.S. Navy), offered a statement of reassuring condescension to the other delegates: "Broad as is the area of the United States," he boasted, "vast as must be its foreign and domestic commerce, its delegation to this Congress has no desire to urge that a prime meridian shall be found within its confines."[22] The point was won, finally (despite the adamant opposition of the French), on the basis of the U.S. argument that "as a matter of economy as well as convenience" Greenwich should be selected because "more than 70 per cent. of all the shipping of the world uses this meridian for purposes of navigation."[23] At the conclusion of the conference, amid accusations from France (and with less zeal, Brazil) that the choice was an expression of "national rivalries" rather than geographic precision, a majority of delegates adopted Greenwich as the official site of the prime meridian and the place where the universal day would begin; and they agreed to pass this resolution on to their individual governments to be put into effect, an outcome none of the predecessor conferences could claim to have achieved (Figure 2).[24]

Exploding Time and National Fictions

Significantly, the 1884 Washington conference took place in the same year in which the Berlin Conference of European powers was convened to determine the future of Africa and to redraw the map of the "dark continent" in accordance with the economic interests of European imperial ambition, thereby orchestrating the infamous "scramble for Africa." Equally significant, 1884 was the year the Fabian Society was established in England, commencing an era when London would attract a host of socialists and anarchists from the Continent who were seek-

FIGURE 2. Composite of views of the Greenwich Observatory. This illustration appeared in *The Graphic* on 8 August 1885, soon after the Washington conference designated Greenwich as the prime meridian. The domestically styled exterior views in the center show the time ball that daily announced Greenwich Mean Time and the public clock that displayed the "real" time to the world. The framing interior views illustrate the elaborate scientific apparatus within for observing the heavens and calculating the precise time, warranting Greenwich as "the home of time" *(J. R. Preston,* The Graphic, *8 August 1885. Courtesy of the Watkinson Library, Trinity College, Hartford, Conn.)*

ing refuge in the more tolerant and democratic state of Britain. As radical movements throughout Europe increased their activity, advocating the overthrow not only of individual governments but of the whole notion of the modern state, a string of anarchist bombings took place in London and other European capitals. Perhaps the most confounding of these incidents was the Greenwich outrage of 1894, when a bomb was exploded on the grounds of the Royal Observatory. Since most of the other attacks of the era had been aimed at government buildings far more significant than this isolated astronomical observatory, we still do not know whether the bombing was intended as a strike on the observatory as a symbol of political power and scientific authority or whether the anarchist responsible for the attack, French radical Martial Bourdin (who was killed in the blast), had detonated the device in that location by accident.[25] As Joseph Conrad stated in a note on his novel that recounted the events of the outrage a decade later, this attack was "a blood-stained inanity of so fatuous a kind that it was impossible to fathom its origin by any reasonable or even unreasonable process of thought."[26]

In *The Secret Agent*'s fictionalized versions of these events, the observatory was selected as a target for this crime because it represented science in its purest form. Significantly, however, the bomb plot is conceived not by the rather innocuous band of international anarchists in Conrad's novel but by a friendly foreign government that wishes to stimulate Britain's repression of radical elements on the principle that, if Britain were to more strongly defend its national sovereignty, all the European states would be more secure. This political intrigue, in Conrad's retelling, exploits the confusing contemporary discourse around the longitude debate and the establishment of time zones to deploy scientific authority as a cover for defending political interests.[27] This context puts in a new light the novel's sometimes bewildering experiments with time and narrative sequence and the dizzying series of narrative gaps, repetitions, and contradictory retellings of incidents.

But Conrad tampers with time in another telling way in this text. The actual bombing of the Greenwich observatory, upon which the novel is based, had taken place in 1894, twelve years before Conrad wrote *The Secret Agent* (1906). He elected to set the events of the story neither in 1906 nor in 1894, however, but nearly another decade earlier, in 1886. Why 1886? The date is noteworthy for Conrad because it marks the year in which he, a Polish immigrant, became a British subject, the time at which a foreign territory officially took on the fiction of home for him. Equally significant, however, is that Conrad's setting of events at this particular time also places them directly after the Washington conference when Greenwich was designated as the prime meridian, at the time when national governments were ratifying the terms of the Washington agreement. An anarchist attack on the prime meridian at that moment could be immediately understood as an attack on the newly acknowledged fictions underlying modern geographical and political mapping and an exploitation of the subversion of national identities to which they gave rise.

Conrad clearly understood that any scientific rationale for selecting Greenwich as the prime meridian was only a front for more radical political considerations relating to the fictions of national identity. In the novel's version of events, the observatory was chosen as a target by the plotters not so much because it had come to represent science in its purest form but because it had come to represent the authority of a new regime of political geography in which national boundaries had become part of the network of the powerful fictions ordering bourgeois society. In seeking to identify the target that would best represent the "sacrosanct fetish" of modern society and stimulate Britain's repression of international anarchist-socialist attacks on the state, the agent provocateur Verloc's foreign diplomat and employer, Mr. Vladimir, determines that the prime meridian is the most appropriate site: "The whole civilized world has heard of Greenwich," he insists to Verloc in his effort to convince the double agent to initiate the outrage that would upset the British more than any other, more obvious site.[28] Vladimir reasons that since "property seems to [the middle classes] an indestructible thing," perhaps the target should be time, the real source of their prosperity (66). "Yes," he concludes, "the blowing up of the first meridian is bound to raise a howl of execration. . . . Go for the first meridian. . . . Nothing better" (68–70).

It is fitting, in this context, that in the world of *The Secret Agent,* so-called anarchists are stimulated into action by the very state governments they seek to overthrow. Mr. Verloc, an informer and double agent for the British police, has once been convicted of spying for one foreign country and is currently doing so for another. Although he takes pride in the fact that he is "a natural-born British subject," he continually boasts of his father's French citizenship and frequently makes mysterious trips between Paris and London, the principal antagonists in the debate over establishing a universal meridian (59).[29] The novel abounds with this kind of confusion about national identity. The assistant commissioner of police who investigates the crime is specifically represented as a foreigner in his own country. He began his career extending British rule as a police administrator "in a tropical colony," where he was "very successful in tracking and breaking up certain nefarious secret societies amongst the natives."[30] When the assistant commissioner returned home, he found that "he himself had become unplaced," we are told, and was constantly mistaken for a foreigner by others; and he was disturbed by the fact that the people he saw in London's public places seemed to have lost "all their national and private characteristics" (152). In this environment where everyone seems to be "denationalized," even this honorable civil servant "seemed to lose some more of his identity" as well as his place in the world (151–152). When pressed about the politics of the crime he had been investigating, the assistant commissioner states, paradoxically, that the bombing was a "domestic" crime even though it had international origins, adding that the bombing was planned "theoretically only, on foreign territory; abroad only by a fiction" (209).

By setting the events of *The Secret Agent* in the year he became a British subject, Conrad links the "theoretical" idea of national identity with the official establishment of London as the center of the globe *and* (at the same time) with the anarchist attack that exposed the political and scientific devices for world order as powerful and necessary fictions. In his account of this critical event in the history of longitude we see the establishment of the prime meridian at the 1884 Washington conference for what it was: a worldwide recognition of the dawn of a postnational (and postcolonial) organization of space, an organization in which the domestic and the foreign were at best powerful theories and fictions. This event signaled an official remapping of the globe that subjected the world to the very transnational political and economic forces that depend, ironically, upon the preservation of a "fiction" of "foreign territory" and a nostalgia for clinging to an idea of the nation as "the home of time." Accordingly, double agents are necessary to Conrad's "simple tale of the nineteenth century," even if they are only pawns in a larger game of political intrigue. Foreign nations conspire to stimulate British nationalism in such a context to preserve the fiction of their own sovereignty, and states sustain the very anarchist elements that would seem to threaten them.

Conrad's interest in these questions is to be expected, since he may be regarded as the most geographical of nineteenth-century British writers. After a career as a merchant seaman, he went on to write fictions set in the Far East, Russia, the South Seas, Latin America, Africa, and, with *The Secret Agent,* the foreign territory of London itself. In the last year of his life, Conrad made explicit the geographical subtext of *The Secret Agent* in an essay he published on the history of geography and European exploration. The essay, "Geography and Some Explorers," first appeared in Britain as part of a new travel serial and was later reprinted in the United States under its proper title in the *National Geographic Magazine* (March 1924). In the course of tracing the history of European geography from its initial "fabulous" period to the days of "geography militant" and finally to the nineteenth-century quest for a "scientific geography," Conrad's essay recounts how his own professional plans evolved; he went from being a cartographer to becoming a novelist. After describing how "the problem of longitude" bewildered the minds and falsified the judgment of so many explorers who had "no means of ascertaining their exact position on the globe," Conrad recounts how his childhood reading of Sir Leopold McClintock's "The Voyage of the *Fox* in the Arctic Seas" (published in the year of Conrad's birth) provoked in him "the discovery of the taste of poring over maps; and revealed to me the existence of a latent devotion to geography which interfered with my devotion (such as it was) to my other schoolwork." "The honest maps of the nineteenth century nourished in me a passionate interest in the truth of geographical facts," he concludes, "and a desire for a precise knowledge which was extended later to other subjects."[31]

Conrad proceeds to describe how he was especially fascinated by contemporary maps of Africa, which the "scientific spirit" of nineteenth-century "honesty" and "precision" had purged of their fabulous monstrosities and speculations, replacing them with the blankness of the unknown (13). That empty space on the nineteenth-century map of Africa, Conrad claims, prompted in him "the idea of bringing it up to date with all the accuracy of which I was capable," compelling him to imagine himself "stepping in the very footprints of geographical discovery" (14). This preoccupation drove Conrad to a life at sea, he explains, and finally brought him to the Africa of the map's blank spaces. But when he arrived there, he discovered how even the most scientific maps reflected not only "hardwon knowledge" but also "the geographical ignorance of its time" (13). Confronted with the spectacle of European imperial policies masquerading as a quest for scientific knowledge, Conrad could no longer recall the "great haunting memory" of his youthful desire to fill the map with truth; he could see only those blank spaces being filled by "the distasteful knowledge of the vilest scramble for loot that ever disfigured the history of human conscience and geographical exploration" (17). We might well conclude that in the moment to which he refers, marked by the Berlin Conference on Africa and the Washington conference on the prime meridian, Conrad the scientific geographer became Conrad the novel-writing critic of empire.

Time Lines and Empty Spaces

In this light, Conrad's treatment of the prime meridian in *The Secret Agent* seems to anticipate uncannily the combination of transnational entrepreneurial extravaganza and monument to British primordial space embodied in the planning for the Millennium Dome nearly a century later. Designed as an exhibition space without a specific plan for an exhibition, the Dome emerged as the perfect (and perfectly empty) signifier for controlling space by asserting control over the terms of time. By taking authority over the narrative of history, the Dome offered itself to the "New Britain" as a symbol of the replacement of territorial domination. In a 1998 *New Yorker* article about the Dome, which was under construction at the time, architecture critic Paul Goldberger referred to it as "The Big Top," emphasizing not only the circuslike, theme-park character planned for the structure but also the fact that it was designed as a geometric form following form rather than as a form following any particular function.[32] Since Britain's millennium project had first been proposed by the Conservative government under John Major as a vague celebration of British private enterprise to take place in Birmingham, most social commentators were certain that Tony Blair's Labour Party would kill the idea when he took office. Once the Labour government was elected, however, Blair surprised everyone, viewing the idea as an opportunity to symbolize the New Britain that formed the central theme of his campaign and as a place to stake a claim for the nation at the cutting edge of modern technology and productivity. The new prime minister instantly became an ardent champion of the project, relocating it to the Greenwich site, on the prime meridian and, coincidentally, in an area of London in need of economic redevelopment. Blair awarded the design commission to architect Richard Rogers, codesigner of the Pompidou Center in Paris, the impressive skyscraper headquarters for Lloyd's insurance company, and a number of visionary architectural schemes for the city of London.

With the Crystal Palace exhibition of 1851 as an explicit model, Rogers conceived of the Millennium Dome as a monument to Britain's significance in the world, if not by its domination of territory then by its control of time. He even referred to himself as the heir, architecturally speaking, of Sir Joseph Paxton, the noted Victorian designer of the controversial Crystal Palace, a design that, when it was first proposed, received a level of ridicule and scorn comparable to that which the Dome has garnered in the modern press.[33] Blair then appointed Minister without Portfolio Peter Mandelson as his "Dome Secretary" to take over arrangements for the project. Mandelson, the grandson of Herbert Morrison, who had organized the 1951 Festival of Britain (itself modeled after the Crystal Palace exhibition of one hundred years earlier), also saw himself in the line of the producers of grand British national events. The only real difficulty was that neither he nor Rogers had any clear plan about what should go into the Dome; they knew only that there should be one. What was to go under the big top remained a mystery for some time.

When Mandelson flew to Disney World for inspiration, therefore, he invoked the wrath of the British populace, who wondered why the symbol of the New Britain should model itself after the quintessential American fantasyland. But if we understand the Dome as a celebration of the simulacrum of national sovereignty, of the nation's ghostly memory rather than its actual presence, Disney World is the perfect reference point for a strategy of celebration. The purpose for the 1851 Great Exhibition was (according to Prince Albert) "the encouragement of arts, manufactures, and commerce" through the display of "products of all quarters of the globe," expressly for the "stimulus of competition and capital" among nations.[34] The Millennium Dome, however (according to Blair), presents an entertainment experience rather than a display of objects or industry, "a *time* for the nation to

FIGURE 3. "All the World Going to See the Great Exhibition of 1851." This George Cruikshank illustration seems to place the Crystal Palace at the prime meridian, a spectacular beacon (like the Dome of the Millennium that will follow) drawing the entire world to its display of commodities. (*From Henry Mayhew and George Cruikshank, 1851; or, The Adventures of Mr. and Mrs. Sandboys, Their Son and Daughter, Who Came Up to London to Enjoy Themselves and See the Great Exhibition [London: D. Bogue, 1851]. Courtesy of the Watkinson Library, Trinity College, Hartford, Conn.)*

come together to be excited, entertained, moved and uplifted."[35] The emphasis upon products and territory in the Great Exhibition—organized as it was as a virtual commodity map of the world, with individual exhibits displayed nation by nation— has been replaced by the Dome's emphasis upon experience and time in a series of entertainments, and upon redefining what it means to be a nation in a world defined by spectacle and simulation (Figure 3).

Like many British critics, Paul Goldberger laments the decision to make the content of the Dome a theme park presiding over primordial time, even as he maintains his admiration for the *form* of the Dome itself; the Dome "wants to be a pure monument," he contends, though "the present government appears to see only wastefulness in that."[36] But even in its inevitably crass celebration of commodities and empty British neojingoism, the Dome is, arguably, a pure monument—to the idea of the nation renewing itself as a nation in the context of the new global capital. Ritualizing the movement through space as a repetition of capitalist exchange expressed in the vague discourse of unspecified nationalistic pride, the

FIGURE 4. London's Dome of the Millennium. This view appeared on 1 January 2000 in the *New York Times* and other newspapers around the world. It was taken at the stroke of midnight as Greenwich recorded the birth of the new millennium. *(Photo: Hugo Philpott. By permission of Agence France-Presse.)*

Dome is fashioned as the great emblem of postnational space, the nation as cybermemory of itself. It seems to be offered as a palimpsestic recollection of the empire upon which the sun never set, of the Great Exhibition, of the Big Ben clock tower, of the prime meridian of the earth, and of the place by which the world still sets its clocks. As the putative home of time, poised astride the prime meridian of longitude, the Millennium Dome presents itself as a giant narrative that retells and revises the story of the United Kingdom as the story of the line of time itself. And the Dome tells this tale at the very moment when the territory of the United Kingdom is less united and less of a kingdom than it has ever been. Indeed, as the title of a more recent *New York Times Magazine* article announced, "The End of Britain" is at hand; Britain is being swallowed up by the European Union and diminished by the independence movements of Scotland and Wales, the gutting of the British constitution in the celebrated "reform" of Parliament through the strategy Blair ominously called "devolution," and the possible loss of its prized national currency.[37]

As the nation celebrated its own passing into history, it built the largest enclosed parcel of space in the world to stake its claim to the less vulnerable precincts of time itself. At the Dome's completion, Goldberger had predicted, it would be "the world's most spectacular empty space," a place that would allow the people of the New Britain to "sustain some illusion that they possess common ground."[38] The ground they shared, of course, was not a familiar place but a remembered time. It was the memory to which the British people cling tenaciously as they enter a strange new world in which they will cash in their pounds and shillings for the universal currency of the Euro and finally abandon the nineteenth-century map of nation-states for the new millennium's arrangement of economic communities. Despite the fact that the world press declared the Dome a spectacular failure as an economic enterprise, and regardless of the repeated efforts by the Blair government to rescue the project from bankruptcy, the Dome itself stands as a national monument to the vanishing nation, the modern translation of Britain's nineteenth-century domination over space into its dominion over time (Figure 4).

In the final months of the Dome's existence as the millennial exhibition, it adopted a new theme by which to define itself. Rather than "Time for a Change," it took as its motto "Essentially British," in an effort to bolster attendance from the "domestic market" and, it would seem, shore up the ruins of the lost Victorian fantasy of a distinctly British nationality. In this modern translation of Britain's nineteenth-century Crystal Palace domination over territory and trade for its pleasure-dome-like dream of dominion over time, the ruins of the old economy of manufacturing are literally being reclaimed for investment by the speculators of the new economy of entertainment and illusion. It seems only fitting, then, that the Dome of the Millennium, this vacant globelike structure poised at the so-called center of the world and offering itself as the so-called home of time, may be best remembered for its starring role in the opening sequence of a 1999 James Bond film, a film so aptly signaling in its title—*The World Is Not Enough*—the insufficiency of space as a category for domination in the virtual world of the new millennium.

Notes

1. Warren Hoge, "Where Time Begins, a Millennium Pleasure Dome," *New York Times,* 28 July 1997, sec. A, p. 4.
2. Ibid.
3. *Financial Times,* 20 December 2000, 2, London edition.
4. See *Observer,* 9 January 2000, 13; *Times* (London), 22 February 2000; and *New York Times,* 18 February 2000, 1.
5. Michael Specter, "The Blunderdome," *New Yorker,* 29 January 2001, 46–51.
6. David Harvey, "Between Space and Time: Reflections on the Geographical Imagination," *Annals of the Association of American Geographers* 80 (1990): 430.
7. J. Brian Harley, "Deconstructing the Map," *Cartographica* 26 (1989): 20.
8. See Dava Sobel, *Longitude: The True Story of a Lone Genius Who Solved the Greatest Scientific Problem of His Time* (New York: Walker and Co., 1995), 1–10.
9. Ibid., 39–40.
10. Ibid., 164.
11. Derek Howse, *Greenwich Time and the Discovery of the Longitude* (New York: Oxford University Press, 1980), 87.
12. Harvey, "Between Space and Time," 427.
13. Howse, *Greenwich Time,* 123.
14. Matthew H. Edney, "Cartographic Culture and Nationalism in the Early United States: Benjamin Vaughn and the Choice for a Prime Meridian, 1811," *Journal of Historical Geography* 20 (1994): 387.

15. Bruno Latour, quoted in David N. Livingstone, "The Science of Knowledge: Contributions towards a Historical Geography of Science," *Environment and Planning D. Society and Space* 13 (1995): 39.

16. S. R. Malin, "The International Prime Meridian Conference," *Journal of Navigation* (London: Royal Institute of Navigation) 12 (1985): 293.

17. Ibid., 203.

18. *Protocols of the Proceedings of the International Conference Held at Washington for the Purpose of Fixing a Prime Meridian and a Universal Day* (Washington, D.C.: Gibson Brothers, 1884), 27.

19. Ibid., 29, 36.

20. Howse, *Greenwich Time,* 140.

21. The establishment of standard time zones in the United States, finally effected by the railroad industry in 1883, was even more vehemently contested in the United States than it was in Britain. For an account of the struggle, see Michael O'Malley, *Keeping Watch: A History of American Time* (New York: Penguin Books, 1991), 99–144.

22. *Protocols of the Proceedings,* 6.

23. Ibid., 40.

24. Ibid., 47–57.

25. For an account of these events, see H. Oliver, *The International Anarchist Movement in Late Victorian London* (London: St. Martin's Press, 1983), 101–109.

26. Joseph Conrad, author's note in *The Secret Agent* (Harmondsworth, Eng.: Penguin Books), 39.

27. According to Mark Conroy, "[T]he all-seeing eye of the Greenwich Observatory is, in one sense, the ultimate political institution, insofar as it grounds and gives form to time; as creator of time, the Observatory itself can be said to occupy a kind of eternity" (*Modernism and Authority: Strategies of Legitimation in Flaubert and Conrad* [Baltimore: Johns Hopkins University Press, 1985], 144).

28. Conrad, *The Secret Agent,* 66, 68.

29. Terry Eagleton argues that each of *The Secret Agent*'s amalgam of genres contributes to an ideological contradiction between the "exotic" and the "domestic," a contradiction that centers on the figure of Verloc (Eagleton, *Against the Grain: Essays, 1975–85* [London: Verso, 1986], 23–32).

30. Conrad, *The Secret Agent,* 116.

31. Joseph Conrad, "Geography and Some Explorers," in *Last Essays* (Garden City, N.Y.: Doubleday, Page, and Co., 1926), 11–12.

32. Paul Goldberger, "The Big Top," *New Yorker,* 27 April and 4 May 1998, 152–159.

33. See Asa Briggs's account of the reaction to the Crystal Palace in *Victorian People: A Reassessment of Persons and Themes, 1851–1867* (Chicago: University of Chicago Press, 1955), 35–37. The popular reaction to Paxton's structure, notably, was much more generous than that of the critics and the press.

34. Quoted in Theodore Martin, *Life of the Prince Consort* (New York: D. Appleton and Company, 1877), 2:205.

35. Quotation attributed to Blair that was posted on the Greenwich Millennium Dome website, <www.londonnet.co.uk/In/guide/about/dome.html>, 1999–2000.

36. Goldberger, "Big Top," 158.

37. Andrew Sullivan, "The End of Britain," *New York Times Magazine,* 21 February 1999, 70, 78.

38. Goldberger, "Big Top," 159.

Mapping the Orient

NON-WESTERN MODERNIZATION, IMPERIALISM, AND THE END OF ROMANTICISM

USSAMA MAKDISI

*I*n a nineteenth century dominated by European colonialism, there were few corners of the globe that were not incorporated directly or indirectly into a culture of modernity. The Ottoman Empire was no exception. Following its defeat at the hands of Russia in 1774 and, again, following Napoleon's invasion of Egypt in 1798, the Ottoman Empire lay at the mercy of modern European imperialism. The once mighty sultanate became the locus of a literature—well plied by scholars in the wake of Edward Said's *Orientalism*—that relegated the empire to an abode of decadence, indolence, and, above all, an Islamic fanaticism that progress had left behind. Although there has been much discussion following Said's own argument about the preponderance and continuity of representations of the Orient, far less attention has been paid to the differences within these representations and to how these differences relate to changing geographies of East and West.[1]

This essay examines how a particular geographical location—the Ottoman Orient—was defined by competing historiographies of timelessness (the unchanging Orient) and earliness (the premodern and developing Orient). These historiographies grew out of a relentless Western encroachment on the Ottoman Empire as well as Ottoman attempts to resist this encroachment. The nineteenth-century Ottoman Empire became the venue for a modernization politics in which a beleaguered group of Ottoman officials made the case for independent reformation of the Ottoman Empire in the face of a majority of European authors, travelers, politicians, and missionaries who insisted on its inevitable subordination to a European civilizing mission. In either case, the Ottoman Empire was configured as a premodern place in relation to Western modernity. The struggle between Ottomans and Europeans about the fate of the premodern empire, however, set the stage for a romantic rediscovery of a timeless Orient. I analyze Richard Burton's *Personal Narrative of a Pilgrimage to Al-Madinah and Meccah* to illustrate how, in the sec-

ond half of the nineteenth century, the advent of non-Western modernization constituted a point of departure for a romantic search for an uncorrupted and timeless Orient. I argue that writers such as Burton and T. E. Lawrence specifically
repudiated the attempts by non-Western governments to modernize independently.
Their search for a romantic Orient ended abruptly during World War I, with the
British-led dismemberment of the Ottoman Empire. The British and French imposition in 1920 of the colonial mandate system on the former provinces of the
Ottoman Empire signaled the final incorporation of the Orient into an imperialist
geography of modernization.

Romantic literature about the Orient was punctuated by two distinct crises.
The first, which scholars of British romanticism such as Saree Makdisi and Nigel
Leask have studied at length, was produced by revolutionary and industrial tur-
moil in late-eighteenth-century and early-nineteenth-century Britain and France.[2]
Writers, poets, and travelers sought to carve out an independent space outside of
European modernization, particularly in the Orient. This first round of romanticism played on a long-established sense that irreconcilable differences separated
the Ottomans from the West—a sense that initially had inspired dread of an Ottoman invasion of Europe but, by the middle of the eighteenth century, had changed
into narratives of exoticism, eroticism, and finally Oriental decadence. This romantic literature contended with, and eventually gave way to, a literature that celebrated rationality and modernization, including that which called for the full-scale
and complete transformation of the Ottoman Empire into a European-dominated
modernity.

The second crisis, which I take up in this essay, emerged from the process
of non-Western modernization of the mid–nineteenth century. The process of modernizing the Ottoman Empire blurred the categories of East and West, especially
after the Ottomans deliberately incorporated themselves into what Johannes Fabian
has called the evolutionary "stream of Time," in which all societies, Eastern and
Western, were located on the same spectrum of progress as a result of modernization.[3] The politics of incorporating the hitherto immutably different Ottomans into
a universalizing discourse of modernization forced the Europeans to consider the
location of the Ottoman Empire in the modern world. What I suggest is that the
beginnings of non-Western modernization represented by the *Tanzimat* reform
movement in the Ottoman Empire, a movement that formally began in 1839 but
whose antecedents were already clear by the end of the eighteenth century, heralded the transformation of the heterogeneous Ottoman Empire into a modern centralized state and inaugurated an independent Ottoman claim to modernity. The
modernization movement in the Ottoman Empire, in turn, inspired a similar development in Egypt, whose rulers, nominally under the suzerainty of the Ottomans,
made their own efforts to build a "Paris on the Nile." These claims directly challenged an imperialist geography of modernization, for the reforming Ottomans (and
Egyptians) insisted that the path to modernity could be traversed without European imperialism. In the context of an overall nineteenth-century moment of European, and particularly British, imperial assertion that divided the world into
modern and premodern, civilized and barbarian, advanced and backward, the

initiative of independent Ottoman modernization unleashed a European prediction, and insistence, that European imperialism should save the premodern inhabitants of the Ottoman Empire. In response to Ottoman modernization, a second round of romanticism emerged, which categorically refused to take seriously the "half-civilized" Ottomans. This second wave of romantic writers sought to preserve a pristine East in the face of non-Western modernization. Far from echoing the earlier romantic literature that arose in opposition to a culture of modernization, however, this new strain of romanticism reflected prevalent nineteenth-century Western conceits about the innate cultural and racial superiority of the West. It depended on a culture of imperialism to make its romanticism possible, and its foray into the timeless Orient was predicated on its refusal to accept the possible dissolution of the temporal boundaries between the Ottomans and the Europeans and, more generally, between East and West, boundaries that had been threatened by autonomous Ottoman and Egyptian efforts at modernization. The challenge of this second burst of romanticism was to swim against the "stream of evolutionary Time," to traverse a temporal as well as a physical gap by venturing eastwards and backwards, but always in order to survive and *return* to a modern West to tell the tale of "authentic" Eastern adventure.

Colonial romanticism did not dispute the inevitability of the incorporation of the Orient into European hegemony. It did, however, challenge a dominant European view of the implications of this incorporation. Most British officials who dealt with the East regarded outside and putatively progressive European intervention to be necessary for the rejuvenation of a stagnant Orient. They understood modernization to mean the transformation of an Orient resistant to, and incapable of, progress on its own into an Orient that might be developed under European tutelage to look ultimately like Europe itself. The abiding tension within British (and European) discussions of Ottoman reform was between a notion of an Orient incapable of independent development—an innately different, irrational, depraved converse of the enlightened and civilized Europe—and a notion of an Orient that must be, for the sake of the political stability of Europe if nothing else, reformed and modernized. Oriental timelessness was, in this particular evolutionary scheme, the antithesis of development. It connoted a backwardness that would inevitably be stamped out by the positivist march of history. For colonial romantics, however, modern colonialism was a method to preserve the timelessness of the Orient from the inauthentic, corrupt, and vile pretense of the "half-civilized" modernizing Ottoman or Egyptian. The timeless Orient was imagined as a corollary of modern development. The desert of Arabia, for example, represented an innately pure and masculine space that could be, indeed had to be, overlaid with a Darwinian framework: the timeless Orient constituted a refuge from the strictures and effeminacy of European civilization, a place where a superior race of British men could reacquaint themselves with their own primordial nature by surviving the harshness of the desert and the savagery of its inferior Arab inhabitants. The timeless Orient, in other words, had an instrumental place in an evolutionary scheme of development: it could make white men better, more capable agents of the modern British Empire.

The Politics of Modernization

When European statesmen, missionaries, and travelers arrived in the nineteenth-century Orient, they encountered an Ottoman Empire that had already undergone several stages of interpretation. The first stage was fear fortified by a sense of awe at the splendor of the late-medieval and early-modern Ottoman court. Roughly speaking, the westward expansion of the Ottomans, which began really with the invasion of the Balkans, was invigorated by the siege and fall of Constantinople in 1453, and culminated in the siege of Vienna in 1683, inspiring dread of what Richard Knolles described as the "present terrour of the world."[4] European kingdoms were pervaded by a fear of the Ottomans, a fear somewhat diminished by the battle of Lepanto in 1571 but conclusively abated only at the beginning of the eighteenth century, when the western frontier of the Ottoman Empire closed. The decline of Ottoman military strength relative to Europe transformed the image of the Ottomans in the West.

From being an object of fear, the Ottoman Empire was gradually annexed into a European literary imagination bent on mapping out an exotic and erotic East. In the process, the Orient was imbued with certain immutable qualities that made it radically and forever different from Europe. It was the celebration or denigration of this difference that was a matter of debate, rather than—as the case would become following the advent of modernization—the eclipse or mitigation of this difference. The advent of the nineteenth century inaugurated a new way of viewing the Orient. In *Orientalism,* Said insists that the Napoleonic invasion of Egypt in 1798 heralded the birth of what he refers to as "modern" Orientalism, in the sense that the Ottoman Empire was incorporated into a project of power that expressed itself through the language of modernization.[5] Rather than representing a historically, religiously, and ontologically separate space, the Orient, as Makdisi notes in *Romantic Imperialism,* was henceforth subordinated into a "diachronic *History* narrated and controlled by Europe."[6]

After a string of catastrophic defeats at the hands of Russia and France at the end of the eighteenth century and at the hands of Muhammad Ali's modern armies in the nineteenth century, the Ottoman Empire embarked upon the *Tanzimat*—a massive project of administrative and cultural reform and modernization. The legendary Janissary corps of the Ottoman army was abolished (and many of its members massacred);[7] new European-style architecture, administration, and uniforms were introduced in both the military and the civilian sectors of society; and, above all, a European discourse of progress was zealously embraced by reforming Ottoman officials who believed that without modernization along European lines the empire would surely perish. The Ottoman Empire sought to define itself (especially after the Crimean War and the 1856 Treaty of Paris) as an equal player on a world stage of civilization. It was a world stage dominated, in Ottoman eyes, by mischievous, but modern, European states, and a world stage in which the Ottoman raison d'être was to "catch up" to a putatively advanced West.

High-ranking Ottoman officials such as Aali Pasha and Fuad Pasha—men who personified the *Tanzimat* because they were trained in new, European-style schools and were at the vanguard of a new, European-style imperial administration—were determined to lay the foundations of an independent and modern Ottoman state that drew on an allegedly rich heritage of tolerance. They insisted that the empire was comparable to modern European nation-states, both in the past and at present, and therefore that there was very little to preclude—from a historical, a moral, or a philosophical standpoint—the Ottoman Empire from becoming a fully fledged member of a modern state system. For example, in his 1855 memorandum on tolerance, which he had delivered to London and Paris, Aali Pasha stated that even a "brief" examination of history would reveal that "the Ottoman Empire passed, in this respect, through the same phases as other countries, and it may even be boldly asserted, without fear of denial, that in the times of darkness and intolerance which more or less oppressed the whole of Europe, it was not in the Ottoman Empire that the conquered minorities had most reason to lament their condition."[8]

Aali Pasha claimed, furthermore, that the Ottoman Empire had "progressed" more than any other country in a short space of time, and he suggested that what the empire needed was the freedom to complete this journey of modernization without political interference from Europe. While he underscored his own commitment to a teleological and evolutionary model of development in which the Ottoman state held, at present, an inferior position, Aali Pasha claimed that modernization was perfectly compatible with the history of the empire. His insistence on a universal discourse of tolerance (for which the Ottomans were avatars) underscored his belief in a shared sense of historical time that linked the Ottomans and Europeans. To Aali Pasha, modernization (signified by a discourse of tolerance) was an outgrowth of Ottoman history rather than a rupture with it. Aali Pasha, in short, suggested a universal history that accepted modernization as its principal dynamic at the same time that he rejected the imperialist implications of modernization articulated by nineteenth-century British writers, missionaries, historians, and diplomats.[9]

The timing of Aali Pasha's memorandum on tolerance was far from fortuitous. He wrote it during the Crimean War to influence directly the outcome of negotiations between Russia, Britain, and France that ultimately led to the 1856 Treaty of Paris. Because European authors had consistently criticized the Ottoman Empire for being fanatical and intolerant, and because Russia had justified its belligerence on the pretext of Ottoman subjugation of Christians, Aali Pasha wanted to "prove" that the Ottoman Empire enjoyed a history of unparalleled tolerance of minorities. From an Ottoman standpoint, one of the most important results of the Treaty of Paris, which brought the Crimean War to a close, was the formal European recognition of the sovereignty of the modern Ottoman state. Ottoman reformers felt that they had arrived as a *nation* on a world stage of modernization—in the aftermath of their commitment to the free trade treaties of 1838; their successful defeat (with British help) of their erstwhile Egyptian vassal and, more importantly, competitor for the title of the Orient's first modernizer, Muhammad Ali, in 1840; and the Crimean War. Henceforth, their duty was to protect and safeguard a very tenuous independence and, concomitantly, to oversee the rapid physical transformation of the empire into a modern nation-state. To wit, provincial administration was radically reformed in 1864, a new secular Ottoman citizenship law was promulgated in 1869, and new urban building codes and municipalities were introduced. At the same time, the Ottoman government also embarked upon a deliberate campaign to present an image of itself as a modern, civilized member of the European colonial club.[10] In this regard, Ottomans and Europeans viewed the temporality of the politics of modernization very differently. The Ottomans prided themselves on their arrival at the doorstep of modernity despite European political and military hostility. Europeans, however, insisted that modernity could not be achieved by a non-Western nation as much as it could be bestowed by benevolent Western imperialism. Arrival in modernity was to be judged by Europeans, not announced by non-Europeans.

Just as British India had seen a debate between Orientalists and Anglicists firmly resolved in favor of the latter (whose position was famously encapsulated in Thomas Macaulay's 1835 Minute on Education), the Ottoman Empire was no longer perceived by the overwhelming majority of Europeans concerned with the Orient to be immutably different and hence irrevocably alien; it was perceived as different (especially in the religious sense) in the sense of inferior. This difference, although an obstacle to so-called improvement and civilization, did not preclude a European-instigated and -dominated drive to save the so-called sick man of Europe—the term was itself highly ambivalent, for it could indicate the incorporation of the hitherto immutably different Ottomans into a European moral and political geography (as the Ottoman reformers believed), or it could underscore the Ottoman Empire's tenuous, and ultimately untenable, position in Europe following the Greek war of independence (as many Europeans thought). The question discussed by those debating Ottoman reform was not whether reform should take place but what shape it would assume. Given the military weakness of the empire, those Ottomans and Europeans (such as David Urquhart) who insisted that the empire might be able to achieve modernity independently were overwhelmed discursively, politically, and militarily by those who argued that it must be subordinated to a European civilizing mission. As Jas Brant, a British official, put it in a memorandum on reform prepared in 1856, the Ottoman Empire "must be reformed, and her positive independence will have to be placed in abeyance until she has learned to administer her own government on an enlightened and equitable system."[11]

From the very outset of the *Tanzimat,* there developed a general Western refusal to acknowledge the sincerity of Ottoman reform. In European eyes, the Ottoman Empire became suspended in a state of perpetual nonarrival; it was deemed to have failed (thus far) to reach the mark of modernity no matter how extensive, how bold, and how unprecedented the *Tanzimat* reforms were. At its simplest, a discourse of unremitting and incorrigible Ottoman intolerance and fanaticism reminded Western readers of the Ottoman Empire's nonarrival at modernity at the same time that the discourse kept open the possibility of the empire's potential (but always distant) arrival in the future, pending a European civilizing mission. In the face of Aali Pasha's memorandum on tolerance, the British government issued several of its own memoranda on reform. What I would like to suggest here, and develop in the final part of this essay, is that this Western refusal to acknowledge the sincerity of Ottoman reform masked an anxiety about the implications of non-Western modernization—in particular, its implications for an imperialist geography of modernization that had neatly divided the world into modern West and premodern East, into advanced and backward, and into master and subject races.

Beginning in the mid–nineteenth century, however, it was as much non-Western modernization as it was a cultural engagement with Western imperialism that defined the work of a new wave of romantic travelers such as Richard Burton and T. E. Lawrence. Western refusal to acknowledge the possibility of Ottoman and "Oriental" modernity pointed to a crisis inherent in modern Western representations of self and other, a crisis embedded in the logic of a supposedly universal culture of modernization that ultimately and theoretically respects no national border, no national identity, and no national representation. Because independent non-Western modernization threatened to homogenize difference at a pace and in a manner unacceptable to a nineteenth-century imperial geography of modernization, European travelers, officials, and missionaries had either to subordinate non-Western modernization to European imperialism—in essence, to continually defer non-Western modernity—or to completely deny its possibility.

Nowhere was this denial more apparent than in a Western discourse that developed in the wake of modernization politics, politics that condemned the "half-civilized" Oriental and rejected out of hand the manifestations of independent non-Western modernization, be it on the shores of the Bosporus or on the banks of the Nile.[12] Jas Brant's memorandum on reform is particularly instructive in this regard:

> The official Turk at Constantinople, both he who has visited Europe, as well as he who may have acquired some notion of the ideas of Europeans from others, know well how to adapt their conversation to gain the good-will of—or, I should rather say, to deceive—any distinguished foreigner who may chance to visit them. They will affect the most liberal ideas and the most enlarged views; they will flatter their guest by praising the institutions of his country, and by their extravagant admiration of the wonderful progress made by Europeans in the arts, they will readily acknowledge the inferiority of their own countrymen; will express the hope that, one day, by imitating the European, they may arrive at the state of civilization to which he has already attained. These praises are accepted with pleasure by the visitor, as an honest tribute to his country's superiority, and he fancies his Turkish acquaintance to be an enlightened and discerning person; but he little suspects that under this specious parade of enlightenment and admiration is concealed the most perfect contempt of the person he has been complimenting, and of all that he has been praising, added to a hatred of everything emanating from Christians, which is further envenomed by a sense of inferiority, wounding most bitterly and offensively his bigotry and his pride.[13]

Brant's opinions on Turkish reform constituted a direct response to Aali Pasha's memorandum on tolerance. Brant was, in effect, warning his own government not only of the impossibility of independent Turkish modernization but of the dangers inherent in any British policy that put any stock in the modernized Turk. "However the modernized Turk be disguised by the polish of a superficial civilization," he continued, to make his point absolutely clear, "he is essentially a wily barbarian, as false as cruel."[14] Brant's emphasis on the modernized Turk's alleged insincerity and scheming marked his own apoplexy at the impertinence of the Ottomans who had yet to submit themselves completely to the supposedly up-lifting colonizing gaze of Europe. Brant's outrage at what he perceived to be Ottoman subterfuge reflected another common European reaction to non-Western modernization, namely, disgust at what was considered to be blind imitation and a vulgarity quintessentially "Levantine," devoid of any authenticity.[15]

Richard Burton and the Inevitability of Empire

At the same time that Brant envisioned an imperialist vision of modernization in which all that was stagnant in the Ottoman Empire was synonymous with Islam—and therefore to be abandoned wholesale—a wave of romantics arrived in the Orient in search of precisely this Islamic and Eastern timelessness. Unlike Brant, who wanted to forcibly lead the Ottomans forward along the stages of Western civilization, travelers such as Richard Burton came to experience and revel in Oriental difference. To be sure, as Edward Said has suggested, Burton's pilgrimage may be understood as an expression of the ascendancy of modern Orientalism, which, at its most basic level, put the considerable erudition and knowledge of Orientalist scholarship in the service of nineteenth-century European imperialism. Burton, for example, pointed out the advantages of controlling Egypt nearly three decades before the British conquest of Egypt in 1882, and his narrative is littered with advice on how to better, more efficiently and authoritatively rule British India.[16] Missing from such analysis, however, is any effort to locate Burton and other romantic travelers as part of a wider reaction to non-Western, and specifically Ottoman (and Egyptian), modernization. Without detracting from Said's reading of Burton, it is important to recognize that post-*Tanzimat* romantic traveler writing marked the apogee of a Western refusal to accept non-Western modernization: the genre's "discovery" of the timeless Orient was predicated on its simultaneous denigration of autonomous non-Western modernization.

Burton's *Personal Narrative of a Pilgrimage to Al-Madinah and Meccah* was a text that affirmed modern imperialism at the same time that it rejected out of hand independent avenues for non-Western modernization. Burton's purpose was not to negate or resist modernization as much as it was to go backwards in the stream of evolutionary time, to traverse it upstream, as it were, and return an invigorated man in order to instruct and upbraid the very civilization from which he initially escaped. Burton embarked upon his great adventure in 1853 at precisely the same time that the Ottoman and Egyptian governments inaugurated their attempts to resist Western imperialism by trying to build their own modern states.[17] For Burton, the world was divided into two authentic experiences: Western civilization and an original Orient inhabited by "real Easterners." On the very first page of Burton's tale, the author wrote that his purpose in undertaking the pilgrimage to Islam's holiest cities in 1853 was to remove "that opprobrium to modern adventure, the huge white blot which in our maps still notes the Eastern and Central regions of Arabia."[18] On the second page, he admitted that because he was "thoroughly tired of "progress" and of "civilisation," he was yearning to see "Moslem inner life in a really Mohammedan country; and longing if truth be told, to set foot on that mysterious spot which no vacation tourist has yet described" (2). Burton, in other words, desired to acquaint himself (and his English readers) with an authentic and as-yet-unknown part of the Orient inhabited by uncorrupted Orientals. From the outset, Burton's desire to embark on adventure for the sake of adventure, to escape from what he saw as a "life of European effeminacy" (141) and "the hypocritical politeness and slavery of civilization" (150), as well as from the "degeneracy and the ill effects of four years' domicile in Europe" (159), was predicated on an idea of a pure and timeless Orient where a civilized man could find himself in a primordial state. As Burton put it,

> Phrenology and physiognomy, be it observed, disappoint you often amongst civilised people, the proper action of whose brain upon the features is impeded by the external pressure of education, accident, example, habit, and necessity. But they are tolerably safe guides when groping your way through the mind of a man in his so-called natural state, a being of impulse, in that chrysalis condition of mental development which is rather instinct than reason. (17)

For Burton, the romance of the "real" Orient lay in the crucial role it could play alongside the science and civilization of the West. The "spoiled child of civilisation" and "the tamest citizen," he claimed, would find "their pulses beat strong, as they look down from their dromedaries upon the glorious Desert" (151). He added,

> It is another illustration of the ancient truth that Nature returns to man, however unworthily he has treated her. And believe me, when once your tastes have conformed to the tranquility of such travel, you will suffer real pain in returning to the turmoil of civilisation. You will anticipate the bustle and the confusion of artificial life, its luxury and its false pleasures, with repugnance. Depressed in spirits, you will for a time after your return feel incapable of mental or bodily exertion. The air of cities will suffocate you, and the care-worn and cadaverous countenances of citizens will haunt you like a vision of judgement. (151)

Because, in Burton's estimation, the real Orient was defined in apposition to Western civilization, it held important lessons and possibilities for unmediated expressions of masculinity, vigor, clarity of thought, and the natural stimulation of senses that were impossible in the West. According to Burton, the Orient, if it was to be authentically experienced, had to be purged of all manifestations of modernization. Burton desired to maintain the East as a pristine refuge, several stages removed from, but at the same time accessible to, the civilized West. Burton, therefore, considered the Orient to be an integral part of a permanently diachronic history controlled by Europe for the benefit of Europe.

Reaching this primordial state, however, necessitated traversing the place and space of the half-civilized Orient. It meant going from civilization to nature, from England to Arabia, by way of an Ottoman Empire in the throes of the *Tanzimat*. Burton's text must be read against the backdrop of an Ottoman Empire and a khedival Egypt in transition. "The land of the Pharaohs is becoming civilised, and unpleasantly so: nothing can be more uncomfortable," he wrote, "than its present middle state, between barbarism and the reverse" (17). Reaching the "real" East entailed rejecting this "middle state" of non-Western modernization. In Burton's eyes, the process of non-Western modernization, not Western modernity, threatened to undo the authentic and premodern Orient that held hitherto untapped moral, spiritual, and material resources for the civilized West. Herein is Burton's contempt for the urbanized Oriental, the Levantine, and the *Tanzimat*. In his discussion of mosque architecture in Cairo, for example, Burton lambasted modern European influences that spoiled the traditional architecture (98). He compared the "genuine Osmanli of past ages, fierce, cold, with a stalwart frame, index of a strong mind" with the "pert and puny modern Turk in pantaloons, frock coat and Fez, ill-dressed, ill-conditioned, and ill-bred, body and soul" (99). He mocked the dragomans (invariably Levantine and Christian) as the "most timid and cringing of men" (128), and he dismissed the *Tanzimat* as "the silliest copy of Europe's folly—bureaucracy and centralization—that the pen of empirical statecraft ever traced" (258).

The trope of the half-civilized resolved a tension, inherent in Burton's narrative, between his criticism of modern civilization and his advocacy of modern imperialism, a tension that Said discusses at length in *Orientalism* and that Patrick Brantlinger refers to as Burton's "double arrogance."[19] Burton rejected non-Western modernization because he saw it as a travesty that reproduced the worst aspects of European civilization. Therefore, Burton felt that it was far more important to maintain the Orient as a pristine refuge to be experienced by what he considered to be real British men. Modern imperialism could accomplish this task in two principal ways. First, it could prevent the course of non-Western modernization, which, according to Burton, was doomed anyway.

> What Milton calls "The solid rule of civil government" has done wonders for the race that nurtured and brought to perfection an idea spontaneous to their organization. The world has yet to learn that the admirable exotic will thrive amongst the country gentlemen of Monomotapa or the ragged nobility of Al-Hijaz. And it requires no prophetic eye to foresee the day when the Wahhabis or the Badawin, rising *en masse,* will rid the land of its feeble conquerors.

"Fate," he added in a footnote to this observation, "has marked upon the Ottoman Empire in Europe *'delenda est':* we are now witnessing the efforts of human energy and ingenuity to avert or to evade the fiat."[20] This constituted Burton's most explicit intervention in the politics of modernization: the "sick man of Europe"—

the Ottoman Empire (which was one manifestation of the East)—was not of Europe but in Europe, from where it inevitably must be ejected.

Second, modern imperialism could clean up the rude and dangerous aspects of Arabia. While Burton romanticized the East, he was always sensitive to the role Western imperialism could play for the mutual benefit of both Arabs and Westerners. Just as Arabs ("real" Easterners, as opposed to modernized city dwellers) contributed to an uninhibited expression of European manhood, proper authoritarian rule would "in one generation" rid Arabia of its bandits, who threatened travelers and pilgrims and prevented them from enjoying the regenerative splendors of the desert. Burton wrote, "By a proper use of the blood feud; by vigorously supporting the weaker against the stronger classes; by regularly defeating every Badawi who earns a name for himself; and, above all, by the exercise of unsparing, unflinching justice, the few thousands of half-naked bandits, who now make the land a fighting field, would soon sink into utter insignificance" (258–259).

By doing both these things—by preventing non-Western modernization and by pacifying (and preserving) Arabia—modern imperialism could maintain a diachronic history, a spatial and temporal segregation that would benefit the civilized West and the timeless East. Burton's ideal was to combine two radically different experiences rather than, as Aali Pasha or Jas Brant sought to do, to homogenize them. Herein lies the paradox that runs through the entire narrative of his journey: his commitment to being a true British man and his desire to maintain the pretense of being the Muslim al-Hajj Abdullah, the Indian *darwish* (literally, a "dervish," "a simple but pious man").

Conclusion

Both Brant and Burton were committed to modern imperialism, but for radically different reasons. One tried to hasten the process of allegedly universal European modernization and the obliteration of cultural difference; the other, to retard it in order to maintain a primordial Eastern refuge at the disposal of Western men. A principal dynamic that informed both their writings, however, was a rejection of any possibility of autonomous non-Western modernization as it was articulated by Ottomans, such as Aali Pasha, who tried to Europeanize (administratively, architecturally, bureaucratically, civilly, and militarily) their empire, and by the khedives of Egypt, who tried to make Egypt modern. In short, by rejecting non-Western modernization as a travesty, as base emulation devoid of sincerity, Burton and Brant effectively dismissed the hopes and aspirations of many hundreds of thousands of Ottomans (and Egyptians) who made increasingly strident claims to modernity. The more modern the Ottoman Empire and Egypt became—the more they began to function and to narrate within a discourse of modernization that threatened to bridge the metaphorical gap inherent in an imperialist geography of modernization—the more intense became European reactions.

Following Brant, Western powers, epitomized by Lord Cromer, effectively the British ruler of Egypt in the decades after 1892, became more committed to imperialism as the only possible way forward for the East.[21] Egypt was occupied by Britain in 1882, and the Ottoman Empire was ultimately destroyed and partitioned into "mandates" by Britain and France in 1920. Yet following Burton, a romantic search for the Eastern premodern became more widespread and urgent in the face of the construction of modern Cairo, Beirut, and Istanbul and the accompanying articulation of anticolonial nationalist movements. At the same time, this search was paradoxically facilitated by the process of non-Western modernization itself. In the second half of the nineteenth century, rail links and steamer services and tourist agencies and hotels brought and catered to an ever larger number of tourists satisfied to do thoroughly managed tours of the "medieval," "authentic," and "unspoiled" East.[22] Modernization, however, also brought in its wake a number of travelers, such as Charles Doughty, who set off on their own into the deserts of the East, not as tourists but as explorers determined to find and experience the "real" and "unmediated" East.[23]

The collapse of the Ottoman Empire following World War I heralded the end of a long nineteenth-century Western romantic involvement with the Orient. It marked, as well, the final incorporation of the East into an imperialist geography of modernization. The Sykes-Picot agreement, a secret 1916 pact between Britain and France to partition the Arab provinces of the Ottoman Empire into British and French spheres of influence—despite prior British promises to aid the Arabs in their quest for independence—anticipated, indeed, laid the groundwork for, the formal imposition of mandates at San Remo in 1920.

Already on the eve of World War I, railroads had utterly changed the map of pilgrimage (a point noted by Doughty in *Arabia Deserta*), but following the conflict, those areas directly subjected to British (Palestine and Iraq) or French (Syria and Lebanon) rule were "opened" up to colonial development. Even Arabia itself—which for so long represented a space for putatively manly pilgrimages for travelers such as Burton—was soon radically transformed by the discovery of oil.[24] The East, in short, no longer represented a site of Western fantasy, a space apart from Western modernity, and a place where one could contest the strictures, norms, and morals of metropolitan culture. Rather, it became an inextricable part of an imperial and strategic landscape.

No figure better represents this final triumph of an imperialist geography than T. E. Lawrence. As an author and a soldier, Lawrence resolves the two apparently contradictory strands of modern imperialism. He is at once the romantic explorer alone in Arabia and the loyal foot soldier of British interests. He is immersed in a search both to know and befriend the Arabs and to lead the Arabs into modernity. "I meant to make a new nation" writes Lawrence, only to add:

> It was evident from the beginning that if we won the war these promises [to the Arabs] would be dead paper, and had I been an honest advisor of the Arabs I would have advised them to go home and not risk their lives fighting for such stuff; but I salved myself with the hope that, by leading these Arabs madly in the final victory I would establish them, with arms

in their hands, in a position so assured (if not dominant) that expediency would counsel to the Great Powers a fair settlement of their claims. In other words, I presumed (seeing no other leader with the will and power) that I would survive the campaigns, and be able to defeat not merely the Turks on the battlefield, but my own country and its allies in the council-chamber. It was an immodest presumption: it is not yet clear if I succeeded: but it is clear that I had no shadow of leave to engage the Arabs, unknow-ing, in such hazard. I risked the fraud, on my conviction that Arab help was necessary to our cheap and speedy victory in the East, and that bet-ter we win and break our word than lose.[25]

This remarkable confession from Lawrence captures the empowering sense of ro-mance that drove him into an inexorable fulfillment of national and imperial duty. The doubt expressed by Lawrence when he admits that "I had no shadow of leave" to deceive the Arabs is expunged by his final statement that the fraud was necessary, and that victory had to come, however high the price.

The price, of course, was a triumph, as the title of his account of the Ara-bian campaign declares. A triumph of what and over whom? A triumph of British imperial interest, certainly, over the Ottomans, but also over the Arabs and, ironi-cally, over an imaginary and romantic Orient. But more important, Lawrence's nar-rative reflected the complete triumph of an imperialist geography over non-Western claims of equality and parity on the world stage of modernization. "At a time when Western Europe was just beginning to climb out of nationality into international-ity, and to rumble with wars far removed from problems of race," Lawrence notes, "Western Asia began to climb out of catholicism into nationalist politics, and to *dream* of wars of self-government and self-sovereignty, instead of for faith or dogma" (45; emphasis added). Perhaps nothing was more symbolic of this triumph than Lawrence's destruction of parts of the Hijaz Railway, that great monument to Ottoman modernization. The scene Lawrence describes of the accompanying mas-sacre of the Turkish soldiers defending it is evocative, as the once proud Turkish officers, now defeated, come groveling to his feet in a desperate effort to save their lives. All Lawrence can say is, "A Turk so broken down was a nasty spectacle: I kicked them off as well as I could with bare feet, and finally broke free" (369). There was no longer any Orient to escape to, only one to govern, to rule, to domi-nate, and to stabilize; there was no longer any Orient to explore, only one to exploit. The Orient was finally and fully mapped.

Notes

1. Edward Said, *Orientalism* (New York: Vintage, 1979). The classic account of the con-tinuity of early Christian polemic against Islam and of the elaboration of a medieval European "cannon" on Islam is Norman Daniel's study, *Islam and the West: The Mak-ing of an Image* (Edinburgh, U.K.: University Press, 1958).
2. Saree Makdisi, *Romantic Imperialism: Universal Empire and the Culture of Moder-nity* (Cambridge: Cambridge University Press, 1998), 9–10; Nigel Leask, *British Ro-*

mantic Writers and the East: Anxieties of Empire (Cambridge: Cambridge University Press, 1992), 10.

3. Johannes Fabian, *Time and the Other* (New York: Columbia University Press, 1983), 17.

4. Richard Knolles, *The Generall Historie of the Turkes, from The first beginning of that Nation to the rising of the Othoman Familie: with all the notable expeditions of the Christian Princes against them Together with The Lives and Conquests of the Othoman Kings and Emperours unto the yeare 1610* (London: Adam Islip, 1610), 1.

5. Said, *Orientalism,* 86–87.

6. Makdisi, *Romantic Imperialism,* 124 (emphasis in original).

7. The suppression of the Janissaries, referred to by Ottoman chroniclers as the "auspicious event," actually occurred in 1826 during the throes of the Greek war of independence but before the *Tanzimat* was formally inaugurated in 1839. However, the suppression should be seen as one in a series of "modernizing" efforts that came to see their fullest form during the *Tanzimat.*

8. "Mémoire transmis à Londres et à Paris par Aali Pasha," May 1855, Archives du Ministère des Affaires Étrangères, Paris, AE MD/T series, Mémoires et Documents, Turquie, 1840–63, vol. 51, no. 9.

9. See Makdisi, *Romantic Imperialism,* 113–115.

10. See Selim Deringil's *The Well-Protected Domains: Ideology and Legitimation of Power in the Ottoman Empire, 1876–1909* (London: I. B. Tauris, 1998) for more on this campaign.

11. Kenneth Bourne and D. Cameron Watt, eds., *British Documents on Foreign Affairs: Reports and Papers from the Foreign Office Confidential Print,* pt. 1, *From the Mid-Nineteenth Century to the First World War,* ser. B, *The Near and Middle East, 1856–1914,* ed. David Gillard, vol. 1, *The Ottoman Empire in the Balkans, 1856–1875* ([Frederick, Md.]: University Publications of America, 1984), 8. Hereafter referred to as *BDFA.*

12. In 1869, of course, the Suez Canal was officially opened under the patronage of Khedive Ismail of Egypt, who desired to make Cairo the "Paris on the Nile."

13. *BDFA,* 11.

14. Ibid.

15. See Douglas Sladen's *Oriental Cairo: The City of the "Arabian Nights"* (Philadelphia: J. B. Lippincott, 1911) for a good example of this reaction. It was precisely during the most active period of non-Western modernization in Egypt that "medieval" Cairo was rediscovered by European tourists and archaeologists.

16. Said, *Orientalism,* 196; see also Makdisi, *Romantic Imperialism,* 120.

17. For the case of Egypt, see Timothy Mitchell, *Colonizing Egypt* (Berkeley: University of California Press, 1991). For the Ottomans, see Erich Zürcher, *Turkey: A Modern History* (London: I. B. Tauris, 1993), 52–74.

18. Richard 0F. Burton, *Personal Narrative of a Pilgrimage to Al-Madinah and Meccah* (New York: Dover Publications, 1964), 1:1.

19. Said, *Orientalism,* 195; Patrick Brantlinger, *Rule of Darkness: British Literature and Imperialism, 1830–1914* (Ithaca, N.Y.: Cornell University Press, 1988), 164.

20. Burton, *Personal Narrative,* 1:259.

21. Evelyn Baring [Earl of Cromer], *Modern Egypt* (London: Macmillan, 1911), 6.

22. See appendix 1 of Sladen, *Oriental Cairo,* 351–353.

23. See Charles M. Doughty, *Travels in Arabia Deserta,* 2 vols. (Cambridge: Cambridge University Press, 1888).

24. See Abdelrahman Munif, *Cities of Salt* (New York: Vintage, 1989), for an account of this process of transformation.

25. T. E. Lawrence, *Seven Pillars of Wisdom: A Triumph* (New York: Anchor Books, 1991), 25–26.

"Water Leaves No Trail"

Mapping Away the Vanishing American in Cooper's Leatherstocking Tales

David C. Lipscomb

$\mathcal{H}$istorical novels construct "biographies of nations," life passages from earlier national selves.[1] In the words of the genre's best-known critic, Georg Lukács, historical novels bring "the past to life as the prehistory of the present."[2] But as a kind of national bildungsroman, the historical novel raises a few questions that the novel of individual formation does not have: namely, Who is going to represent the earlier national self? and Where is this past to be located?

In the heyday of the genre—the early nineteenth century—new tools for reading time on geographic space helped historical novelists answer these questions. New theories about the development of "savage" peoples in distant lands and the extinction of plant species buried in rocks—together with the popularity of picturesque ruins—meant that the Bible was no longer the only place to read the passage of historical time. Geographic space now also told time. Indeed, the invention of the chronometer in the eighteenth century had made it not only romantic but also practical to read time across geographic space, a time-telling approach made even more irresistible when Harrison's clock became the accepted method for calculating longitude, as Ronald Thomas has shown in this volume. For historical novelists, these new time-telling tools meant that the movement from national past to national present could be rendered as a journey across geographic space within the manageable frame of a hero's lifetime—that is, if the novelist could find the right setting.

Just as philosophical and scientific developments had increased the temporal potential of a novel's setting, early-nineteenth-century political realities—including imperialism—had already inscribed some settings with particular problems and associations that a historical novelist could not ignore. Geography was an especially complex problem for the historical novelist James Fenimore Cooper, writing his fictional biographies of the American Republic when the nation was still young, geographically undefined, and haunted by the "removal" of the land's "savage" occupants. But Cooper was a gifted mythographer of national history who knew not only the tools for reading time on geographic space but also the particular history written on the map of the American frontier. By refashioning contemporary time-telling tools to fit the American political landscape, Cooper developed an imaginative geography of fluvial forgetfulness and a hero who was as much a path-eraser as a pathfinder.

The most famous of the new time-telling tools, the chronometer, made it possible to measure longitude from the difference between the local time and the meridian time at the Royal Observatory Greenwich in London. Although there were several acknowledged prime meridians in the late eighteenth and early nineteenth centuries and Greenwich would not officially become the international standard meridian until 1884, it was by convention the meridian used most often by navigators. By the 1820s, even school-age children were learning to see London as time zero on a map of the world. American readers of William Woodbridge's popular school atlas, for example, learned not only specific longitudinal points but also how to measure longitude. Along the bottom of Woodbridge's "Moral and Political Chart of the Inhabited World" (Figure 1)—a chart that appeared in at least six editions of *Woodbridge's School Atlas* in the 1820s and 1830s—the cartographer supplemented the more common numerical degree notations with the descriptive phrases "Hours before noon" and "Hours after noon in London."

Woodbridge's "Moral and Political Chart," however, teaches more than one method of telling time on the map of the world. The chart and the accompanying text of the school atlas show students how to measure cultural time in terms of "degrees of civilization" away from London, as well as how to measure longitude in terms of hours away from London. Translating the Scottish Enlightenment's stage theory of social development into a five-tiered system of shading, Woodbridge divides the world into "Savage," "Barbarous," "Half Civilized," "Civilized," and "Enlightened" countries.[3] The "Savage" countries of the world are marked with the darkest shade, whereas the spaces of the "Enlightened" nations are marked only with beams of light radiating outward from their centers. To reinforce this imaginative geography, the text of the atlas contains study questions, including the following: "What parts of the Western Continent are in the Savage state? What of the Eastern? Is any Part of Europe in the Savage state? What countries are Barbarous? Are any of the countries of America Barbarous? What countries of the world are Half Civilized?"[4] Looking back at the chart, a student learns that the western and northern sections of North America, the northern sections of Asia, and the very centers of Africa and South America remain in the "Savage" state; according to the chart, many parts of Asia and Africa have barely emerged from the earliest state of civilization to a "Barbarous" state; meanwhile, the countries of South Asia are "Half Civilized." In the advertisement that accompanies the various editions of *Woodbridge's School Atlas,* the author claims that his "Moral and Political Chart" is "unlike any ever executed before" in the way it attempts "to present the great features of Physical, Political, and Statistical Geography *to the eye,* with as much clearness as circumstances admit, and thus produce the most distinct and perma-

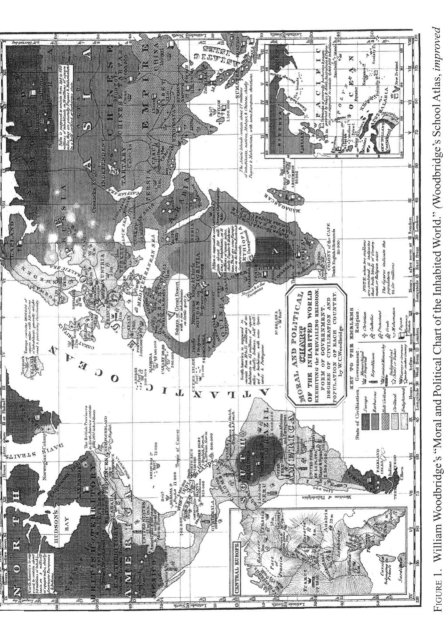

FIGURE 1. William Woodbridge's "Moral and Political Chart of the Inhabited World." (Woodbridge's School Atlas, *improved ed. [Hartford, Conn.: Oliver D. Cook, 1830], Library of Congress, Map Division.*)

nent impression on the mind" (italics in the original). In reference to his particular method of shading, Woodbridge's claim for originality appears accurate. Yet his chart is just one among many late-eighteenth-century and early-nineteenth-century cultural productions that attempted to spatialize the Scottish Enlightenment's stages of development *to the eye,* or at least to the eye of the mind.[5]

Edmund Burke gives perhaps the most poetic description of the new method of telling time enabled by the stage theory of development. Writing in 1788 to congratulate the Scottish Enlightenment historian William Robertson on his *History of America,* Burke declared Enlightenment stage theory superior to traditional history because one could now read history synchronically across the globe.

> I have always thought with you, that we possess at this time very great advantages towards the knowledge of human Nature. We need no longer go to History to trace it in all its stages and periods. History from its comparative youth, is but a poor instructour. But now the Great Map of Mankind is unrolld at once; and there is no state of Gradation of barbarism, and no mode of refinement which we have not at the same instant under our View. The very different Civility of Europe and of China; The barbarism of Tartary, and of arabia. The Savage State of North America, and of New Zealand.[6]

Burke's description of the "Great Map of Mankind" applies to William Woodbridge's map just as much as it applies to Robertson's *History:* both works bring each "state" of social development "at the same instant under our View." In both cases, readers are given an imaginative geography that lets them measure *when* they are, in relation to those peoples living in different stages of social development.

Like the cartographer Woodbridge and the historian Robertson, novelists of the Romantic period offered their own "great maps of mankind." In the Jacobin romances of Elizabeth Inchbald and Robert Bage, noble savage heroes from Sierra Leone and the Sioux territory in the United States become time travelers as they journey from locations of "savagery" to England, where the purity of their primitive education brings out the corruption of advanced society; and in the national tales of Sydney Owenson (Lady Morgan) and Maria Edgeworth, English heroes, usually absentee landlords, travel backwards in time away from London toward the west coast of Ireland, where they find versions of what England was in simpler and more primitive stages of the civilization process.

In 1814, Sir Walter Scott's *Waverley* launched the genre of the historical novel, which takes as its central theme the analogy between growing up and becoming civilized. When Edward Waverley marches with the rebellious Jacobite army from the Highlands of Scotland across the Lowlands and into England, his geographic movement signals a host of temporal movements, including his personal growth from youth to maturity, the novel's transformation from romance to realism, and the nation's movement from savagery to civilization.[7] Scott is especially careful that his readers see his island map of social evolution in relation to its larger-scale model, what Burke had called the "Great Map of Mankind." When the Highlanders first march into the Lowlands, Scott's narrator says that they "conveyed to the south-country Lowlanders as much surprise as if an invasion of African Negroes or Esquimaux Indians had issued forth from the northern mountains of their native country."[8] Although Scott was not the first British writer to read the British island's tripartite topography (England, Scottish Lowlands, Scottish Highlands) in terms of Scottish Enlightenment stages of development,[9] Scott perhaps did more than any other writer to make British readers see the nation's map as an analogue for the nation's biography. Most importantly, Scott enabled his readers to measure their nation's progress in terms of their own geographic distance from a past located in the Highlands.

As Saree Makdisi points out, it is an *imaginary* past that Scott maps out for his readers: "[*Waverley's*] imaginary map of the Highlands is not, strictly speaking, a map of the past, but rather of a possible past, an imaginary past that is forever spatially (and temporally) different and distinct. It is a past that can never become present because it cannot be modernized and remain identical to itself— it is necessarily anti-modern."[10] *Waverley,* like many of Scott's subsequent novels, ends with a marriage between the heiress of a Lowland Scottish estate and the English hero; but the novel's Highland characters take no part in the happy endings. Savagery, romance, and the symbolic youth of both the hero and the nation are all securely placed in the imaginary past, on the other side of the Highland Line.

In the United States, no Highland Line set one region apart to provide American imitators of Scott, such as James Fenimore Cooper, with a locus for the nation's symbolic past. "[I]n Western Europe," George Dekker notes, "the stages [of social development] were separated from each other by long periods of time or by formidable topographical barriers like the Scottish Highland Line." So in translating Scott's imaginative geography of Enlightenment stages to American soil, the novelist faced a cartographic problem. "In nineteenth-century America," Dekker continues, "the stages were often thought of as being distributed across American space like colors across a spectrum."[11] In addition, the American representatives of the "savage" stage could not be associated with primarily one geographic region, as could the Scottish Highlanders; instead, Native Americans had been displaced (and were still being displaced) from *every* geographic region.

Perhaps nothing illustrates the problem better than a famous American map from the middle of the eighteenth century, Lewis Evans's "A General Map of the Middle British Colonies"[12] Originally printed in 1755 at the beginning of the French and Indian War and then continuously revised and reprinted for the next fifty years, Evans's map contains an enormous amount of information about the history of territorial dispossession in America. Evans attempts to show the layeredness of American space. In the original edition, the letters and symbols that denote British settlements are set alongside and sometimes atop blue italic crosshatched capital letters that spell out the names of Native American peoples who formerly inhabited the settlement areas; some areas contain white italic capitals that spell out the names of extinct Native American peoples (Figure 2). In the 1771 edition printed by Carrington Bowles in London (Figure 3), the italic capitals of Native American names are appended with "*an S*" or "*ex S*" to indicate

FIGURE 2. Detail from Lewis Evans's "A General Map of the Middle British Colonies in America," 1755. *(Library of Congress, Map Division.)*

Native American "ancient seats" and "those that are extinct," respectively, as the "Explanation" from this edition makes clear (Figure 4). Starting just above Philadelphia and running along the Delaware River, for example, the italic letters *LENE LENAPE an S* mark the area as the former home of the people later known as the Delaware; and the letters *WAMPONOACS an S* extend across southeastern Massachusetts, indicating the territory that this people occupied before King Phillip's War (Figure 5).

On a map such as this, how and where was Cooper to construct his narratives about the nation's coming of age, its passage from savagery to civilization? In his first Leatherstocking Tale, *The Pioneers* (1823), Cooper offered a solution similar to Evans's map.[13] He limited his geographic scope to one upstate New York settlement, Templeton—a fictionalized version of his own family's settlement at Cooperstown. Starting from a present in the 1790s, he traced the land's passage backwards from its Republican owner (Temple) to its pre–Revolutionary War Loyalist owner (Major Effingham), who had in turn been granted the land from the local Delaware tribe. This last link in the chain of property transfers is especially intricate: it turns out that Major Effingham had saved the life of John Mohegan (elsewhere called Chingachgook), and so Effingham is adopted into the Delaware nation as "the son of Mohegan" and is therefore granted the land. The adoption involves no racial "mingling" (a constant source of mystery and tension in the novel, as Major Effingham's grandson, Oliver Edwards, is rumored to have "Indian blood"), but it does secure the land as an *inheritance*. Thanks to the marriage—between Loyalist (Oliver Edwards Effingham) and Republican (Elizabeth

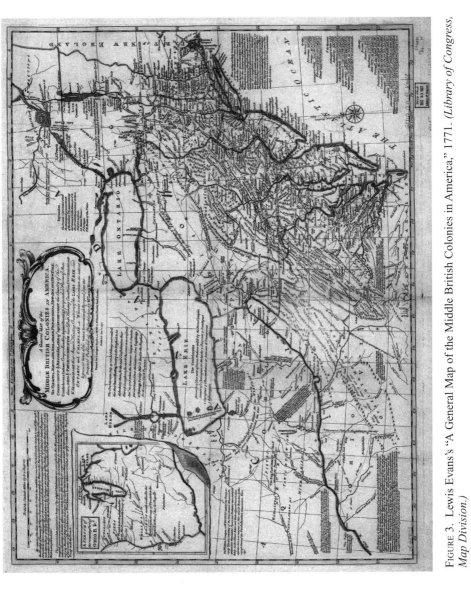

FIGURE 3. Lewis Evans's "A General Map of the Middle British Colonies in America," 1771. *(Library of Congress, Map Division.)*

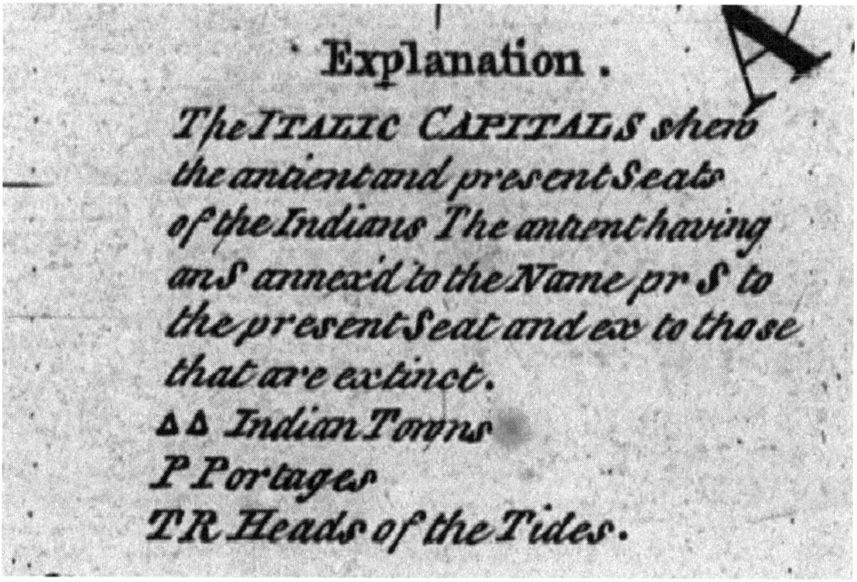

FIGURE 4. Detail from Lewis Evans's "A General Map of the Middle British Colonies in America," 1771. *(Library of Congress, Map Division.)*

Temple)—that closes the novel, the land will continue to function as an inheritance after the novel ends. This convoluted and implausible formulation attempts to write American history as biography, one in which the land could be imagined as inherited from adoptive American ancestors, the land's first Americans.

But the attempt fails, and not simply because the adoption lacks credibility. *The Pioneers* asks too much of a reader's imagination, spatially as well as historically. By crowding onto the same space characters from different stages of history, Cooper, like Lewis Evans before him, was trying to overcome a problem that Freud, in *Civilization and Its Discontents,* would declare unsolvable. "If we want to represent historical sequence in spatial terms," Freud says, "we can only do it by juxtaposition in space: the same space cannot have two different contents." After considering whether or not the different historical periods of Rome could be represented in one image, Freud concludes that "only in the mind is a preservation of all earlier stages alongside of the final form possible, and we are not in a position to represent this phenomenon in pictorial terms."[14] The early editions of Evans's map survive as exceptions to Freud's rule. Yet only the early editions are exceptions. In the later revised editions of Evans's map, the layers of American Indian settlements gradually disappear.[15]

By the Jacksonian Age, the era of "Indian removal," Anglo-Americans would have no use for a landscape cluttered with Lewis Evans's "ancient seats"—and neither would Cooper. Although the geographic setting of *The Pioneers* offered no discrete and distant space where Cooper could put the past, it did provide him with a place to eliminate it: Lake Otsego, the source of the Susquehanna River. In

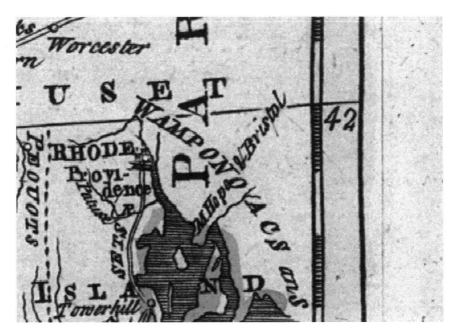

FIGURE 5. Detail from "A General Map of the Middle British Colonies in America," 1771. *(Library of Congress, Map Division.)*

The Pioneers, the breakup of the ice on Lake Otsego works as an organic analogue for the historical dispossession of the Native Americans. His description of this "natural" process in *The Pioneers* is worth quoting at length:

> For a week, the dark covering of the Otsego was left to the undisturbed possession of two eagles, who alighted on the centre of its field, and sat eyeing their undisputed territory. During the presence of these monarchs of the air, the flocks of migrating birds avoided crossing the plain of ice, by turning into the hills, apparently seeking the protection of the forests, while the white and bald heads of the tenants of the lake were turned upward, with a look of contempt. But the time had come, when even these kings of birds were to be dispossessed. An opening had been gradually increasing, at the lower extremity of the lake, and around the dark spot where the current of the river prevented the formation of ice, during even the coldest weather; and the fresh southerly wind, that now breathed freely upon the valley, made an impression on the waters. . . . At each step the power of the winds and the waves increased, until, after a struggle of a few hours, the turbulent little billows succeeded in setting the whole field in motion, when it was driven beyond the reach of the eye, with a rapidity, that was as magical as the change produced in the scene by the expulsion of the lingering remnant of winter. Just as the last sheet of agitated ice was disappearing in the distance, the eagles rose, and soared with a

wide sweep above the clouds, while the waves tossed their little caps of snow into the air, as if rioting in their release from a thralldom of five months' duration.[16]

This little fable of seasonal dispossession is set on a fluid space, where the normative state for both the territory and its occupants (eagles and migratory birds) is *motion*. The ice can impede that motion only temporarily, and each creature is in its turn dispossessed. When Cooper returns to this water fable at the end of the novel, he identifies the Delaware John Mohegan with the ice itself. Moments before his death, Mohegan declares that his own passage is natural: "Where is the ice that covered the great spring?" he asks. "It is melted, and gone with the waters. John has lived till all his people have left him for the land of the spirits; his time has come, and he is ready" (402). His time has come, but his people have already "gone with the waters." How, exactly, did they go? Returning to the description of the breaking of the ice, we can see that it was "a struggle of a few hours," a clearing operation that drove the ice "beyond the reach of the eye" in order to "release from a thralldom" the spring's water and set nature's course in motion.

In each of the subsequent Leatherstocking Tales, Cooper gives his readers just such a clearing operation. After *The Pioneers,* Cooper drops his attempt to write national history as an inheritance plot, as well as his attempt to represent several stages of historical development in one place. Instead, he sets the remaining Leatherstocking Tales at the moment of frontier struggle, the moment in history that Cooper calls "rescuing the region from the savage state."[17] Through plots that depict "the expulsion of the lingering remnant," his novels symbolically transform the richly layered surface of American terrain into "virgin land."

Instead of a regeneration through inheritance, Cooper's subsequent Leatherstocking Tales offer American audiences what Richard Slotkin has called "regeneration through violence."[18] At the "healing waters" of Balston spa in *The Last of the Mohicans* (1826), for example, Natty Bumppo and his two Mohican companions release from thralldom the white woman who has been taken captive by the villainous Magua and his Huron followers; in a brief struggle, they kill five Hurons and chase Magua from the scene. To make sure readers understand exactly what has taken place, Cooper points directly at the topographical analogue. After the Hurons have been cleared away, Natty Bumppo leads the newly reunited band to the spring, where, he notes, "the knaves have trodden in the clay, and deformed the cleanliness of the place, as though they were brute beasts, instead of human men." But not to worry. The Huron footprints in the clay are not permanent. "Cleanliness" is quickly restored after Natty Bumppo, Chingachgook, and Uncas "commenced throwing aside the dried leaves, and opening the blue clay, out of which a clear and sparkling spring of bright, glancing water, quickly bubbled."[19] Just in case the reader has not yet made the connection between the two clearing operations, Cooper closes the chapter with a description of what, exactly, Natty and his companions have accomplished: "The whole party moved swiftly through the narrow path, towards the north, leaving the healing waters to mingle unheeded with the adjacent brook, and the bodies of the dead to fester on the neighboring mount" (123). This passage echoes an earlier description in *Last of the Mohicans* that is placed directly after Natty Bumppo, Chingachgook, and Uncas have killed at least six Hurons at Glens Falls: "The uproar which had so lately echoed through the vaults of the forest was gone, leaving the rush of the waters to swell and sink on the currents of the air, in the unmingled sweetness of nature" (81). In both scenes the natural course of mingling waters is left unmingled with the course of man. What has just taken place in the course of human history is a clearing operation: nothing more than that. The characters in the novel do not participate in the subsequent stages of history. The "uproar" and all the characters who made the uproar are gone from the site when the flow of history they have enabled continues along its course. The characters exit, "gone with the waters."

Occasionally, Natty Bumppo is tempted to remain for the next stage of history. At the end of *The Deerslayer* (1841), after the Leatherstocking hero has helped clear the Lake Otsego area of Hurons, the heroine, Judith Hutter, tries to convince him to stay. "This lake will soon be entirely deserted," she says, "and this, too, at a moment when it will be a more secure dwelling place than ever."[20] But Natty Bumppo does not take the bait. Whereas the typical Waverley novel ends in marriage, the typical Leatherstocking Tale, after *The Pioneers,* ends in massacre. Once Natty Bumppo can leave a space "entirely deserted" for the benefit of a future in which "it will be a more secure dwelling place than ever," he will move on. "I do not mean to pass this-a-way, ag'in," he tells the disappointed Judith Hutter, "for to my mind no Huron moccasin will leave its print on the leaves of this forest, until their traditions have forgotten to tell their young men of their disgrace and rout" (539). The Pathfinder tracks down the Native American presence and eliminates it; at novel's end, he is satisfied knowing that the land will suffer no moccasin print to muddy the next stage of development. In Cooper's novels, the violence of frontier clearance is a discrete moment, discontinuous with the history that follows.

What is perhaps most extraordinary in Cooper's novels is that this stage of history leaves no trace: no footprints, no inheritance, nothing. One day after Natty Bumppo and British troops massacre the Huron village at the end of *The Deerslayer,* the event itself is erased from the fluid landscape: "When the sun rose on the following morning, every sign of hostility and alarm had vanished from the basin of the Glimmerglass [Lake Otsego]. The frightful event of the preceding evening had left no impression on the placid sheet" (523). The Hurons are cleared away and so too is the force that has done the clearing. Fifteen years later, when Natty Bumppo returns to the lake, he notes that more clearing has been done. All traces of the novel's events have now been, or are in the process of being, washed away. Not only are there no traces of that "frightful event," but now the shallow-water graves of two main characters who died violently during the novel have disappeared. Even the floating log fortress that had been the center of action for the main characters is now caught in the final stages of the "blotting" process: "The palisades were rotting, as were the piles, and it was evident that a few more recurrences of winter, a few more gales and tempests, would sweep all into the lake, and blot the building from the face of that magnificent solitude. The graves could not be found. Either the elements had obliterated their traces, or time had caused those who looked for them, to forget their position" (546). The novel's various events, frightful or otherwise, leave "no impression," no print of any kind. By novel's end, the geographical setting is suddenly virgin territory as the earth is repeatedly cleansed by the waters that make it new again and again.

After the massacre of a Sioux village at the end of *The Prairie* (1827), the focus is once again on the obliteration of traces with the next dawn:

> The day dawned, the following morning, on a more tranquil scene. The work of blood had entirely ceased, and as the sun arose its light was shed on a broad expanse of quiet and solitude. Here and there, little flocks of ravenous birds were sailing and screaming above those spots where some heavy-footed Teton had met his death, but every other sign of the recent combat had passed away. The river was to be traced far through the endless meadows by its serpentine and smoking bed, and the little silvery clouds of vapor. . . . The Prairie was, like the heavens after the passage of the gust, soft, calm, and soothing.[21]

Like the passage of a gust, the traces of characters and events vanish from the landscape, leaving no trace. How very different is Cooper's traceless American landscape from the landscape mapped by Lewis Evans.

Yet how is such a traceless landscape possible? The answer lies in the element that is common to all the above scenes of violence in the Leatherstocking Tales: water. Cooper turns Enlightenment developmental history into national history by staging the historical dispossession of the "savage state" on or alongside American watercourses. On Cooper's fluvial landscapes, history becomes natural history. The flow of water, often a metaphor for history itself, has a particular quality in Cooper's fiction: historical *erasure*. Important watercourses set the stage and *clean up the stage* in every Leatherstocking tale: the Hudson (*Last of the Mohicans*), the Missouri (*The Prairie*), the St. Lawrence (*The Pathfinder*), and the Susquehanna (*The Pioneers* and *The Deerslayer*). In fact, Cooper's novels never move far away from water.

Cooper's hero tells the reader why. "Water," notes the Pathfinder, "is the only thing in natur that will thoroughly wash out a trail."[22] As Natty Bumppo tells the reader again and again, "water leaves no trail." Natty Bumppo repeats these words in every novel after the first, and in *The Pathfinder* they become a kind of mantra as he repeats them at least four times.[23] The Pathfinder is, in fact, an expert path-eraser. He knows what to do to make sure no trace is left behind. When on the run in *The Prairie,* Natty Bumppo admonishes his fellow characters, "[I]t is needful to wash our trail in water."[24] Nothing else works quite as well. Water is the only "scene" that can thoroughly wash out the "sign" and so resolve the conflict between "the scene of nature" and "the sign of man" that Tony Tanner says perplexes American fiction.[25]

In Cooper's novels, water functions as a chronotope, but its function is far different from the chronotopes, such as the road and the Gothic castle, that Bakhtin discusses in his famous essay on the subject, "Forms of Time and of the Chronotope in the Novel." In the Spanish picaresque novel, Bakhtin says, "the road had been profoundly, intensely etched by the flow of historical time, by the traces and signs of time's passage, by markers of that era."[26] Far from being etched by historical time, the rivers and other watercourses in Cooper's novels wash history away.

But they do not wash it away until after Cooper's narratives have recorded the history that is to be erased. Cooper's narratives often recall the very process of erasure. After Natty Bumppo kills one Iroquois at the "Great Falls" in *Last of the Mohicans,* Cooper describes the disappearance of the body as if the author were wielding a slow-motion camera: "[T]he body parted the waters, like lead, when the element closed above it, in its ceaseless velocity, and every vestige of the unhappy Huron was lost for ever."[27] With careful detail, Cooper remembers the loss of the trace. In other words, Cooper's fluvial chronotope enables him to recall the "first" Americans and simultaneously forget them, their "ancient seats," and the process that unseated them.

By remembering and forgetting the first stage of historical development, Cooper invites his readers to do the same. He maps the New World with bodies of water being the only forms of inscription on the earth. Cooper's early readers thus had the advantage of Woodbridge's map without the disadvantages of Evans's; they could measure how far the nation had come from the "Savage" stage of civilization without having to see themselves atop a terrain etched with the history of earlier stages. In the Jacksonian Age, Roy Harvey Pearce says, Americans needed the Native American as a symbol of the past even while they were eliminating him from the landscape: "[T]he Indian was the remnant of a savage past away from which civilized men had struggled to grow. To study him was to study the past. . . . To kill him was to kill the past."[28] To wash away his footprints, Pearce might have added, was to wash away the past. Cooper's imaginative geography of fluvial forgetfulness was enormously useful. In his second annual message (6 December 1830), President Andrew Jackson admonished the nation to practice "true philanthropy" by accepting the inevitable loss of the Native Americans: "To follow to the tomb the last of his race and to tread on the graves of extinct nations excite melancholy reflections. But true philanthropy reconciles the mind to these vicissitudes as it does to the extinction of one generation to make room for another."[29] In an age of active government "removal," true philanthropy turns out to be reconciling the mind with the process of removal, a process Jackson says is as natural as the passing of one generation "to make room" for the next. James Fenimore Cooper was thus an exemplary true philanthropist. His fictions reconcile the memory of Native Americans with their loss from the scene; he renders the process of making room for the next stage of civilization as natural as the flow of a river.

While Cooper's particular imaginative geography helped him meet a specific cultural need, his pattern of remembering and forgetting is typical of nineteenth-century national fictions. "[T]he essence of a nation," according to Ernest Renan, "is that all individuals have many things in common, and also that they have forgotten many things."[30] Benedict Anderson, revising Renan, says that remembering national tragedies that one is simultaneously obliged to forget is a vital characteristic of national genealogies. In fact, Anderson sees Cooper's *The Pathfinder* as typical of such genealogies:

> In 1840, in the midst of a brutal eight-year war against the Seminoles in Florida, James Fenimore Cooper published *The Pathfinder,* the fourth of his five, hugely popular, Leatherstocking Tales. Central to this novel (and to all but the first of its companions) is what Leslie Fiedler called the "austere, almost inarticulate, but unquestioned love" binding the "white" woodsman Natty Bumppo and the noble Delaware chieftain Chingachgook ("Chicago"!). Yet the Renanesque setting for their bloodbrotherhood is not the murderous 1830s but the last forgotten/remembered years of British imperial rule.[31]

The last years of British imperial rule are also the forgotten/remembered moment of frontier clearance, the moment of "rescuing the region from the savage state" (in *The Prairie,* which takes place along the Missouri and its tributaries, the moment of frontier clearance is the time of the Lewis and Clark expedition). Only in this Renanesque stage of history is a familial relationship between a white man and a red man possible. Unlike the adoptive father-son bond between Major Effingham and Chingachgook in the first Leatherstocking Tale, the blood-brother bond between Natty Bumppo and Chingachgook suggests no generational continuity: no land could be inherited and no name could be left to clutter the terrain. Their bond is isolated to the first stage of American history.

Cooper's landscape makes certain that there is a disconnect between stages. In the words of the noble Tuscarora woman Dew of June, explaining the limits of her own knowledge to the white heroine in *The Pathfinder,* "Water got no trail— red man can't follow."[32] But white settlers can follow, secure to remember/forget the red man on Cooper's imaginative geography.

Although at odds with historical reality, Cooper's geography is remarkably consistent with at least one aspect of contemporary cartography—the erasure of Native American names nearly everywhere on the map except American rivers. As the spaces where Lewis Evans had once recorded Native American seats became increasingly crowded with the names of white settlements, the names of waterways—such as the Missouri and the Allegheny—continued to recall the Native American presence. In her 1834 poem entitled "Indian Names," Lydia Huntley Sigourney reminds readers how the rivers on American maps got their names:

> Ye say, they have all passed away,
>   That noble race and brave,
> That their light canoes have vanished
>   From off the crested wave;
> That mid the forests where they roamed
>   There rings no hunter's shout;
> But their name is on your waters,
>   Ye may not wash it out.[33]

While the poem's tone is defiant, its meaning is also ironic. The names are registered where no deeds or titles can make the names marketable or inheritable. The presence of American Indian names poses no threat to property if the names are mapped on water. Why wash them out? They can, in fact, be useful. Upon the rivers depicted on nineteenth-century American maps, Native Americans can be read and simultaneously washed away, both remembered and forgotten. As Cooper's readers know, water leaves no trail.

Notes

1. I am borrowing the phrase "biography of nations" from Benedict Anderson, *Imagined Communities: Reflections on the Origin and Spread of Nationalism,* rev. ed. (London: Verso, 1991), 204.
2. Georg Lukács, *The Historical Novel,* trans. Hannah and Stanley Mitchell (Lincoln: University of Nebraska Press, 1983), 53.
3. The best work on the Scottish Enlightenment's stage theory of social development remains Ronald Meek's *Social Science and the Ignoble Savage* (Cambridge: Cambridge University Press, 1976). See also Johannes Fabian's *Time and the Other: How Anthropology Makes Its Object* (New York: Columbia University Press, 1971); and Reinhart Kosellek's *Future's Past,* trans. Keith Tribe (Boston: MIT Press, 1985), for detailed studies of the ways that social evolutionary time is written on geographic space in the eighteenth and nineteenth centuries.
4. William Channing Woodbridge, *Woodbridge's School Atlas,* improved edition (Hartford, Conn.: Oliver D. Cooke and Co., 1831), n.p.
5. A year after Woodbridge's "Moral and Political Chart" first appeared, a variation on it was printed in London: "Clark's Chart of the World" (1822). Like Woodbridge's "Moral and Political Chart," "Clark's Chart" uses a five-part taxonomic system, but it employs color-coding instead of shading to delineate the states of civilization. As a result, the north of Asia, the centers of South America and Africa, and the top half of North America are all colored pink to indicate a "Savage" state; most of Africa and Asia is colored blue for "Barbarous" territory; India, China, and Japan are beige to indicate a "Half Civilized" state; the coastal regions of South America are light green to signal a "Civilized" state; and finally, two spaces are bright yellow—Europe and the east coast of North America, indicating that they are "Enlightened" ("Clark's Chart" is reproduced in Peter Whitfield, *The Image of the World: Twenty Centuries of World Maps* [San Francisco: Pomegranate Artbooks, 1994], 116–117). James Wyld's 1815 "Chart of the World Shewing the Religion, Population and Civilization of Each Country" is, I believe, the earliest map to order space according to the Scottish Enlightenment stages of development. In the key to his chart, Wyld explains his system for ordering geographic complexity: "The degrees of civilization are expressed by Roman Figures. The 1rst degree comprehending the Savage and proceeding up the scale to V, which includes the most civilized nations as England, France, &c." (Peter Barber and Christopher Board, eds., *Tales from the Map Room: Fact and Fiction about Maps and Their Makers* [London: BBC Books, 1993], 18–19).
6. Quoted in P. J. Marshall and Glyndwr Williams, *The Great Map of Mankind: British Perceptions of the World in the Age of Enlightenment* (London: Dent, 1982), ii.
7. Focusing on Edward Waverley's initial journey from England to the Highlands, Franco Moretti notes that the hero "travels backward through the various stages of social development described by the Scottish Enlightenment: the age of Trade, of Agriculture,

of Herding (the pretext for seeing the Highlands is a cattle raid), and finally of Hunting (the essence of Highland culture is embodied in Fergus' ritualized hunting party—which also coincides with the beginning of the rebellion)" (*Atlas of the European Novel, 1800–1900* [London: Verso, 1998], 38). Scott's use of Scottish Enlightenment theory, including his translation of geographic space into social evolutionary time, has received much attention in Scott criticism. See, for example, George Dekker, *The American Historical Romance* (Cambridge: Cambridge University Press, 1987), esp. 29–78; Avrom Fleishman, *The English Historical Novel: Walter Scott to Virginia Woolf* (Baltimore: Johns Hopkins Press, 1971), esp. 37–51; Duncan Forbes, "The Rationalism of Sir Walter Scott," *Cambridge Journal* 7, no. 1 (1953): 20–35; Peter Garside, "Scott and the Philosophical Historians," *Journal of the History of Ideas* 36 (1975): 497–512; Saree Makdisi, "Colonial Space and the Colonization of Time in Scott's *Waverley*," *Studies in Romanticism* 34, no. 2 (summer 1995): 155–187; and Katie Trumpener, *Bardic Nationalism: The Romantic Novel and the British Empire* (Princeton, N.J.: Princeton University Press, 1997).

8. Sir Walter Scott, *Waverley; or, 'Tis Sixty Years Since,* ed. Claire Lamont (Oxford: Oxford University Press, 1986), 214.

9. See Trumpener, *Bardic Nationalism,* 69.

10. Makdisi, "Colonial Space," 171.

11. Dekker, *American Historical Romance,* 81.

12. Lewis Evans, "A General Map of the Middle British Colonies in America" (London: Carrington Bowles, 1771), Library of Congress, Map Division, G3710 1771.E8.

13. James F. Cooper, *The Pioneers,* ed. James Franklin Beard (Albany: State University of New York Press, 1980).

14. Sigmund Freud, *Civilization and Its Discontents,* trans. James Strachey (New York: W. W. Norton, 1961), 19–20. For a discussion of Freud's indebtedness to Scott and the stage theory of history, see Martin Meisel's "Waverley, Freud, and Topographical Metaphor," *University of Toronto Quarterly* 48, no. 3 (spring 1979): 226–244.

15. Henry N. Stevens, *Lewis Evans, His Map of the Middle British Colonies in America; a comparative account of the ten different editions published between 1755 and 1807* (London: Henry Stevens, Son, and Stiles, 1905), 50.

16. Cooper, *The Pioneers,* 243.

17. James F. Cooper, *The Deerslayer,* ed. James Franklin Beard (Albany: State University of New York Press, 1987), 15.

18. Richard Slotkin, *Regeneration through Violence* (Hanover, N.H.: University Press of New England, 1973), 5.

19. James F. Cooper, *The Last of the Mohicans,* ed. James Franklin Beard (Albany: State University of New York Press, 1983), 119.

20. Cooper, *The Deerslayer,* 539.

21. James F. Cooper, *The Prairie,* ed. James Elliott (Albany: State University of New York Press, 1985), 341.

22. James F. Cooper, *The Pathfinder,* ed. Richard Dilworth Rust (Albany: State University of New York Press, 1980), 32.

23. Natty Bumppo's evocations of water's traceless quality are usually occasioned by a route decision made in the middle of a chase. The master hunter and guide of the woods, when faced with a decision between a route over water and one through the woods, will choose the water. To mention one example, in *The Last of the Mohicans,* Uncas and Chingachgook, Natty's Mohican companions, lobby for a forest route in pursuit of the villainous Magua and the captive Munro daughters, but Natty convinces them

to canoe the length of Lake George, noting that because they are themselves being pursued, they must travel "in a manner that should leave no trail" (199).

24. Cooper, *The Prairie,* 259.

25. The first chapter of Tony Tanner's *Scenes of Nature, Signs of Men* (Cambridge: Cambridge University Press, 1987) focuses on the Leatherstocking Tales.

26. Mikhail M. Bakhtin, *The Dialogic Imagination,* trans. Caryl Emerson and Michael Holquist (Austin: University of Texas Press, 1981), 244.

27. Cooper, *Last of the Mohicans,* 75.

28. Roy Harvey Pearce, *Savagism and Civilization: A Study of the Indian and the American Mind* (Berkeley: University of California Press, 1988), 49.

29. Quoted in ibid., 57.

30. Ernest Renan, "What Is a Nation," trans. Martin Thom, in *Nation and Narration,* ed. Homi K. Bhabha (London: Routledge, 1990), 11.

31. Anderson, *Imagined Communities,* 202–203.

32. Cooper, *The Pathfinder,* 347.

33. Lydia Huntley Sigourney, *Illustrated Poems* (New York: Leavett & Allen Brothers, 1869), 237–238.

Commodities and Exchanges

Mapping Enterprise

Cartography and Commodification at the 1893 World's Columbian Exposition

Diane Dillon

*T*he first visualizations of the World's Columbian Exposition, the spectacular world's fair staged in 1893 on 686 acres of reclaimed swampland in Chicago's Jackson Park, took the form of maps. In November 1890, the members of the fair's consulting board—architects Daniel H. Burnham and John W. Root, landscape architects Frederick Law Olmsted and Henry S. Codman, and engineer Abraham Gottlieb—met to devise a general plan. The team examined sketches Codman had made on the site while working out Olmsted's ideas for a formal architectural court framed by waterways and balanced by the natural landscape of a wooded island (Figure 1). They then produced the first large-scale rendering of the entire design, rapidly drawn on brown paper "mostly with a pencil in the hands of Mr. Root." The board presented this map, accompanied by similarly sketchy written specifications, to the fair's governing boards, who approved it as the official plan of the exposition.[1]

This rough sketch, now lost, established several precedents that would be followed throughout the exposition. First, and most obviously, the sketch set forth the working plan for the layout of the buildings and grounds; over the next three years the designers would adjust and refine the scheme, but the basic configuration remained unchanged. Second, the sketch established the centrality of maps to the fair's production and representation. Throughout the planning and construction, maps served as a primary medium through which the fair-makers comprehended and shaped the spatial features of the exposition. The use of site plans by architects and builders was hardly unusual, but the complexity of the project, the physical challenges presented by the landscape, the large number of players involved, and the time pressure made these evolving visualizations all the more crucial. Maps of the grounds were, at the same time, well suited for promoting the fair because they outlined the entire scheme for the public at large. The visual appeal of the exposition, so vital to this promotion, depended less on the self-contained aesthetics of individual buildings or landscape features than on the

FIGURE 1. West End Court of Honor, World's Columbian Exposition. *(Photo: C. D. Arnold. Courtesy of Special Collections, Chicago Public Library.)*

exposition's status as a *Gesamtkunstwerk,* or synthesis of the arts. Burnham and Root understood this when they accepted their appointment as consulting architects who were "not to participate in the work as designers of any of the buildings" but to shape the "plan of the Exposition as a whole."[2] Third, the sketch, which represented the joint efforts of the men on the board, stressed the corporate nature of the work of designing and producing the fair.[3] Subsequent maps would index the fair's larger social relations, mediating between designers, administrators, engineers, laborers, exhibitors, and tourists. More broadly, the sketch established cartography as a fundamental way of seeing and knowing at the exposition, not only for its creators and actual fair-goers but also for a much larger public of entrepreneurs and consumers.

In this essay I will examine the web of connections between maps and the fair in three cartographic genres: maps produced by the fair-makers to construct and promote the exposition; maps disseminated by exhibitors to guide visitors to their displays; and maps distributed by real estate developers to advertise property in conjunction with the fair. The first two genres comprise maps of the exposition itself, and the third consists of city maps highlighting the fair site. The construction and real estate maps illustrate the ways maps were used as practical tools to shape space and sell places. These genres, along with the exhibitors' maps, also show how particular locations—the fair site and the developers' neighborhoods—in turn prompted the production of additional maps, which served as advertisements and souvenirs. By encompassing plans inspired by the site along with sites developed using maps, and by tracing the links between the literal building of the fair and its larger imaginative life, these genres highlight complementary aspects of the dynamic relationship between map and place.

Regarded together, these maps document the production of space at the exposition, charting the ways, as Henri Lefebvre has argued, capitalist space assumes a special and powerful reality of its own, a reality resembling—and intimately connected to—that of commodities, money, and capital.[4] Although the appearance and use of the fair maps were related to each of these three different forms, the maps' material and metaphorical connections to money were the most pervasive. Much as money is the root form of representation in capitalist society, maps were, literally and figuratively, at the base of representation at the exposition. For fair-makers and fair-goers, maps served functions similar to those of money: the maps facilitated measure and comparison, competition and circulation, accumulation and investment. The parallel encompassed form as well as function, as the maps echoed money's multifaceted, protean character. Many were printed on two sides, incorporating several different maps and views; over time they met evolving demands, from acting as working documents for the construction of the fair to serving as souvenirs for the preservation of its memory.

At every stage, the maps evoked money's faculty for generating illusions, in Marx's words, "for turning an *image* into *reality* and reality into a mere *image.*"[5] Like exchanges of commodities and money in the marketplace, the fair existed as an imaginary construct before and after it existed as built forms in space, and it concealed the social character of the labor that produced it. The maps shaped and recorded these complex fictions, negotiating between the illusions fundamental to capitalism and those particular to the exposition's visual culture. Cartography was a medium through which ideas about selling and owning could literally and figuratively be mapped onto ways of seeing and knowing. In making these connections visible, the maps dramatized the fair's self-reflexivity, enabling us to glimpse the relationships between the production and consumption of the exposition itself and the goods it displayed, between its promotional image and its built reality, between the anticipation, experience, and memory of its contours.

Maps for Construction and Promotion

Maps were as integral to the construction process as they were to the design process. On 22 October 1891, Burnham, as chief of construction, centralized the fair's mapmaking within the Bureau of Surveys and Grades, a unit of the construction department led by Chief Engineer John W. Alvord. Alvord's corps of surveyors, engineers, and draftsmen had prepared a complete set of topographical maps of the grounds in April 1891. As the construction progressed, they maintained a working map, recording the location not only of all buildings but also of temporary features such as railroad tracks, piles of lumber and other building materials, and the contractors' construction shacks. In November, Burnham stipulated that this map be updated monthly and distributed to all exposition departments. Supervised by the chief draftsman, Hermann Heinze, these progress maps, constructed

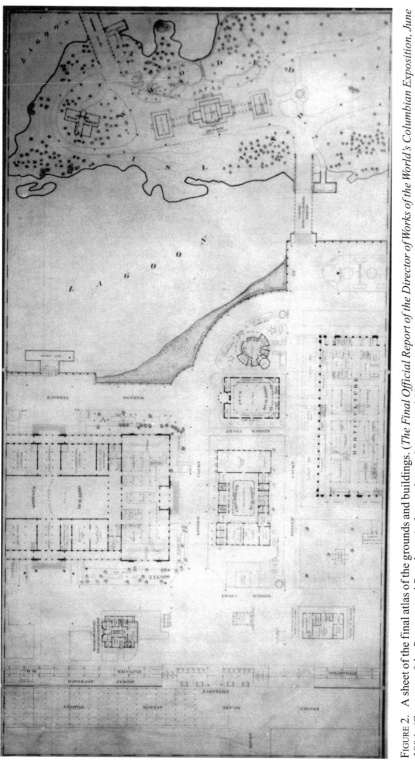

FIGURE 2. A sheet of the final atlas of the grounds and buildings. (*The Final Official Report of the Director of Works of the World's Columbian Exposition, June 1894. (Courtesy of the Ryerson and Burnham Archives, the Art Institute of Chicago.)*

on a scale of one hundred feet to one inch, summarized the developing configuration of the buildings and grounds.[6]

To guide the detailed construction work, the bureau prepared two more-specialized atlases. The first plotted the exact location of the buildings and the intervening distances, and the second traced the network of utilities underground. Constructed on a scale of forty feet to the inch, each sheet showed five hundred square feet of the park. As Alvord noted, the importance of the first atlas to all the fair's departments became so great that "it became necessary to send out regular bulletins of the changes which occurred upon it." He added, "Many of the sheets had to be redrawn, owing to the wear and tear consequent to their constant use." In late 1892 the bureau began a third atlas, to facilitate the installation of exhibits. Prepared on an even grander scale—twenty feet to the inch, with each sheet showing a one-thousand-by-five-hundred-foot section of the park—these maps presented detailed floor plans for each building and indicated the location of every display (Figure 2).[7]

Maps conceived to build the fair were soon pressed into a second use: promotion. Because the maps were so vividly detailed, it was logical that the fair-makers would turn to them to explain and advertise the exposition. During the early stages of construction, the Bureau of Surveys and Grades distributed small blueprints of the ground plan to tourists who came to see the work in progress, and the bureau subsequently printed and disseminated larger engraved maps by the thousands to department chiefs and distinguished visitors. The demand for maps became so great that Alvord hired a custodian of records to take charge of their inventory and circulation.[8]

The most aggressive distributor of maps, however, was Moses P. Handy, head of the Department of Publicity and Promotion. Handy's department sent out a staggering volume of press releases to editors around the world, many accompanied by maps and views of the fair. These materials were often published without any editorial changes or comments—in the guise of objective reporting—in newspapers, architectural periodicals, general interest magazines, and guidebooks. Distributed for promotion but packaged as information, the maps and notices consistently blurred the line between news and advertising. The popular association of cartography with scientific documentation further enhanced the maps' veneer of objectivity.

If the usefulness of the construction atlases depended on their precise measurements and accurate data, the promotional value of all the fair maps turned on their ability to shade the truth. From the first sketch of the grounds to the installation atlas, the construction maps grew progressively more exact. The maps used as advertising, in contrast, incorporated an increasing array of complex illusions. This trend began with the engraving of the construction maps for press releases, when many of their precise details were sacrificed to make them legible in a small format. These plans were also appealingly suggestive in more conceptual ways. When viewers contemplated maps before the fair was completed (or before they saw it), the images invited them to imagine what the exposition would be like in its final form. This mental leap was shortened when the maps were presented alongside

FIGURE 3. Daniel Burnham explaining the large map to a congressional party visiting the fairgrounds, 22 February 1892. *(From* The Illustrated World's Fair *2, no. 9 [March 1892].* Courtesy of Special Collections, Chicago Public Library.)

bird's-eye views and renderings of the individual buildings, which provided a sense of the physiognomy of the structures. Double-sided promotional brochures often incorporated all three genres, offering a breathtakingly comprehensive sense of the architecture and overall plan.

The fair-makers relied on the power of these imaginative leaps when they lobbied Congress for an appropriation to bolster the fair's finances. Handy bombarded Washington with promotional images, which soon lined the corridors of government offices and hotels.[9] The directors, realizing that members of Congress would be more likely to appropriate the funds if they saw the work in progress firsthand, invited the legislators to visit the exposition grounds on 22 February 1892. Even while the politicians were on the site, however, maps and views were deployed as instruments of persuasion, to focus their attention on the wonders that lay ahead. The tour route was outlined in red on a small map distributed to all the legislators and the press. The longest stop was inside the Woman's Building, the most nearly completed hall, which was fitted out as a gallery lined with maps and views of the rest of the fair as it would appear. The centerpiece was a giant, multicolored plan of the grounds, elaborately festooned with flags, specially prepared for the occasion by the Bureau of Surveys and Grades (Figure 3). When Burnham addressed the group, he identified buildings and sites on the map as he rattled off their dimensions and described the overall scheme, all with an eye toward rationalizing its enormous cost. After the presentation, Burnham led the delegates to the roof of the building, from which they could see the plan unfolding in three dimensions. The legislators were ultimately convinced, and they granted the fair $2.5 million.

The construction maps fostered still more illusions when they were given away or sold as keepsakes. Official souvenir editions, available in black and white or color, in formats ranging from fourteen-by-fifteen-inch sheets to thirty-eight-by-forty-inch wall maps, were marketed at newsstands and from Heinze's downtown office. Early versions guided preopening tourists during their visits, fueling anticipation of the completed exposition. Later editions led visitors through the finished fair and aided their recollections, shaping and preserving memories of their experiences. A promotional line on one iteration—"Send one to your friends!"—indicated that the intended audience included individuals who may not have visited the fair. For these consumers, the maps created a fantasy of having been there, serving as tangible tokens of an imaginary experience. Although the functions of the three incarnations of the construction maps—as working plans, publicity tools, and souvenirs—were complementary, each incarnation was conditioned by, and in turn inflected, the other two.

On another level, one's reading of the construction maps was complicated by the deceptive appearance of the structures they represented, an appearance that made the maps, as it were, inherently idealized projections of an idealized world. The fair's gleaming beaux arts palaces were not genuine marble structures but temporary compositions formed with staff, a mixture of plaster and hemp, coated with paint. The illusion encompassed function as well as form: the grand halls were not royal palaces or civic institutions, as their style and scale might suggest, but showcases for the latest commodities. The elegant exteriors worked to counter Chicago's reputation as a city where money was pursued at the expense of high culture and refinement, while the displays inside simultaneously furthered commercial goals.

Cartography and Commodities

Much as the fair-makers used maps to package and sell the exposition as a commodity, fair exhibitors distributed more-specialized maps of the grounds to draw visitors to the displays of their own commodities. The Libbey Glass Company marked the location of its model factory in bright red ink on one map, adding advertising copy and an engraving of the building. Similarly, the Old Times Distillery highlighted its site and included an image of its miniature plant on a vest-pocket map. The cover featured an additional promotional gimmick, inviting visitors to come to the exhibit to register their guesses as to the number of admission tickets that would be sold during the run of the fair. The best guesser would receive "fifteen barrels of Kentucky's famous Old Times Sour Mash Whiskey." The Lemp Brewing Company, the American Biscuit Manufacturing Company, and Aultman, Miller & Company likewise distributed small, easily wielded maps to direct visitors to their exhibits (Figure 4).

FIGURE 4. "Aultman, Miller & Co.'s Buckeye Map of the World's Columbian Exposition." *(Chicago Historical Society.)*

FIGURE 5. Advertisement for Aultman, Miller & Co.'s pavilion in the Agriculture Building, printed on the reverse of their "Buckeye Map." *(Chicago Historical Society.)*

These maps presented a new twist on the economic mapping that had begun to proliferate in the 1840s and 1850s. The earlier maps charted the distribution of agricultural and manufactured products on national, European, and global scales. Some marked the points of origin for the goods, while others mapped their directional flow from production locations to markets. The first world's fair, staged in 1851 at the Crystal Palace in London, inspired August Petermann, a German cartographer recently relocated to London, to produce a large panel incorporating eighteen small maps titled "Geographical View of the Great Exhibition of 1851 Shewing at One View the Relative & Territorial Distribution of the Various Localities from Whence the Raw Materials & Manufacturers Contributed to the Exposition Have Been Severally Supplied."[10] In light of this tradition, the maps made for the 1893 fair are distinctive in their more minute scale, tight focus, and obvious advertising purpose.

Many of the firms that distributed advertising maps may have been prompted to make an extra effort to point fair-goers to their displays by their out-of-the-way locations on the grounds. The Libbey Glass model factory was situated on the outlying Midway Plaisance rather than within Jackson Park, and the Old Times Distillery was wedged into the far southern corner of the grounds. Aultman, Miller's reapers were displayed in the annex behind the Agriculture Building (Figure 5), and the Lemp Brewing and American Biscuit exhibits were housed in the building's second-floor gallery. Over the course of the fair, numerous exhibitors complained that displays in the upper galleries attracted fewer visitors than those on the main floors of the buildings.[11]

Studebaker Brothers, makers of wheels and carriages, issued an elaborate, two-sided map to direct viewers to their locations at the fair and beyond. One side featured a city map highlighting the site of the fair and the company's Chicago headquarters along with a view of its wagon works in South Bend, Indiana, thereby offering a concise summary of the geographic extent of Studebaker's operations. The fair map on the reverse targeted the company's exhibits around the grounds, juxtaposed with an engraving of the company's downtown building. Studebaker's fair displays were more prominently placed than those of the firms mentioned above, but the company had good reason to call attention to the location of its downtown showroom. In the mid–1880s, the firm had moved into a new Michigan Avenue building, whose elegant architecture was roundly praised.[12] But the move took the showroom away from the main cluster of carriage makers on Wabash Avenue and thus off the path typically beaten by shoppers.

Several manufacturers recognized the limited value of displaying their products passively and arranged for them to be demonstrated in action at the fair. Henry R. Worthington distributed a map highlighting the many places at the fair where visitors could witness his pumps and machinery in operation, and Wyckoff, Seamans & Benedict mapped the placement of its Remington typewriters in use and on exhibit throughout the grounds. Worthington had agreed to furnish pumps for steam power and fire protection to the exposition for $1—their estimated value was $150,000—in exchange for the promotional value of having them in use on the grounds. Similarly, Wyckoff, Seamans & Benedict provided typewriters for the convenience of the public, the press, and exposition officials. In exchange, the company was permitted to charge a fee for work turned out by its typists and to advertise its Remington Standard Typewriter as "the official writing machine of the World's Columbian Exposition."[13]

Despite their active use throughout the fair, these machines might easily have been overlooked by fair-goers. Worthington's main pumping station was tucked behind the annex to the Machinery Building, and most of his engines in the exhibits supplied power for the machines of other manufacturers, which were the featured items in those displays. The typewriters likewise blended into the offices where they were used. Both firms relied on the maps to convince fair-goers that their machines were worth notice, using clever graphics to make their utilitarian devices more eye-catching on the maps than they were in actuality. Worthington marked the installation of his machines inside buildings with red flags and the use of his marine pumps with tiny boats on the waterways. Wyckoff, Seamans & Benedict featured an engraving of the company's latest model on the outside of its map and used big red stars to designate its work stations as well as the site of the Remington exhibit in the Manufactures Hall. Both maps also stressed quantity: Worthington incorporated a numbered list of his various devices, and Wyckoff, Seamans & Benedict proclaimed that more than 370 Remingtons were in use on the grounds. The aggressive accumulation of pumps and typewriters represented on these two maps can be interpreted as a metaphor for the accumulation of money in the form of capital, the driving force behind the fair and bourgeois society in general.

These advertising maps contributed to the fair's atmosphere of commercial competition in multiple ways. The maps typically led visitors not only to the displays of the promoted goods but to those of their rivals as well, as most of the exhibits were grouped by type. The advertisements thus facilitated, perhaps inadvertently, comparison of different brands of beer, biscuits, or reapers. And because the maps could also be consulted by fair-goers to guide them to still other exhibits, the maps enabled visitors to compare the advertised goods to the broader array of commodities on view throughout the exposition.

Much as individual displays competed for the attention of fair-goers, the exhibitors' maps rivaled one another. In addition, the advertising maps vied with the wide array of maps offered to fair-bound tourists on trains, in depots, and at hotels as well as with those bound into guidebooks and printed in newspapers and magazines. A variety of promotional circulars incorporating small maps and floor plans pinpointing specific displays offered still more competition.

At the same time, mapmakers vied with one another for the business of exhibitors. The exposition carved a new niche in the map trade, a niche that engravers and printers around the country rushed to fill. Some exhibitors, such as Wyckoff, Seamans & Benedict, had their maps engraved and printed in the cities where they were headquartered. But most exhibitors patronized Chicago firms. Libbey Glass customized a souvenir edition of the construction department map, which Heinze (who held the copyright) sold to advertisers as the only map compiled from "actual official data." American Biscuit turned to the Globe Lithographic Company to print its flyers, while other exhibitors patronized competing engravers such as George H. Benedict & Co., J. Manz & Co., and A. Zeese & Co. The largest group of advertisers, however, took their business to Rand, McNally & Co., who had been producing customized vest-pocket maps since at least the 1880s.

Rand, McNally came to dominate the market for fair maps—and Chicago mapmaking in general—through efficient production and aggressive marketing. The company printed maps from wax engravings (a process less costly than lithography)[14] and developed a few standardized templates that could be easily adapted and combined by advertisers. The fair maps distributed by Studebaker, Worthington, and the Old Times Distillery were all versions of Rand, McNally's "Indexed Standard Guide Map of the World's Columbian Exposition." The maps are not exactly alike—the company issued updated versions to reflect the changes in the ground plan as the fair was constructed—but the designs of the map and indexes were clearly derived from the same model. The dimensions of the sheets varied from 14 by 10 1/2 inches to 15 1/4 by 19 1/2 inches, but all were designed to fold into a pocket size, a format Rand, McNally had been promoting since the 1870s.[15] After the ground plan was more firmly set, the company issued a "New Indexed Miniature Guide Map" of the grounds (9 3/4 by 7 1/2 inches) and a "Letter Size Map of Chicago and the World's Fair" (10 5/8 by 8 1/2 inches). These compact maps could be printed on the reverse of flyers (as was the miniature guide of Aultman, Miller), bound into small guidebooks, or combined with other maps.

In 1892 Rand, McNally developed an even tinier (5 1/2 by 2 7/8 inches) format for maps of the fair, the central portion of Chicago, and the larger metropolitan area, which could readily serve as insets on larger maps or be printed on cards. Around the same time the company also introduced a "new and concise" 28-by-20-inch sheet map of Chicago, highlighting the fair site and the transportation routes to it.

Rand, McNally emphasized the status of its maps as commodities by insuring the recognizability of its brand. Unlike competing mapmakers, it consistently included its name on every map. Because Rand, McNally's maps were so standardized and ubiquitous, they were readily identifiable even when they were re-titled by advertisers who relegated the maker's name to a tiny credit line along the bottom edge. The company's maps were also distinguished by their clean design and typography, making them more legible than many competitors' maps featuring more-crowded layouts and fussier typefaces. Furthermore, the firm typically printed its fair maps in shades of bright orange or yellow, enhancing their identification as Rand, McNally products. Much as the advertising maps served to promote the fair along with specific commodities, Rand, McNally's line of innovative, adaptable maps doubled as advertisements for the firm itself. Each customized map that went into circulation represented one more Rand, McNally product on the market, one more promotion for the company.

All the advertising maps underscored the illusions of the fair's marketplace through a series of deceptions and substitutions. Fair-goers were tantalized by the merchandise the maps led them to, but they were denied the satisfaction of acquiring it on the spot because the fair's official rules prohibited exhibitors from conducting sales from displays on the grounds.[16] This policy ensured that the visual coherence of the exhibits would not be disturbed, and it enabled foreigners to show their wares without paying the stiff protective tariff levied on goods imported for sale. In this context, the maps acted as surrogate acquisitions, tangible commodities visitors could instantly possess and take home. Moreover, the use value of the maps made it more likely that fair-goers would hang on to them, whereas they might discard a less serviceable trade card or flyer.

Through another set of displacements, the maps obscured the labor that produced the advertised goods and defined their utility. The maps' stress on travel, on getting the fair-goers to the advertised location, overshadowed any indication of the human manufacture of the commodities themselves. This overshadowing was emphasized by the frequent incorporation of exterior views of showrooms and factories. Like the maps themselves, these architectural icons helped visitors locate the sites where they could peruse the goods, but the icons effectively replaced not only the labor that took place inside the buildings but also the products themselves, which were less frequently represented.

The advertisement on the reverse of Aultman, Miller's map did depict its Buckeye harvesters but showed them inside its luxurious fair pavilion rather than on a farm. The elegance of the structure made utilitarian products more desirable, as did Studebaker's Chicago building. Like the fair's artistic architecture and landscaping, the design of the pavilion at once masked and refined its commercial goals.

Appointed with rich green curtains, plush carpet runners, a stuffed moose head, and stylishly clad fair-goers, the setting evoked domestic comfort and relaxation. The lavishness of the pavilion was matched by its reproduction in the form of a carefully printed, full-color lithograph. The advertisement implied that Buckeye reapers could help farmers prosper, enabling them to afford the same luxuries that their urban and suburban contemporaries enjoyed. But the actual work of producing the tools or reaping the harvest was nowhere apparent. The practical value of the map to the fair-goer and the display value of the pavilion effectively replaced the use value of the farm implements, clouding their connections to productive labor.

The series of displacements and illusions presented in the maps highlight the role of visibility and perception in the process Marx described as commodity fetishism. Material objects are fetishized when they appear in the marketplace, where the human labor that made the commodities is masked by their appearance as value. The social relations of production become conflated with the material form of the objects, whose value then seems inherent to them as things.[17] The maps show how this deception was perpetuated over the course of the marketing process: the appearance of human labor was replaced by that of the commodity itself, and the commodity was replaced by the site of its production or sale, which in turn was replaced by the map. As visual representations and marketing tools, the maps not only illustrated the course of commodity fetishism but also facilitated its operation.

Mapping Real Estate

Chicago real estate developers also adapted maps to advertise their properties in conjunction with the exposition. Like the maps promoting other local enterprises, these sheets neatly diagrammed the economic connections between the fair and the city of Chicago, as the property agents similarly counted on the power of the exposition to whet consumers' desires. The gratifying purchases that were prohibited on the grounds could be realized in nearby subdivisions, where developers facilitated easy acquisitions. But like the construction plans, these maps built—and were built on—fictions. Aiming to capitalize on the prestige of the exposition, brokers fabricated ties to it from whole cloth, basing their optimism about the fair's salutary influence on the property market on predictions that proved equally illusional.

The land boom of 1889–1890 coincided precisely with boosters' efforts to secure the fair for Chicago.[18] The local newspapers, along with trade journals such as *The Real Estate and Building Journal* and *The Economist,* reinforced investors' high expectations. In 1889 the *Journal* predicted that "the attendance here of such a vast concourse of people will be a benefit to the city beyond estimate" and went on to stress that the benefit would be most marked for the side of the city selected for the exposition grounds.[19] Residents and developers who were invested in different sections of the city lobbied for the selection of a site in their districts, but the South Siders advocating Jackson Park campaigned the most energetically. *The*

Economist acknowledged the gossip about a "real estate ring" operating "in connection with the Jackson Park site" but quickly came to the defense of the officials, denying that there was any evidence "that the directors of the corporation have been influenced in the slightest degree by real estate speculation in their deliberations respecting the site."[20] After the choice of site was announced, the *Journal* predicted that Jackson Park property would "now begin to look up" and anxious investors would be found "exploring for choice lots in the southern district."[21] These optimistic forecasts about the land market echoed the publicists' promotion of the fair itself. In both cases, writers and publishers were quick to substitute press releases for objective reporting, projected visions for material realities.

Enterprising brokers modified city maps to yoke their South Side properties to the star of the world's fair. W. H. Shepard & Company adapted a Rand, McNally map of the southern portion of Chicago, visually linking the company's headquarters to the exposition site by marking both in distinctive purple ink. A competing brokerage, Eggleston, Mallette and Brownell, called attention to its subdivisions southwest of the fair by adding a photograph and promotional copy describing the company's namesake community of Eggleston to maps of the city (some printed by Rand, McNally; others by George Cram), along with red lines pointing directly to Eggleston and its twin community, Auburn Park. The firm highlighted the fair site in the same shade of red, drawing a visual connection between its properties and the exposition.

These maps were a key element in the developers' promotional campaigns, working in conjunction with articles and advertisements in guidebooks, newspapers, and trade journals. Well before the fair, property and mortgage brokers habitually bought inexpensive printed maps in bulk from local cartographers to distribute to potential clients. The mapmakers frequently overprinted the sheets with the name of the firm and a few lines of promotional text. The incorporation of the exposition tie-in made this advertising strategy more focused and sophisticated.

Eggleston, Mallette and Brownell forged close ties to several guidebook publishers (who likewise expanded their product line in anticipation of the fair) by arranging for their maps to be bound into several editions of *Moran's Dictionary of Chicago, Rand, McNally & Co.'s Handy Guide to Chicago and the World's Columbian Exposition,* and John J. Flinn's *Standard Guide to Chicago.* Flinn included extensive descriptions of Auburn Park and Eggleston in two of his 1893 guides. In these guides, the presence of passages that had appeared verbatim in the *Real Estate and Building Journal* in 1889 indicated that the text had been supplied by the developer.[22] Eggleston, Mallette and Brownell stressed the up-and-coming location of its subdivisions in an advertisement in the *Journal,* noting that because both communities were "within walking distance of the world's fair site, there has been a boom in this location." The *Journal* confirmed this upswing in articles the following month, noting that the new transportation lines to be constructed for the fair would "insure the continued supremacy of this section of the city in development, growth, and activity" and that this would mean "higher prices for real estate, especially for business purposes and for fine residences." Another article pointed out that Auburn Park was "within easy reaching distance of the

World's Fair, yet far enough away from the great Exposition to suffer none of the discomforts that [would] surely result from the presence of the crowd of visitors that [would] centre there."[23]

Samuel Eberly Gross, one of the most famous Chicago real estate entrepreneurs of the period—and surely the most skillful promoter—likewise distributed maps linking his subdivisions to the fair, adapting a standard sheet printed by Chicago mapmaker Rufus Blanchard in 1891. The density of information and the multiplicity of targeted sites on this map underscore the diversity of Gross's enterprise. The locations printed in red identify the wide-ranging urban and suburban communities in which he had a stake. Gross also linked his developments to the exposition by marking its site in the same color, pushing the tie-in a step further by including an inset map of the fairgrounds at the upper right. At the lower left, Gross added photographs of two elaborate Victorian dwellings, noting that he offered "Houses and cottages for sale on easy monthly payments" (Figure 6). In picturing these dwellings, Gross underscored his current emphasis on middle-class communities, in contrast to the working-class subdivisions filled with modest cottages he had developed in the early 1880s. He claimed to provide four hundred different house options, and he enhanced several of his newer subdivisions by building train stations, schools, and meeting halls.[24] Gross maintained affordable prices by buying mass-produced building materials in bulk and relying on formulaic plans— the same mechanisms of standardization that enabled Rand, McNally to offer such inexpensive maps.

Although Gross's map promoted his interests throughout the metropolitan area, in other contexts he stressed the proximity of his South Side developments of Brookdale, Dauphin Park, and Calumet Heights to the fair. He informed readers of Flinn's *Standard Guide* that Brookdale was but a three-minute walk from the fair and assured *Chicago Tribune* readers that they could reach the fair from Calumet Heights in five to eight minutes by the Calumet Electric Road. In 1891 he ran a full-page advertisement in the *Real Estate and Building Journal* featuring a bird's-eye view of Dauphin Park with the exposition visible beyond it. The headline proclaimed that the fair was "only 5 minutes ride from Ideal Dauphin Park." An article in the same issue reinforced the message, noting that "Dauphin Park is so near the World's Fair site that it will surely reap a great benefit from the thousands of mechanics who will be employed there and who will desire homes convenient to their work; also from the great armies of visitors that will come to see the exhibits and surrounding territory."[25]

In distributing maps linking their enterprises to the fair, developers endeavored to capture the attention of the hordes of tourists expected in 1893, hoping to convince at least some of them to stay. In 1891 *The Economist* speculated that the population of Chicago would be increased "through the decision of many individuals who come as visitors to make this their permanent place of abode."[26] The *Journal* reiterated this belief the next year, proclaiming that "[t]he great throng of visitors who for the first time come to see the glories of Chicago will be induced to invest here." The *Journal* pointedly eyed the pocketbooks of wealthy foreigners, predicting that they would "invest several millions of dollars in Chicago real

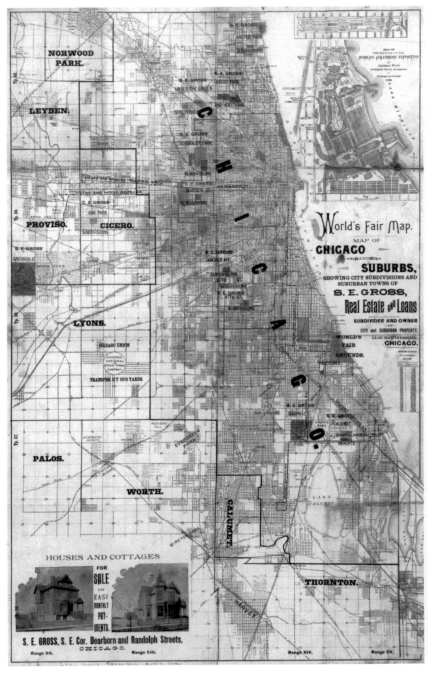

FIGURE 6. "World's Fair Map. Map of Chicago and Suburbs," showing the real estate development of S. E. Gross *(Rufus Blanchard, mapmaker. Chicago Historical Society.)*

estate before their return to their native countries" and that some of them would be "so captivated with the World's Fair city" that they would remain permanently.[27]

Real estate entrepreneurs also sought to reap profits by providing temporary housing during the fair. The developers of the Mecca apartment block, constructed in 1891–1892 on the corner of Dearborn and Thirty-fourth Streets,[28] converted their new building into a hotel for the exposition season, advertising the hotel's amenities on a double-sided Rand, McNally map. One side featured interior and exterior views of the building, a menu for the hotel's restaurant, and a card map of the city pinpointing the hotel and the fair. The text lauded the hotel's modern conveniences and desirable location "midway between the World's Fair Grounds and the business and amusement portions of the city," a location that was easily accessible by nearby cable car and elevated rail lines. The map of the fairgrounds on the reverse followed the typical pattern by including the dimensions as well as the names of the major buildings, and the developers added their own promotions— "To the Mecca Hotel, 20 Minutes" and "L Trains leave this station for Mecca Hotel every 3 minutes"—on top of the outlines of several buildings. By inserting these advertisements directly into the space of the fair, the proprietors gave their hotel an imaginary presence on the grounds, fortifying its invented link to the exposition. The emphasis on the convenient location also made the Mecca seem more competitive with hotels situated just a few blocks from the fair, such as the Park Gate and the Raymond & Whitcomb Grand. These hotels touted their proximity by targeting their sites on maps showing the exposition and its immediate neighborhood.

Those who monitored the pulse of Chicago's real estate market looked to the world's fair to explain both the highs and the lows on the graph of the city's property values. By the time real estate entrepreneurs issued their first fair maps in 1891, the explosion in property values was already over. The maps and fair-related advertisements were thus part of the developers' campaign to reignite the boom and perpetuate the belief that the fair would spur the market.

The trajectory of the real estate trade between 1890 and 1893 reveals that the exposition's influence on the market was but an illusion, despite the boosters' efforts. The *Chicago Tribune* admitted that "the market is at a complete standstill" in December 1890, but remained optimistic that it would revive the following spring.[29] The real estate press remained confident throughout 1891 that South Side land values would rise, and the promotional zeal of the fair-makers and brokers remained undampened.[30] Harmon Spruance, chief of the fair's bureau of subscriptions, hawked shares of exposition stock in the summer of that year by telling local entrepreneurs "how the fair is sending up the price of their real estate and increasing their business."[31] Eggleston, Mallette and Brownell assured readers of the September 1891 issue of *The Illustrated World's Fair* that "the time to buy is now."[32]

By the middle of 1892, however, analysts started to see the fair as having a depressive influence on the market. Waxing belief that the fair had brought unsustainable inflated prices reportedly prompted investors to hold back until after the exposition closed.[33] Observers began to regard the enormous expense of the fair—from preparing the grounds and buildings to constructing hotels and related enterprises—as diverting money away from real estate investment.[34] As the nationwide economic panic that began in the spring of 1893 settled in, the purse strings of fair-goers tightened. Just after the exposition opened in May, several brokers optimistically reported "a good demand for property" from fair visitors, but by September the *Real Estate and Building Journal* admitted that well-to-do tourists and local residents alike were buying "everything and anything but real estate." After the exposition closed, the *Journal* described the White City as a holiday distraction that made it "impossible to get the attention or the interest of the people fixed on anything in the ordinary line of business."[35]

Maps, Money, Illusions

How might we make sense of the broad range of the exposition's cartography, of its complex illusions and its intimacy with money? As advertisements, the maps seem to have brought mixed success. Though effective in promoting the fair itself, they failed to kindle the real estate market in the surrounding area. The direct result of the advertising maps for specific exhibitors at the fair is harder to gauge, though Rand, McNally's expanded production of customized maps for corporate clients into the twentieth century indicates that many firms continued to believe in the promotional value of cartography.[36] The equivocal payoff of the maps matched that of the exposition itself, which likewise did not generate the financial returns that the organizers had predicted. Moreover, as unfulfilled financial promises, these advertisements were part of the fair's larger culture of illusion, contributing to the dynamic tension between the fair's visual and economic imperatives.

Although the fair-makers used elegant architecture, landscaping, and public art to cover the exposition's monetary ambitions with a veneer of aesthetic and civic refinement, the maps distracted from these financial aims in a complementary way, by imparting information and stressing their own practicality. In conveying knowledge the maps evoked the higher realm signaled by the art while referring simultaneously to pursuits outside the market, where objects were valued for their usefulness rather than for their exchange value. The abundance of educational cartography throughout the buildings, from relief maps of individual states and nations to school globes to historic maps tracing Columbus's voyages, would have encouraged fair-goers to regard maps as sources of enlightenment. The genre of exposition maps was added to this collection when a version of Heinze's souvenir map, made conspicuous by its twelve-foot-square dimensions, was installed in the Transportation Building.

The ability of the maps to convey complex fictions while simultaneously maintaining an association with truth made them suited to multiple purposes at the fair. That ability not only enabled them to serve as construction documents, marketing tools, visitors' guides, and souvenirs but also secured their resemblance to commodity forms in general and money in particular. As Marx explained, the commodity leads a double life, existing first in a natural form, manifesting its particular use value, and then assuming a value form, expressing its economic properties. As the commodity moves through the market, this initial bifurcation is followed by a series of further doublings: the act of exchange is split into purchase and sale; money plays additional dual roles, first, in facilitating measure and circulation and, later, in serving as the embodiment of wealth and as the more general material representative of wealth.[37]

At each stage, the metamorphosis from one form to the next occurs first in the imagination. In Marx's words, "[C]ommodities are transformed into bars [of gold] in the head and in speech before they are exchanged for one another." The appearance and value of the fair were also figured mentally—in the heads of mapmakers and map viewers—before any earth was moved or any symbolic transactions were conducted in its space. After a commodity assumes this double existence in the mind, Marx continued, "the idea proceeds . . . to the point where the commodity appears double in real exchange: as a natural product on one side, an exchange value on the other. I.e., the commodity's exchange value obtains a material existence separate from the commodity." When it is transformed into money, "all connection with the natural form of the product is stripped away";[38] these natural properties, embodying the commodity's use value, are displaced by a mental abstraction, which in turn is realized objectively and symbolically—in representation—as money.

The multiple functions of the maps likened them to the money metals in particular. Gold and silver demonstrated their own specific use values as commodities (for filling teeth or providing the raw material for luxury goods) before assuming a formal use value in facilitating the circulation of other commodities.[39] Similarly, the use value of the fair maps could be found both in the geographic guidance they offered and in the exchanges they promoted. The double-sided format of many of the maps underscored their ability not only to embody both use value and exchange value but also to perform multiple roles: to act as guides to Chicago and the fair, as advertisements for exhibited products and mapmakers, as literal and metaphorical images of space and money.

The illusions produced and reflected in the maps were also doubled, drawing together the deceptions fundamental to commodities and money in the market with the idealized forms that shaped the fair's visual culture, from its mock palaces to its model factories and ersatz retailing. Much as money lubricates exchanges between sellers and buyers, the maps negotiated the transactions between these two genres of illusion at the fair.

If the maps' dual functions likened them to gold and silver, in their appearance and substance they more closely resembled paper money. Much as paper currency served, according to nineteenth-century money theorist David Wells, as a mere "promise to deliver a commodity," the advertising maps acted as promises of the goods fair-goers could obtain only after they left the grounds. Wells further maintained that the paper bills could never be money; they could only represent it: paper "could not possibly be the commodity or the thing itself, any more than a shadow could be the substance, or the picture of a horse a horse, or the smell of a good dinner the same as the dinner itself."[40] Similarly, the maps offered representations of the fairgrounds and of the exhibited wares in place of the material objects. In this way the maps replicated the larger structure of the fair, where both use value and exchange value were eclipsed by a more purely representational value. The key to the visual culture of the exposition was to be found less in the glamour of the commodity-in-the-market than in the sheen of the commodity-on-display.[41] The fair resembled the Parisian arcades that captivated Walter Benjamin as well as the colorful, animated show windows of American department stores; each of these environments oriented visitors toward visual rather than material satisfaction, conditioning them to "[l]ook at everything; touch nothing" and to find pleasure most of all in the spectacle itself.[42]

The fair maps evoked money not only in the multiplicity of their uses but also in the specific functions they performed in facilitating measurement and circulation. As representation, money works by persuading its owners that it stands for a particular measure of value, such as one dollar. Much as money calibrates the relative value of commodities to lubricate their exchange, the construction atlases measured the fair's landscape and built forms. The role of the installation atlas came still closer to that of money, as it gauged the specific spaces where commodities would be exhibited. Similarly, the real estate maps measured distances between the advertised properties and the fair, aiming to convince potential investors that the exposition's proximity would enhance land values.

The attention to accuracy implied by measurement also entered into the maps' representation of circulation. The widely distributed progress maps furthered the exchange of correct information among officials, resembling money in the way that it expedites the circulation of commodities. More subtly, the maps dismantled—or at least reconfigured—some of the illusions produced in the marketplace. If the exchange process sundered the natural properties and uses of commodities from their existence as values in circulation, the advertising maps reunited these aspects. Visually and ideologically, the maps sutured usefulness to exchange by substituting their own use value for that of the products and substituting the circulation of consumers for the exchange of goods and money. By replacing the transfer of commodities with the movement of people, the maps emphasized the social character of the economic process. This human face is precisely what is typically hidden by the forms of commodities and money, which make social relations between people assume "the fantastic form of a relation between things."[43]

At the same time, the maps' metaphorization of circulation perpetuated other deceptions. Maps of Chicago traced the surface grid of the city, and plans of the exposition outlined the superficial configuration of its buildings and grounds, each burying the social and economic foundations of the constructions they represented. This aspect of the maps evokes Marx's description of circulation's character as "pure semblance," as "the phenomena of a process taking place behind it." In this way, the "simple forms of exchange value and money latently contain the opposition between capital and labor."[44]

The most basic deception, shared by all the fair maps, lay in their shrouding of the exposition's origins in human labor. This deception becomes readily apparent when we compare the construction maps to the hundreds of photographs taken by the official photographer, C. D. Arnold, to document the progressive stages of the fair. The landscape pictured by the photographs is filled with workers, whose movement of dirt and joining of beams literally shape the grounds and buildings. In contrast, the lagoons and palaces outlined on the maps seem to have formed themselves: all evidence of human agency is erased. The progress maps indicate that construction is ongoing by locating stacks of lumber and temporary railroad tracks, but the maps offer no clues as to who has wielded or will wield these materials and tools. The maps distributed as publicity mystified the production of the fair still further by excluding all signs of the building process and offering a seemingly definitive image of the fair while it was still in the early stages of construction. This obfuscation was echoed in the advertising maps, which displaced the actual production and utility of the goods they hawked. The ground plan of the exposition worked in a similar way, screening off most of its industrial operations (including the Worthington Pump House and Old Times Distillery) behind a decorative colonnade at the south end of the grounds.

The maps produced in conjunction with the exposition promoted illusions about consumption as well as about production. They whetted consumers' desires to own a piece of what they saw, from merchandise to real estate to the fair itself. But ironically, the peculiarities of the market undermined the acquisition of these items. The fair's prohibition on direct sales put the displayed goods out of reach, and the depressed economy discouraged the purchase of property or exposition stock. In this climate, the maps acted as ready surrogates; usually inexpensive or free, widely available and readily portable, they were easy to own. Moreover, they served as metaphorical reminders of the economic activities that had come to a halt during the depression.

The worth of the maps as guides, advertisements, and commodities was extended by their value as souvenirs. The exposition was a temporary construction; shortly after it closed, few of its monumental forms remained in Jackson Park. They were outlasted by the seemingly ephemeral maps. In the years since the fair, the value of the maps has shifted. Like souvenir coins, the maps are now prized not for their usefulness, facilitation of exchange, or promotional value but as rarities and objects of memory. Mementos that initially confirmed fair-goers' experiences (asserting "I was there") have metamorphosed into documents of a vanished landscape, testifying to what was where on the grounds in 1893. As with coins and paper bills, which we accept as substitutes for bars of gold, we have come to take the documentary value of the exposition maps on faith. A closer look, however, reveals that the maps incorporated the fair's fantasies along with its realities, often blurring the boundaries between them.

Notes

The research for this essay, and for the larger study of the exposition of which it forms a part, was supported by a National Endowment for the Humanities Fellowship at the Newberry Library and a J. Paul Getty Postdoctoral Fellowship in the History of Art and Humanities. For invaluable assistance with this work I would like to thank Patrick Ashley, Jean Jansen, Robert Karrow, Lesley Martin, Rob Medina, Patrick Morris, Elizabeth Patterson, Steven Peters, Timothy Samuelson, Jeanne Solensky, Andrea Telli, Julie Thomas, Emily Clark Victorson, Bronwen Wilson, and Mary Woolever. I am deeply indebted to Daniel Bluestone, Carl Smith, and Alan Wallach for their careful readings of the text, their stimulating questions, and their wise suggestions.

1. Frederick Law Olmsted to Board of Directors, World's Columbian Exposition, 17 October 1892, Frederick Law Olmsted Papers, Library of Congress; Daniel H. Burnham, "Report of the Director of Works," in *Final Official Report of the Director of Works of the World's Columbian Exposition,* by Daniel Burnham (New York: Garland Publishing, 1989), 1:38.

2. Burnham, "Report of the Director of Works," 1:2.

3. Olmsted to Board of Directors, 17 October 1892; Burnham, "Report of the Director of Works," 1:39.

4. Henri Lefebvre, *The Production of Space* (Oxford: Blackwell Publishers, 1974), 26

5. Karl Marx, "The Power of Money in Bourgeois Society," in *The Economic and Philosophic Manuscripts of 1844,* ed. Dirk J. Struik (New York: International Publishers, 1964), 168.

6. For the centralization of mapmaking, see John W. Alvord, "Report of Engineer, Surveys and Grades," in *Final Official Report of the Director of Works of the World's Columbian Exposition,* by Daniel H. Burnham (New York: Garland Publishing, 1989), 1:91. On the topographical maps, see "Reports of Work Done," *Chicago Tribune,* 22 April 1891. For the progress maps, see "Power for Mr. Clowry: Exposition Notes," *Chicago Tribune,* 12 November 1891; and Alvord, "Report of Engineer," 1:91.

7. Alvord, "Report of Engineer," 1:91.

8. Ibid., 1:91–93.

9. "Washington Is Well Billed," *Chicago Tribune,* 25 December 1891.

10. Arthur H. Robinson, *Early Thematic Mapping in the History of Cartography* (Chicago: University of Chicago Press, 1982), 140–154.

11. "Plaint of Gallery Exhibiters," *Chicago Tribune,* 18 June 1893.

12. Perry R. Duis, "'Where Is Athens Now?' The Fine Arts Building 1898 to 1918," *Chicago History* 6, no. 2 (summer 1977): 66.

13. "Money Saved for the Fair," *Chicago Tribune,* 26 May 1892; "Typewriters at the World's Fair," *Chicago Times,* 5 August 1892.

14. Cynthia H. Peters, "Rand, McNally in the Nineteenth Century: Reaching for a National Market," in *Chicago Mapmakers,* ed. Michael P. Conzen (Chicago: Chicago Historical Society, 1984), 70–71.

15. Ibid., 68.

16. *Classification and Rules: Department of Manufactures* (Chicago: World's Columbian Exposition, n.d.), 8–11. The rules issued for other departments contain the same language. Over the course of the fair, many foreign exhibitors protested against this rule, and violations were not uncommon. Toward the end of the exposition, the managers made arrangements for the foreign exhibitors to sell the extra merchandise they had imported with their exhibits, but deliveries could not be made until after the close of the fair. See "Want to Sell Goods," *Chicago Tribune,* 13 July 1893; and "Exhibitors May Sell Goods," *Daily Columbian,* 14 October 1893.

17. Karl Marx, *Capital: A Critique of Political Economy,* trans. Ben Fowkes (New York: Vintage Books, 1977), 1:163–177.

18. In addition to the fair, the extension of new cable and rail lines and the development

of new manufacturing districts also stimulated the South Side real estate market; see Homer Hoyt, *One Hundred Years of Land Values in Chicago* (Chicago: University of Chicago Press, 1938), 161–171. Chicago was selected by Congress as the site of the fair after a competition with New York, Washington, D.C., and St. Louis. On the boosters' efforts to secure the fair for Chicago, see "History of the World's Columbian Exposition" and "They Got the Fair," *Chicago Inter-Ocean,* 1 November 1893.

19. "The Agent's Outlook," *Real Estate and Building Journal* [hereafter, *RE&BJ*] 31, no. 50 (14 December 1889): 876.

20. "The World's Fair," *Economist* 4, no. 7 (16 August 1890): 250

21. *RE&BJ* 32, no. 37 (13 September 1890): 807–808.

22. John J. Flinn, *The Standard Guide to Chicago, World's Fair Edition* (Chicago: Standard Guide Co., 1893), 412, 416; John J. Flinn, *Chicago, The Marvelous City of the West* (Chicago: Standard Guide Co., 1893), 442, 447–448; "Eggleston and Auburn Park," *RE&BJ* 31, no. 39 (28 September 1889): 658.

23. "Eggleston and Auburn Park" (advertisement), *RE&BJ* 33, no. 15 (11 April 1891, suppl.): 11; "South Side Real Estate," *RE&BJ* 33, no. 20 (16 May 1891): 689; "Three-Quarters of a Million Deal at Auburn Park," *RE&BJ* 33, no. 21 (23 May 1891): 724.

24. Patrick Ashley and Emily Clark, "The Merchant Prince of Cornville," *Chicago History* 21, no. 3 (December 1892): 10.

25. Flinn, *Standard Guide to Chicago,* 414; "The Opening Sale and Excursion to Calumet Heights" (advertisement), *Chicago Tribune,* 24 May 1891; "On the Route of the 'L' and Electric Roads" (advertisement), *RE&BJ* 33, no. 15 (11 April 1891, suppl.): 5; "Southside Real Estate: Extensive Purchases Being Made," *RE&BJ* 33, no. 15 (11 April 1891, suppl.): 3.

26. "The World's Fair," *Economist* 3, no. 9 (1 March 1890): 228.

27. "The Real Estate Market," *RE&BJ* 34, no. 46 (12 November 1892): 1451; "Purchase of Real Estate by Aliens," *RE&BJ* 35, no. 7 (18 February 1893): 204.

28. "The Mecca," *RE&BJ* 33, no. 37 (12 September 1891): 1321.

29. "Chicago Real Estate," *Chicago Tribune,* 7 December 1890; Hoyt, *One Hundred Years,* 174.

30. "The Financial Situation" and "The Real Estate Market," *RE&BJ* 33, no. 18 (2 May 1891): 609, 612; "Agent's Budget," *RE&BJ* 33, no. 47 (21 November 1891): 1679.

31. "Money for the Fair," *Chicago Herald,* 19 July 1891.

32. *The Illustrated World's Fair* I, no. 3 (September 1891): 21.

33. "Forces That Operate in the Real Estate Market," *RE&BJ* 34, no. 25 (18 June 1892): 780; "Auction Sales—A Prophecy," *RE&BJ* 34, no. 32 (6 August 1892): 1004; "The Real Estate Market," *RE&BJ* 34, no. 44 (29 October 1892): 1387; "The Real Estate Market," *RE&BJ* 34, no. 45 (5 November 1892): 1419; "The Real Estate Market," *RE&BJ* 35, no. 20 (20 May 1893): 619–620; "South Side Residence Property," *Economist* 8, no. 7 (14 August 1892): 234; "Various Real Estate Matters," *Economist* 8, no. 15 (8 October 1892): 512; "Demand for Rentals," *Chicago Daily News,* 9 April 1892, morning edition; "Big Buyers Hold Back," *Chicago Daily News,* 30 April 1892, morning edition. At the end of 1892, Gross predicted that "the real estate business may be rather quiet for the next few years" (S. E. Gross to Mother, 17 December 1892, Samuel Eberly Gross Papers, Chicago Historical Society, Chicago, Illinois).

34. "The Real Estate Market," *RE&BJ* 34, no. 46 (12 November 1892): 1451; "Current Comment," *RE&BJ* 34, no. 52 1/2 (31 December 1892): 1678.

35. "The Real Estate Market: New Notes," *RE&BJ* 35, no. 20 (20 May 1893): 620; "The Real Estate Market," *RE&BJ* 35, no. 37 (16 September 1893): 1118; "The Real Estate

Market," *RE&BJ* 35, no. 44 (4 November 1893): 1286; *The Economist* echoed this analysis, noting that the fair was, "when in actual operation, rather a hurt to the market, because it diverted the minds of the people from trade, produced the feeling that there might be a collapse after the exhibition was over, and in general threw business out of its ordinary channels." See "Realty and Public Improvements in Chicago," *Economist* annual no. (1 January 1894): 19.

36. Illustrations of these customized products are collected in box 7, Photographs and Scrapbooks, Rand, McNally Collection, Newberry Library, Chicago, Illinois.

37. Karl Marx, *Grundrisse: Foundations of the Critique of Political Economy,* trans. Martin Nicolaus (New York: Penguin, 1973), 148, 187, 221.

38. Ibid., 142–147.

39. Marx, *Capital,* 1:194.

40. David Wells, *Robinson Crusoe's Money* (New York: Harper & Brothers, 1876), 57.

41. Walter Benjamin outlines this evolution in the visual culture of expositions in *The Arcades Project,* trans. Howard Eiland and Kevin McLaughlin (Cambridge, Mass.: Harvard University Press, 1999), 171–202 and passim. See also Susan Buck-Morss, *The Dialectics of Seeing: Walter Benjamin and the Arcades Project* (Cambridge, Mass.: MIT Press, 1990), 81–82.

42. Benjamin, *Arcades Project,* 201. See also Buck-Morss, *Dialectics of Seeing,* 85.

43. Marx, *Capital,* 1:165.

44. Marx, *Grundrisse,* 255, 248.

"A Typically English Brew"

TEA DRINKING, TOURISM, AND IMPERIALISM IN VICTORIAN ENGLAND

JULIE FROMER

$\mathcal{T}$ea is generally regarded as England's "national beverage," a title granted to the drink in the eighteenth century and maintained today.[1] By the early nineteenth century, tea had become a necessary part of English daily life and was consumed by all classes of society. An English national identity had coalesced around the icon of tea, and the rhetoric of the tea industry paid considerable attention to England's relationships with the suppliers of this crucial commodity. Victorian histories of tea include images and descriptions of China to convince readers that drinking tea was a way of touring foreign lands and the British Empire from the domestic comfort of their own homes. But, by emphasizing the foreign origins of English-consumed tea, nineteenth-century tea histories suggest the potential dangers of "consuming" the Orient—anxieties of ingestion, the threat of pollution, and frighteningly permeable political and cultural boundaries. In response, histories of tea offer strategies for reaffirming English physical, political, and cultural boundaries in order to reconstitute English identity. They celebrate tea's ability to unify English social classes, they honor renowned English tea drinkers, and they illustrate the important technological innovations that allowed English tea retailers to protect consumers from questionable Chinese trade practices. Ultimately, however, tea histories reveal that the British tea industry's central strategy for procuring safe, secure sources of tea was to transform tea from a foreign commodity into a product of the British Empire.

During the seventeenth and eighteenth centuries, all European imports of tea came from China, but the early-nineteenth-century discovery and cultivation of tea in British-controlled regions of India resulted in a precipitous decline in the amount of China tea imported to England. English imports of tea continued to increase throughout the nineteenth century, as they had from tea's first introduction to England in the 1650s; but more and more tea came from the British colonies of Assam and Ceylon.[2] Late-nineteenth-century publications sought to establish that tea was not only consumed by tea drinkers in the cultural center of British power but also produced by British planters and, therefore, that tea originated from an outpost of that cultural center. By encouraging tea drinkers to envision themselves as contributing to the growth of British naval, economic, and colonial power, the tea industry helped to construct the image of England as an imperial nation.

My analysis of tea in Victorian England is based upon three histories of tea published between 1850 and 1890. Robert Fortune's *A Journey to the Tea Countries of China; Including Sung-Lo and the Bohea Hills; with a Short Notice of the East India Company's Tea Plantations in the Himalaya Mountains* (1852) provides a firsthand account of his East India Company–sponsored adventures in China.[3] Fortune's purpose in China was to gain access to Chinese secrets of cultivating tea and to control the production of tea by transporting the knowledge and plants he had collected to British-controlled territory in India. Samuel Day's *Tea: Its Mystery and History* (1878) offers a more historical look at tea drinking in England; in addition to describing exotic Chinese traditions of tea drinking, Day's treatise creates the image of tea as quintessentially English. At the same time, however, Day's text emphasizes the cultural anxieties created by consuming Chinese tea and basing English national identity upon a product grown and manufactured outside of British knowledge and control. Day's response to the potential threat of adulterated Chinese tea is to emphasize the purity and hygienic quality guaranteed by the retailer and linked to the brand name of Horniman's Pure Tea. Arthur Reade's *Tea and Tea Drinking* (1884) presents a more scientific account of the history of tea in England, weighing the opinions of tea's historical opponents and advocates. Reade continues Day's celebration of tea's English and Chinese background; and he acknowledges the predominance of Chinese tea on the market, but he is optimistic about the future of Indian tea to provide a safe, secure source for the English national beverage.

Tea Drinking as Tourism

Robert Fortune, Samuel Day, and Arthur Reade's publications discussing the merits of tea and popularizing its history present tea drinking as a form of tourism. All three texts emphasize the cultural otherness of China, and they suggest that the English consumers of tea could capture the exoticism of the Orient in their teacups. Day and Reade's books include numerous illustrations of Chinese tea plantations, focusing the reader's gaze on the familiar domestic beverage being cultivated in faraway, exotic locales. Robert Fortune's *Journey to the Tea Countries of China,* "with map and illustrations," is essentially a travelogue of his adventures in the unknown, forbidden interior of China.[4] Fortune reveals the secrets of China through panoramic scenes:

> As we went over the passes we always rested while on the highest point, from which we obtained a view, not only of the valley through which we

had come, but also of that to which we were going. The long trains of coolies laden with chests of tea and other produce, and with the mountain chairs of travellers, presented a busy and curious scene, as they toiled up the mountain side, or were seen winding their way through the valleys. These were views of "China and the Chinese" as they are seen in everyday life.[5]

According to Fortune, scenes of Chinese everyday life make the perfect picture for English readers contemplating the exotic East. The work of Chinese laborers—carrying tea toward the coast for export to England—is framed within the mountain pass, a vantage point on Fortune's journey from which he can look behind and before him, surveying all of China. The words "China and the Chinese" are put into quotation marks, so that, just as the scenes are enclosed by the mountain passes, the representation of these scenes is set aside, set off, and highlighted within an English text as an exotic tableau offered to the reader's gaze.

Through descriptions of the exotic landscapes surrounding the origin and cultivation of tea destined for English tea tables, nineteenth-century tea authors invited their English readers to experience the mystique of China in every cup of tea. According to anthropologist Pierre Van den Berghe's definition, the central element of tourism is exoticism: "The tourist comes to see and experience something he cannot duplicate at home."[6] By offering imaginary tours of China, nineteenth-century tea histories emphasize elements that marked tea-producing regions as geographically and culturally different from England. The crucial distinction between tourism and tea drinking, however, is that the English tea drinker could and did duplicate this exoticism at home, through the process and ritual of consuming tea. Day, Reade, and Fortune, in their nineteenth-century histories of tea, present the tea drinker as participating in a type of armchair tourism, vicariously witnessing and participating in Chinese traditions, traveling to distant lands via the economic and naval power of British trading companies.

The nineteenth-century custom of drinking tea becomes increasingly meaningful when viewed as a method of tourism, in which the tourist seeks, according to social anthropologist Erik Cohen, "appreciation of that which is different, strange or novel in comparison with what the traveller is acquainted with in his own cultural world." Identifying acts of tourism according to a tourist's relationship to a social or religious center, Cohen emphasizes the process of comparing exotic elements of cultural experiences with a tourist's home culture. According to Cohen, "Every society possesses a 'centre,' which is the charismatic nexus of its supreme, ultimate, moral values."[7] Fortune's text establishes England as an offstage center, the cultural foundation of his own viewpoint and the judgments he makes about Chinese customs. Containing the scenes he describes within English linguistic and textual strategies, he implicitly compares Chinese and English cultural norms. Fortune has given up the literal security of the boundaries of his nation, and he replaces them with the rhetorical boundaries of his written text, emphasizing the cultural power of the narrative gaze.

Arthur Reade's text, however, occupies a different rhetorical location—firmly within England. Reversing the imagery of Fortune's scenes of Chinese life framed by English rhetoric, Arthur Reade's *Tea and Tea Drinking* (1884) frames an icon of Englishness within a flourish of exotic Orientalism. The frontispiece of *Tea and Tea Drinking* depicts a recognizably English teacup (complete with handle, saucer, and teaspoon) within Asian-style lettering and cherry blossoms, creating a central English image surrounded by Oriental ornamentation. Rather than containing scenes of exoticism within English conventions, Reade portrays England as the center of his text, framed by the exotic locales of the empire and the world. In tea drinking, the message of the power of the original center, English culture, is reinforced dramatically by the fact that, in truth, the tea drinker never leaves that center. He or she is able to travel metaphorically, experiencing the wonders of the world that have been transported back to that center by its powerful economic and political forces. Thus, Cohen's schema reveals that nineteenth-century histories of tea offered an opportunity for "recreational" tourism through tea drinking, a recreation that "performs a serious 'function'—it restitutes the individual to his society and its values, which, despite the pressures they generate, constitute the centre of his world."[8] Cohen's definition of recreational tourism suggests that traveling for pleasure has an important societal role—allowing the tourist to witness other cultures in the process of impressing the tourist with the power and values of his or her own culture.

Nineteenth-century histories of tea thus perform the cultural work of investing the tea drinker with an important sense of English national identity. Inderpal Grewal has argued for the similar function of other vicarious forms of tourism within England, such as guidebooks, travelogues, and museums. According to Grewal, creating the impression of a national culture depends upon the representation of the unification of classes within English society.[9] Nineteenth-century histories of tea celebrate the ability of tea to unify English social classes, and Fortune, Day, and Reade emphasize this utopian vision of a classless national culture by contrasting the English consumer of tea with its Asian producer. Robert Fortune, Samuel Day, and Arthur Reade all include illustrations in which Chinese landscapes provide the backdrop for Asian workers in large bamboo hats and pigtails, bending over the tea plants. The captions for these illustrations emphasize the "great care" and the expertise needed to pluck, weigh, dry, and fire the tea leaves: "Immense care is necessary in order to protect the delicate young leaf from injury. . . . Every leaf has to be plucked separately."[10] The descriptions of the labor necessary to produce the tea would have helped to create an intensified appreciation of tea, a consideration of the amount of delicate handwork that went into each teaspoonful of leaves. At the same time, these depictions of Chinese workers laboring to produce British tea inspired a certain class identification, a separation between the manufacturers and the drinkers of tea.

While working to cultivate and harvest tea reduced the Chinese tea gatherers to poor laborers, the act of drinking tea helped to unite all classes of English society and to raise them to a cultural level above the Chinese workers. A popular English advertisement affirms that, in China, "[t]ea . . . is used at meals and upon all visits and entertainments in private families, and in the palaces of grandees."[11]

By offering a contrasting picture of those who drank tea versus those who worked to produce tea, Victorian authors created a rigid class hierarchy within the Asian society they portrayed. The reader, the English tea drinker, was encouraged to identify with the upper-class tea drinkers in this hierarchy. Emphasizing the cross-class nature of tea drinking in England, the rhetoric of these publications helped build a coherent English national identity that crossed class lines, as opposed to the foreign, exotic, hierarchical structure of Chinese society.

Images of Asian "grandees" create a sense of the elite nature of tea drinking, but invoking English royalty allows Day and Reade to bring their tours of Asia back to the cultural center of England. Each author constructs a specifically English tradition of tea consumption around a collection of famous tea drinkers. Reade, in *Tea and Tea Drinking,* proclaims that "the influence of Royalty did more than anything else to make tea-drinking fashionable."[12] Popular legend specifically attributed this influence to "Catherine . . . , queen of Charles II, who had tasted the beverage in Portugal, and grew enamored with the same, [and] brought it into fashion in this country."[13] Subsequent queens followed the tradition established during the seventeenth century, becoming "worshippers at the shrine of tea"; Queen Victoria's first command after her accession to the throne was said to have been "Bring me a cup of tea and the *Times.*"[14] Nineteenth-century sources also include examples of tea-drinking authors and poets, including Pope, Cowper, Byron, and Samuel Johnson. Literary and royal tea drinkers are not the only heroes in Day and Reade's texts—leaders of commercial industry also figure largely in their histories and, thus, in their constructions of English identity. Although tea was produced and manufactured in an exotic, distant country, it was imported to England and marketed there by famous English merchants, such as Thomas Garraway and Thomas Twining. By focusing on such well-known English men and women and highlighting their taste for tea, nineteenth-century authors contributed to the tradition of tea as a national, English beverage: "It is not, possibly, too great an assumption that there must exist something about Tea specially suitable to the English constitution and climate."[15]

Anxieties of Adulteration

Basing a national identity on a product manufactured thousands of miles away, however, caused anxiety within English texts on tea. The very act that asserted British power and control, the act of importing tea into England and enabling the English consumer to tour the world from his or her own armchair, also contained the possibility of weakening and polluting that power. The process of consumption, of taking tea into the English body, involved permeating the boundaries of that body and allowing potentially dangerous substances to invade it. Samuel Day, writing about Horniman's Pure Tea in 1878, argues that the largest threat to the English tea drinker was the false coloration and adulteration of green tea by Chinese manufacturers, who intentionally deluded "English fools" with poisonous substances. According to Day, "The Green Teas sold in England are usually artificially coloured in order to enamour the eye of the unsuspecting purchaser.

The principal medium employed in effecting this result is none other than Prussian blue, a deadly poison." Day contends that the beneficial commercial competition that followed the 1834 dissolution of the East India Company's China monopoly effectively ended English smuggling and tea adulteration. "Yet," he warns, "there is no protection against what is done in China."[16]

Nineteenth-century concerns about the adulteration of food were not limited to tea, but the status of tea as a product imported from a country over which Britain had no economic or military control grants the fears of tea adulteration special consideration. According to Jack Goody's study of the cultural significance of consumption patterns in *Cooking, Cuisine, and Class,* "Adulteration is a feature of the growth of urban . . . or rural society that is divorced from primary production."[17] The problem of adulterated tea presented a more exaggerated case of the gap between production and consumption. Not only was production carried on thousands of miles away from consumers, but the producers maintained strict secrecy about their methods of cultivation and manufacture, preventing the English from observing and maintaining the quality of tea exported from China.[18] Politically, China had maintained a defiant sense of itself, staving off European foreign powers and influence. Despite numerous military losses to the British, the Chinese retained economic and political control of their own boundaries and, therefore, of their tea exports. Chinese officials continued to refuse British merchants access to the Chinese interior, where the tea plantations were located. According to nineteenth-century histories of tea, English tea consumers were vulnerable to the practices of Chinese tea manufacturers because the British could not monitor the production of tea.

The anxieties of adulteration and pollution evident in Day's text resonate on both individual and political levels. *In Purity and Danger: An Analysis of Concepts of Pollution and Taboo,* anthropologist Mary Douglas explores the significance of pollution in relation to cultural taboos concerning food and eating. Douglas argues that "the processes of ingestion portray political absorption."[19] The act of consuming, according to this model, creates permeable boundaries between political entities. In *Tea: Its Mystery and History,* Samuel Day's fears of adulterated China tea echo his fears of a world polluted by the breakdown of Chinese political and physical boundaries. As China became more and more accessible to foreign trade, through trade negotiations and armed conflicts, those boundaries suddenly lost their ability to maintain cultural and racial distinctions:

> Who could have thought that the Tea trade was destined to become one of the most important branches of our commerce, and not only so, but to occasion several wars, lead to the extension of our Eastern possessions, and precipitate the great Chinese exodus, which threatens such important results to the Pacific States of America, to Australia, the Polynesian Islands, and possibly to the world at large?[20]

According to Day, although Britain's power to import tea from China symbolized one of the great achievements for English culture, the success of the tea trade threatened to literally dilute that power within the world. Not only would English consumers be polluted by adulterated, poisoned Chinese tea, but the world at large was in danger, according to Day, of being polluted by the disintegration of Chinese boundaries and the influx of previously isolated Chinese individuals into the rest of the world.

Strategies of Strengthening English Boundaries

Samuel Day's answer to the problem of permeable political boundaries focuses on English commercial opportunities, illustrating his commitment to the free trade capitalism that followed the dissolution of the East India Company's China monopoly. Day argues that the British government, including perhaps the East India Company, had failed to metaphorically maintain the borders between China and England, allowing adulterated tea to be sold to unsuspecting English tea drinkers. Speaking on behalf of Horniman's Pure Tea, Day advocates that English consumers should wield their buying power to protect themselves, choosing the purest, highest quality tea from the most reputable tea merchants.

Horniman's Pure Tea depended upon a new Victorian innovation in tea sales—prepackaged tea. Previously, all tea had been sold in bulk form, blended and packaged by local grocers for individual customers. Horniman's message, according to Denys Forrest, a twentieth-century tea historian, was that "the consumer buying a packet of Horniman's tea in its foil-lined paper wrapping was getting a hygienically protected, uniformly weighed quantity of unadulterated leaf." Placing concerns about hygiene within an imperial context, Anne McClintock argues that late-nineteenth-century packaging innovations encouraged brand recognition and, perhaps more important, signified Victorian interest in sanitizing products that had come from the "dirty" empire and that had been handled by tradesmen.[21] The introduction of individually wrapped packages of Horniman's Pure Tea, like the brands of soap that McClintock discusses, reaffirmed a physical barrier between Chinese tea and English tea drinkers. Since the boundaries separating the Chinese and the English were finally beginning to falter, as more Chinese ports opened to foreign trade and Chinese exports of tea continued to increase, British tea merchants erected new boundaries closer to home—"sealed packets," paper packaging, and certifications of purity.[22] Packaging innovations helped further the construction of tea as the English national beverage, increasing the distance between the dangerous Chinese producer and the certifiably hygienic English tea dealer.

Samuel Day, writing in the 1870s, was situated on the cusp of the shift from Chinese to Indian tea imports. China tea sales reached their peak in 1879, the year after *Tea: Its Mystery and History* was published; from that year onward, tea imports from China fell steadily, even as total amounts of tea imported to England continued to increase.[23] Arthur Reade's *Tea and Tea Drinking,* looking at the tea industry as a whole rather than from the viewpoint of a single brand of tea, proposes a different method of ensuring the protection of English bodies and identities from potentially adulterated Chinese tea. Whereas Day's solution erects boundaries closer to home, between the consumer and the tea that he or she purchased, Reade prefers a more global boundary—expanding the British Empire in order to encompass more of the world and bringing tea production into the British sphere of influence. Reade suggests that the cultivation of Indian tea will permanently solve the threat to English identity posed by Chinese tea:

> The tea plant, although cultivated in various parts of the East, is probably indigenous to China; but is now grown extensively in India. In consequence of the poorness of the quality of the tea imported by the East India Company, and the necessity of avoiding an entire dependence upon China, the Bengal Government appointed in 1834 a committee for the purpose of submitting a plan for the introduction and cultivation of the tea-plant; and a visit to the frontier station of Upper Assam ended in a determination on the part of Government to cultivate tea in that region. In 1840 the "Assam Company" was formed, and it is claimed for them that they possess the largest tea plantation in the world. . . . Every year thousands of acres are being brought under cultivation, and in a short time it seems likely that we shall be independent of China for our supplies of tea.[24]

Like Day, Reade exhibits an awareness of Chinese practices of adulteration, but Reade posits a more aggressive solution to the possibility of pollution. According to Reade, English consumption of Chinese tea formed the basis of commercial and political dependency, a relationship that weakened Britain's international position. He emphasizes the need to avoid "an entire dependence upon China" and enthusiastically champions the goal of finally becoming "independent of China for our supplies of tea." For Reade, tea has become an essential part of the colonizing process; his history of British colonization of Assam is integrally linked to the need for British sources of tea.

The cultivation of tea in India, on British-ruled soil, allowed the British to maintain control over the entire process of tea production, from the initial planting through tea plucking and drying to the final exportation to England. As tea imports from the colonies in India and Ceylon increased, cultural reliance on tea as part of the English national identity acquired imperialistic overtones: "A large quantity of tea is now imported from this island [Ceylon], and new plantations, it is reported, are being made every month; day by day more of the primeval forest goes down before the axe of the pioneer, and before another quarter of a century has passed it is anticipated that the teas of our Indian empire will become the most valuable of its products."[25] Tea was no longer an exotic commodity imported to England from uncertain, malevolent, foreign sources; instead, tea had become a product exported from within the British Empire. Reade asserts possession over Indian teas, the teas of "our Indian empire," and he equates tea production with Victorian pride in national and technological progress. He looks forward to a time when, far from relying on unscrupulous Chinese manufacturers for adulterated, poisoned tea, Britain will be producing tea for English consumption and for exportation abroad.

Conclusion

According to Emiko Ohnuki-Tierney's anthropological investigation of the correlation between eating habits and identity, "Food tells not only how people live but also how they think of themselves in relation to others."[26] Conceptually, both tourism and the literal consumption of food involve the interruption of boundaries between the self and the other, introducing the self to new experiences and ingesting what are essentially foreign elements. By the early nineteenth century, however, tea had become a central part of physical existence and social interaction. Rather than initiating comparisons between self and other, as tourism tends to do, tea had become unquestionably absorbed into the cultural center of English life. Victorian tea histories exhibit anxiety over the extent to which tea drinkers had become inured to tea's foreign origins and to the blurred distinction between self and other that had accompanied the adoption of tea as a daily necessity. Although tea had become known as the "national beverage" during the eighteenth century, owing to the fact that English consumption of tea far outstripped that of other European nations, tea nevertheless remained a foreign import and contained potentially dangerous implications of dependency and pollution. But during the course of the nineteenth century, as Britain began producing tea for its own consumption within its "Indian empire," the significance of tea's label as "the national beverage" acquired new meaning. Consuming tea became a method of absorbing British imperialism, of literally and physically participating in the vital circulation of goods maintained by the British Empire. According to nineteenth-century tea histories, tea constituted English national identity both metaphorically and bodily, contributing to the continued strength of England and its people.

Notes

The words of the title, "A Typically English Brew," are taken from a description of Twining's English Breakfast Tea, a "blend of Indian and Ceylon teas creating a typically English brew," printed on hundred-gram tins (Twining & Co., Ltd., London).

1. Many late-eighteenth- and nineteenth-century treatises claim that tea is the national beverage; for example, see Samuel Phillips Day, *Tea: Its Mystery and History* (London: Simpkin, Marshall & Co. [sponsored by Messrs. Horniman's & Co.], 1878); and Arthur K. Reade, *Tea and Tea Drinking* (London: Sampson Low, Marston, Searle, & Rivington, 1884). There are only a few relatively recent scholarly histories of tea; for two examples that celebrate tea's English heritage, see Edward Bramah, *Tea and Coffee: A Modern View of Three Hundred Years of Tradition* (London: Hutchinson of London, 1972); and Denys Forrest, *Tea for the British: The Social and Economic History of a Famous Trade* (London: Chatto & Windus, 1973).

2. According to Arup Kumar Dutta's *Cha Garam! The Tea Story* (Guwahati, India: Paloma Publications, 1992), 79, the amount of tea imported to Great Britain from China dropped from more than 113 million pounds to 80 million pounds between 1885 and 1888. Over the same three years, combined imports of tea from India and Ceylon rose from approximately 69 million pounds to more than 104 million pounds.

3. Robert Fortune, *A Journey to the Tea Countries of China; Including Sung-Lo and the Bohea Hills; with a Short Notice of the East India Company's Tea Plantations in the Himalaya Mountains* (London: John Murray, 1852).

4. Ibid., title page.

5. Ibid., 204.

6. Pierre L. Van den Berghe, "Tourism as Ethnic Relations: A Case Study of Cuzco, Peru," *Ethnic and Racial Studies* 3, no. 4 (October 1980): 377.

7. Erik Cohen, "A Phenomenology of Tourist Experiences," *Sociology, Journal of the British Sociological Association* 13, no. 2 (May 1979): 182–183, 180.

8. Ibid., 185.

9. Inderpal Grewal, *Home and Harem: Nation, Gender, Empire, and the Cultures of Travel* (Durham, N.C.: Duke University Press, 1996), esp. the introduction and chaps. 1–4. A number of recent Victorian scholars have investigated the ways in which nineteenth-century texts constructed a unified, middle-class audience. For example, see Mary Poovey, *Making a Social Body: British Cultural Formation, 1830–1864* (Chicago: University of Chicago Press, 1995); and Anne McClintock, *Imperial Leather: Race, Gender, and Sexuality in the Colonial Contest* (New York: Routledge, 1995), esp. chap. 3.

10. Day, *Tea: Its Mystery*, 44.

11. Ibid., 33; Reade, *Tea and Tea Drinking*, 3.

12. Reade, *Tea and Tea Drinking*, 6.

13. Day, *Tea: Its Mystery*, 37.

14. David Crole, *Tea: A Text Book of Tea Planting and Manufacture; Comprising Chapters on the History and Development of the Industry, the Cultivation of the Plant, the Preparation of the Leaf for the Market, the Botany and Chemistry of Tea, etc., etc., with Some Account of the Laws Affecting Labour in Tea Gardens in Assam and Elsewhere* (London: Crosby Lockwood and Son, 1897), 2.

15. Day, *Tea: Its Mystery*, 61.

16. Ibid., 46–47, 57.

17. Jack Goody, *Cooking, Cuisine, and Class: A Study in Comparative Sociology* (Cambridge: Cambridge University Press, 1982), 171.

18. See Bramah, *Tea and Coffee*, 86; and Forrest, *Tea for the British*, 41.

19. Mary Douglas, *Purity and Danger: An Analysis of Concepts of Pollution and Taboo* (London: Routledge and Kegan Paul, 1966), 4.

20. Day, *Tea: Its Mystery*, 49.

21. Forrest, *Tea for the British*, 132; McClintock, *Imperial Leather*, 210–211.

22. Goody, *Cooking*, 173.

23. Forrest, *Tea for the British*, 162.

24. Reade, *Tea and Tea Drinking*, 19–20.

25. Ibid., 21.

26. Emiko Ohnuki-Tierney, *Rice as Self: Japanese Identities through Time* (Princeton, N.J.: Princeton University Press, 1993), 3.

Cultural Ecologies of the Coast

SPACE AS THE EDGE OF CULTURAL PRACTICE IN MARY KINGSLEY'S *TRAVELS IN WEST AFRICA*

JULES LAW

> *Nor indeed do I recommend African forest life to any one. Unless you are interested in it and fall under its charm, it is the most awful life in death imaginable. It is like being shut up in a library whose books you cannot read, all the while tormented, terrified, and bored.*
> Mary H. Kingsley, *Travels in West Africa*

Texts and Maps

In a famous, virtually modernist image, Mary Kingsley compares the African forest to an illegible text—specifically to a "library whose books you cannot read."[1] The textual metaphor has been welcomed by literary critics whose primary interest has been to analyze the way Kingsley's travelogue, like other nineteenth-century documents, produced Africa as a "text." The principal tool in such contemporary analyses has itself been rhetorical; that is, critics have concerned themselves with the way various, even competing, literary genres and conventions mingled in those European texts that in turn produced Africa as a text for various interests and purposes. Specifically, such notable critics as Mary Louise Pratt, Sara Mills, Anne McClintock, and Karen Lawrence have examined the triangulated set of relations between gender, genre, and imperialism.[2] Their studies reveal how representations of non-European cultures and places by European writers were heavily inflected by a highly gendered set of tropes (for example, "monarch of all I survey") and genres (adventure story, ethnography, journal, domestic tale, and so on).

The interest in genre has produced important insights into the relation of gendered subjectivity to political subjecthood. Victorian women's texts on Africa were in various and complicated ways both parodic of, and complicit in, the aggrandizing and eugenicist public discourses dominated by male politicians, explorers, entrepreneurs, and novelists. One might then quite justifiably characterize Kingsley's travelogue itself as a battleground containing two competing discourses, variously characterizable as relativist and ethnocentric, comic and epic, idiosyncratic and empirical. If Kingsley's text is a site for competing discursive schemes, however, it is also, perhaps less obviously, a site for competing mapping schemes. In this paper I would like to consider Kingsley's *Travels* in the light of recent work on geography and, while not ignoring issues of rhetorical mediation, to examine the tension between two significantly differentiated conceptions of space in her text. We might call these conceptions the distributive and the migratory. My final aim will be to reintroduce the variable of gender into the discussion and, in doing so, to subsume the distributive and migratory schemes into the more general idea of space as the "edge of cultural practice."

"I was a trade product," announces Kingsley halfway through her narrative, when beset by the revelation that the European traders in the French Congo regard her journey in a commercial rather than a "philosophical" light.[3] Taken together with Kingsley's pervasive textual metaphors and her constant interest in questions of translation, this figure foregrounds the central role played by routes of sexual and commodity exchange in outlining the imaginary geography of Africa for fin de siècle Britain. But if the space of Africa was *defined* in terms of overlapping distributive and exchange systems, it was nonetheless *imagined* principally in terms of the peripheries of those systems. Networks of distribution and exchange were imagined to fill up, or trace out, the African continent. Thus Kingsley writes endlessly of distribution: the distribution of disease, the distribution of shirts, the distribution of wives, the distribution of flora and fauna, of languages, and of sexual and cultural practices. My goal here will be to trace the relationships between these systems of distribution and their imagined borders, between the entropic tendency of cultural systems and the putatively tropic finitude of representational space. My argument is that fin de siècle writers identified something like an ecological system at work in Africa, in which the ebb, flow, development, and devolution of cultural systems (such as language, kinship, and economic exchange) merged imperceptibly with patterns of geology and geography. Furthermore, I want to argue that these systems were understood as visible principally at their edges or limit points.

Africa as Grid

Late-nineteenth-century Europeans employed four distinct imaginary geographical grids to elaborate Africa as a comprehensible system of differences. First among these grids was the ideologically freighted distinction between the East and West. It was certainly no coincidence that William Booth's 1890 scandalogue, *In Darkest England,* a London remapping of Stanley's *In Darkest Africa,* turned on a distinction between London's civilized West End and its primitive East End. For all its reputation as the "white man's grave," West Africa still retained all the connotations of "Western-ness" in Victorian cultural teleology. Whereas the colonization of Africa's east coast bespoke a classical, Mediterranean, even Asiatic history, with successive waves of Arabic and Persian colonizers (as Rider Haggard's 1880s and 1890s novels constantly reminded his readers), the west coast was a series of modern Euro-national departments. Superimposed over this grid, with ease, was a second geocultural grid: north and south. Tipped on its side, "Westward the course of civilization" became "Northward the origins of civilization." Whether the "civilizing" influence on sub-Saharan Africa was seen to be European (as on the west coast) or Arabic (as on the east coast), there was a consensus that culture penetrated Africa from the north and that the further south one went, the more primitive conditions became. To read Africa either from west to east *or* from north to south was still to read it teleologically.

However, the third grid, that of interior and exterior, presented complications. Joseph Conrad's Marlow might treat the scheme teleologically, equating "the farthest point of navigation [upstream]" with "the culminating point of [his] experience,"[4] but as we shall see the scheme did not apply equally on all the perimeters of the continent and was fraught with internal contradictions as well. Kingsley, for instance, begins her narrative by noting that the eminent geographer Alfred Wallace had identified the interior of West Africa as an unmapped region beckoning to the aspiring zoologist. Yet Wallace's own characterization of West Africa (in *The Geographical Distribution of Animals*) is as internally contradictory as his characterization of East Africa. In contrast to the monotonousness of East Africa, writes Wallace, West Africa is zoologically vital and diverse. He asserts this despite admitting that we know little of its interior; in fact, paradoxically, it is the unknown character of the interior that marks it out as a site of interest: "This extensive and luxuriant district has only been explored zoologically in the neighborhood of the . . . coast. Much, no doubt, remains to be done in the interior, yet its main features are sufficiently well known, and most of its characteristic types of animal life have, no doubt, been discovered."[5] The passage remains poised between an unknown interior and a sufficient exterior, and the passage's fundamental ambivalence about the distinct ontological status of the interior is casual rather than sublime. This ambivalence certainly harmonizes with the deadpan style of Kingsley's narrative, which represents Africa as a constantly shifting set of economies rather than as an ever receding, ever immanent revelation. And this strategy points in turn toward a fourth grid, the most significant for Kingsley's text: Africa as a distributive or circulatory grid.

Distributive Space

Travels in West Africa recruits Victorian anthropology to a view of Africa as the outline or limit point of a varied set of self-regulating economies, principally of women, commodities, and disease, but also, subtending all these, of language. Kingsley's constant interest in linguistic practices and her pervasive linguistic analogizing provide a crucial insight into her sense of space. For without the example of language one might gather the impression that space, in Kingsley's view, was simply the phenomenon opened up by systems of social distribution rather than an independent entity upon which practices and patterns were inscribed. Instead, we will see that in many ways an analogy obtains in Kingsley's text between the relation of language to its objects (as it obtained in Kingsley's pre-Saussurean conception) and the relation of culture to geography. Both language and culture are imagined as extending over, reflecting, and imposing patterns on some independent entity that is imagined, variously, as the "mind," the objects, or the space of Africa.

Africa, to Kingsley, is first and foremost a place of linguistic confusion. African languages, she complains, fail adequately to distinguish between genders, or even between animate and inanimate objects.[6] This failure at the level of the signifier is reflected in the economy, where the ubiquitous barter principle is constantly undermined by the shifting values of objects from one transaction to the next (203–204). The alleged lack of stable signifiers—the fact that the "same" thing is called by different names—is perhaps Kingsley's greatest frustration with the cultures and space of Africa. But she is careful to distinguish between language and its referents. It is a virtual axiom of her text that African languages are inadequate instruments for rendering the sophistication of "the native mind":

> No one who has been on the Coast can fail to recognise how inferior the native language is to the native mind behind it. (431)

> I will not go into the subject of African languages here, but only remark of them that although they are elaborate enough to produce, for their users, nearly every shade of erroneous statement, they are not, save perhaps M'pongwe, elaborate enough to enable a native to state his exact thought. . . . In all cases I feel sure the African's intelligence is far ahead of his language. (504)

Of course it is not only the African "mind" that is inadequately reflected in language but also, apparently, African space. And this failure is not attributable exclusively to the intrinsic liabilities of specific African languages. Beyond these liabilities is the sheer problem of multiple linguistic communities: "Geographical research in this region is fraught with difficulty, I find, owing to different tribes calling one and the same place by different names" (237). Not only native linguistic practices but European ones as well seem to play this same trick on the cartographic impulse:

> The various divisions of the West Coast of Africa are very perplexing to a new comer. Starting from Sierra Leone coming south you first pass the Grain Coast, *which is also called* the Pepper or Kru Coast, or the Liberian Coast. Next comes the Ivory Coast, *also known as* the Half Jack Coast, or the Bristol Coast. Then comes the Gold Coast; then the old Slave Coast, *now called* the Popos. . . . *In addition to these names* you will hear the Timber Ports, and the Win'ard and Leeward Ports referred to, and it perplexes one when one finds a port, say Axim, referred to by one competent authority, *i.e.* a sea-captain, as a Win'ard port, by the next as a Timber, by the next as a Gold Coast port. (75; my emphasis)

Fortunately for Kingsley, this proliferation of signifiers is not critical; she is interested more in cultural than in physical geography. Although at points in *Travels* she goes into fairly detailed accounts of the various topographical and logistical networks that divide up African space (discussing, for instance, the respective merits of rivers, roads, footpaths, and rail lines in facilitating trade [for example, 634–641]), her sense of space tends to be dominated more by the outlines of cultural systems than of physical systems. Excepting on the coast—where the ubiquitous denominator of "disease" appears to be absolutely determining—the significant obstacles to, and facilitators of, travel in West Africa, in Kingsley's mind, are cultural. How far one can travel, how fast, how safely, how easily—all these are dictated by the reach and interaction of cultural systems. "[A]ll trade in West Africa follows definite routes," she writes (314); but these routes are marked not so much by topographical markers as by kinship systems. Traders who wish to avoid being poisoned by envious or hostile villages quite literally map out their routes with kinship ties:

> [A]lthough the village cannot safely kill him [the trader], and take all his goods, they can still safely let him die of a disease, and take part of them, passing on sufficient stuff to the other villages to keep them quiet. Now the most prevalent disease in the African bush comes out of the cooking pot, and so to make what goes into the cooking pot—which is the important point, for earthen pots do not in themselves breed poison—safe and wholesome, you have got to have some one who is devoted to your health to attend to the cooking affairs, and who can do this like a wife? *So you have a wife—one in each village up the whole of your route. I know myself of one gentleman whose wives stretch over 300 miles of country,* with a good wife base in a Coast town as well. (315; my emphasis)

Cultural and physical geography run very close in this scheme, in which the map of wives presumably traces over some previously existing map of trade routes, which in turn may have been dictated by either topography or social relations. But Kingsley does not depict space and culture as conceptually indistinguishable from each other in these instances. Instead, and this is my point here, Kingsley imagines cultural patterns extending over space much in the same way she imagines language extending over its objects. Just as objects are conceived here as having an existence independent of and prior to language (which, in this prestructuralist scheme, may be more or less adequate to its referents), so too does physical space possess an ontological, if not epistemological, priority, as the theater of human meanings and actions. The analogy is not entirely symmetrical, since Kingsley does not bemoan the failure of the physical environment to secure its own reflection in culture the way she bemoans the failure of thoughts to determine language. Yet even if space is not credited with determining culture in Kingsley's text, as the postmodern critique of geography points out, passive and determining space are equally Kantian concepts, equally rationalized. What is important is that landscape be imagined in the abstract as pure extension, over which particular patterns (whether economic, linguistic, or sexual) may be traced. And this is precisely how Kingsley appears to imagine African space. One thrust of her text is an emphasis on the utter monotony and frustratingly undifferentiated nature of African topography

and culture: its swamps, rivers, and forests extend in emphatically nonsublime re-
petitiveness; its languages and barter systems are hopelessly muddled by the lack
of any stable system of differentiation; and its kinship practices—polygamous,
matrilineal, orally recorded, and undermined by contravening principles of chattel
slavery—make any consistent pattern of allegiances and obligations impossible to
track. All in all, writes Kingsley, "It is like being shut up in a library whose books
you cannot read, all the while tormented, terrified, and bored" (102). But Kingsley's
library-of-Babel metaphor captures precisely her paradoxical understanding of African space, in which rational differentiation on the one hand and category confusion on the other testify equally to a conception of space as pure extension. It is
across this particular conception of undifferentiated space that Kingsley imagines
the distinctive fetish- and kinship-practices of various African tribes distributing
themselves. And distribution is precisely the phenomenon that defines the reach
of a culture in Kingsley's view, whether it is the distribution of goods or, as she
puts it, of wives.

My point here is to distinguish a certain impulse in Kingsley's text—in which
cultural geography is isomorphic with, yet categorically distinct from, physical geography—from a more radical conception of space as something essentially in flux.
Whereas the former conception treats space as the static theater of distribution,
the latter conception treats physical geography as dynamic. Though Kingsley's
sense of landscape as mapped out by social processes of distribution represents a
thoroughly social conception of space, it does not unsettle her sense of the stability of that space. This is not quite the case when she turns from the stabilizing
economies of distribution within particular cultures to the more unsettling phenomenon of migration.

Migratory Space

> *Social and spatial structures are dialectically intertwined in social life,*
> *not just mapped one onto the other as categorical projections.*
> Edward W. Soja, *Postmodern Geographies*

If Kingsley apprehends African space in terms of the peripheries of distributive
systems, then nowhere is this apprehension more acute than on the periphery of
the continent itself: the coast. The important work of Paul Gilroy, Joseph Roach,
and others on trans- and circum-Atlantic cultures has heightened our awareness
of the ways in which the African coast functioned as a contact zone, not in the
simple binary sense of confrontation or exchange but in the sense of a circulatory
site where cultures were appropriated, hybridized, and performed.[7] Kingsley shares
this sense of the coast as a dynamic site—far from the monotonous, impenetrable,
indecipherable exterior mythologized by Conrad. For Kingsley, the coast is a place
both of cultural and of physical flux, with this difference from the interior: that
physical space seems to be much more actively intertwined with human lives. But
when Kingsley looks at the coast she does not see hybrid cultures; rather she sees
a temporal succession of communities seemingly impelled toward the sea by some

evolutionary force deep within the continent. It is the task of the African coast, in Kingsley's ecological view, to give shape and meaning to this flow of populations.

Not surprisingly, given the governing tropes of her era, Kingsley sees Africa as a place of excess; but is it an excess of vitality or of decay? An extreme of production or negation? The undecidability is managed, both literally and physically, by the coast. At first glance this indeterminacy would seem to cast Kingsley's spatial schemes in a most conventional and ideologically transparent light. The African coast is represented as a zone virtually constituted by the encounter of vitality and death. Central Africa, she repeatedly emphasizes, is fringed with malarial swamps, which Kingsley views as a kind of natural population-control device. Over and over again her text depicts the "vital" Fan tribe of interior East Africa migrating seaward, where they presumably will replace, and follow the fate of, the "lethargic," "dying-out coast tribes":

> These Fans are a great tribe that have, in the memory of living men, made their appearance in the regions known to white men, in a state of migration seawards, and are a bright, active, energetic sort of African, who by their pugnacious and predatory conduct do much to make one cease to regret and deplore the sloth and lethargy of the rest of the West Coast tribes.[8]

> The Igalwa are one of the dying-out coast tribes. . . . [T]hey are, so far, undisturbed by the Fan invasion, and laze their lives away like lotus-eaters. (226–227)

> It always seems to me a wonder we have so many traces of early man as we have, when one sees here in Africa how one tribe sweeps out another tribe that goes like the foam of a broken wave into the *Ewigkeit* before it, leaving nothing, after the lapse of a century, to show it ever existed. (400)

> [T]he South-West Coast tribes have all migrated from a region in the northeast that seems to be perpetually throwing off tribe after tribe, which all come west, and die out in the swamp-lands of the West Coast. (459)

The rhetorically baroque tenor of these passages barely covers over the unresolved contradiction in Kingsley's apprehension of the African diaspora. She characterizes Africa's ostensibly centrifugal demographic currents alternately as evolutionary and devolutionary. At one moment the logic impelling migration toward the coast is Nordau-esque degeneration; at the next it is some sort of entropic vitality. In either case, African cultures and peoples are held to be under the sway of massive historical-environmental forces that far transcend the individual or even the culture. We shall see momentarily how different this is from the picture Kingsley gives of the subsumption of European individuals in the collective project of colonialism.

To Kingsley, then, the coast is paradoxically a space both of birth and of death, a space where ascendant cultures triumph over those that have met their evolutionarily prescribed demise. At times the ideological investments in such a

picture become particularly transparent. Witness the barely veiled allegory of colonization in her description of the making of Africa on the coastal delta:

> At corners here and there from the river face you can see the land being
> made from the waters. A mud-bank forms off it, a mangrove seed lights
> on it, and the thing's done. . . . [T]his pioneer mangrove grows. . . . He gets
> joined by a few more bold spirits and they struggle on together . . . but
> they always die before they attain any considerable height. Still even in
> death they collect . . . so that these pioneer mangrove heroes may be said
> to have laid down their lives to make that mud-bank fit for colonisation,
> for the time gradually comes when other mangroves can and do colonise
> on it, and flourish, extending their territory steadily. . . . Soon the salt wa-
> ters are shut right out, *the mangrove dies,* and that bit of *Africa is made.*
> (90–91; my emphasis)

Here the zero-sum (and largely phantasmatic) internecine struggle of African tribes
(predatory Fans versus lotus-eating Igalwas) is replaced by the progressive corporate individualism of European "pioneers," each of whom gives way to the next in
orderly and heroic fashion. In an almost perfect chiasmus, African corpses become
the fraying edge of the continent, while European corpses "make" the new land.

And yet, abstracting from the rhetoric and from a political allegory that we
have no right to ignore, we can see in this passage an instructive and quite literal
instance of what Edward Soja terms "the restless formation and reformation of
geographical landscapes."[9] It is crucial to emphasize how literally Kingsley takes
the idea that the space of Africa is actually forged at its coastal periphery. The
description is hardly an exceptional one in *Travels in West Africa,* which consistently depicts the coast as the zone where Africa is physically produced. Just as it
is the teleological horizon for endless "waves" of African cultures, so does the coast
function as the terminal point for a variety of environmental phenomena that tear
Africa apart in the interior. Kingsley is fascinated with the constant rush of torna-
does and rivers from the interior to the coast, and she gives extensive accounts of
the way in which whole swatches of African landscape are uprooted in the interior and deposited in the coastal deltas.[10] These reciprocal patterns of cultural and
ecological space are more difficult to disentangle than the mutual tracing-over of
rivers, trade routes, and wives outlined above. For here, the physical ground of
social space is literally in flux; the continent appears literally to expand and contract along with its populations. The African landscape is thus conceived of neither as the passive medium nor as the active determinant of social dynamics but
as a phenomenon complexly bound up with cultural space. This mutual imbrication is reflected in the relationship between language and geography, a relationship that on the coast is no longer simply a case of mapping but rather one of
profound isomorphism. The notorious proliferation of names for coastal towns and
rivers (noted above) turns out to reflect actual proliferations within the landscape.
The coastal deltas, Kingsley repeatedly notes, are characterized by the dispersal
of great rivers into multiple channels and estuaries: "Gaboon is the finest harbour
on the western side of the continent, and was thought for many years to be what it

looks like, namely, the mouth of a great river. Of late years, however, it has been found to be merely one of those great tidal estuaries like Bonny—that go thirty or forty miles inland and then end in a series of small rivers" (104). Not only linguistic signifiers but also physical signifiers multiply on the coast. Thus it is no longer a case of language being inadequate to its objects, as in the earlier analogy *language:objects = culture:space*. Instead, the conditions of language, social behavior, and physical space appear to mirror one another perfectly; in short, to be indistinguishable. This is particularly the case with the crucial category of the "coast," which, as the site where destruction and renewal meet, turns out in Kingsley's scheme to have identical cultural and physical functions.

Of course Kingsley's fascination with the literal *making of Africa* on the coastal delta must ultimately be resituated or reread. The rationalizing account of the coast as a zone where evolutionary dead ends are converted to productive use is, as we have noted, a classic imperialist creation-myth, and the indigenous peoples have their own versions, as Kingsley is well aware. Nominally in support of her thesis that the coast is a zone of negation, perilously colonized by life, Kingsley cites the native countermyth, in which death is introduced to the verge of the African continent from beyond and to the interior from the coast:

> Mr. Ibea and myself agree . . . that there is something inimical to human life, black or white, in the immediate Coast region of West and South-West Africa, as far down as Congo: and the interior tribes also join us in our opinion. Many times have I, and others, been told by interior tribes that there is a certain air which comes from the sea that kills men—that is just their way of putting it. . . . I [suspect] that it comes from the rotting, reeking swamp land and lagoons, and not from the sea. (409)

> The visitation of these maladies [smallpox and sleeping sickness], indeed of all maladies in West Africa, take the form of epidemics, and seem periodic. . . . The natives all along the Coast from Calabar to the South will tell you: "It is when the crabs come up the river," which means when the crayfish come down the rivers; but that is just their artless, unobservant way of putting things. (401–402)

The native myths virtually reverse the flow of rivers and of populations as perceived by the Europeans; the entropic space envisioned by Kingsley becomes a contracting, centripetal space. It is not clear to what extent Kingsley senses the political thrust of this alternative picture; what is clear is that the countermyths confirm her sense of the inextricability of populations and space. Communities are defined by their movements across landscape, while landscape itself is experienced as the movement of communities.

The Space of Fetish

Kingsley's conception of a contact zone, then, is more sequential than organic, more diachronic than synchronic; I have called it migratory. Yet if cultures do not so much jostle with as displace one another in this scheme, the spaces where they meet still retain a special status. Despite her concern that the Fans and other interior tribes might suffer the same fate as Africa's coastal peoples, Kingsley is fascinated with the figure of an African continent whose exterior fringes are charged with the evolutionary task of eliminating excess population. "Population" is indeed, perversely, the master trope of Kingsley's text, with its endless accounts of communities swelling to excess and sweeping toward the coast, where they are destined to do evolutionary battle with contending tribes on a terrain that is literally being made and unmade in some constantly ebbing and flowing no-man's-land. Countering this thrust is the periodic contraction of population due to "internal" practices such as infanticide and wife-sacrifice: "[O]ur European Government," Kingsley notes, "puts a stop to the action of those causes which used to keep the native population down, intertribal wars, sacrifices, &c., &c; *and to the deportation of surplus population in the form of slaves,* so unless means of support are devised for 'the indigenous ones,' as Mrs. Gault calls them, Africa will have us to thank for some smart attacks of famine" (121; my emphasis). The chilling reference to the Atlantic slave trade as a population-control mechanism helps to clarify the subtle epistemological distinction between the (post)modern paradigm of the contact zone and the more traditional binary scheme of interior and exterior; for if slavery is sufficiently comprehended in ethical terms by a binary scheme (freedom and captivity, good and evil, interior and exterior), that scheme fails nonetheless to account for the economies that hold the coastal margin in place, giving it contour and meaning. But here too we need to look beyond the two paradigms that we have traced in incipient form throughout Kingsley's text—distributory space, in which culture is always mapped out over some ostensibly more material schema, and migratory space, in which the continent and its peoples become interchangeable—in order to see how economies, cultures, and space become visible in relation to one another.

In her book *Imperial Leather,* Anne McClintock argues that the commodities produced by late-Victorian economies required, in some sense, a colonial horizon to activate their status as fetishes.[11] Such a requirement does not mean at all that the quotidian commodity was reenchanted under the gaze of colonial subjects, for whom it had an occult status. Rather, under the pressure of the colonial encounter, the ritual nature and intensity of the beliefs underpinning rationalization and commodification became themselves exposed. Thus could a Victorian writer like Kingsley imagine a future in which the African shore would be littered with the rejected magical instruments of science and progress (for example, infant feeding bottles, imported by European missionaries but unused by Africans, which sit idly in warehouses on the coast).[12] In McClintock's scheme, the zone of cultural contact is a fetish zone precisely because it is the site where objects are endowed simultaneously, and thus uncannily, with incommensurate meanings. I would like to adapt and revise this scheme slightly and suggest that Africa's contours become defined, in the fin de siècle imagination, by outworn fetishes—in other words, as the borderland where fetishes start to lose their ritual function. Here, then, one must adopt a somewhat non-Freudian and non-ethnological model of fetish. The model of fetish I want to employ here follows the theory of totems developed by early structural-functionalists such as Bronislaw Malinowski and A. R. Radcliffe-Brown.[13] In this system fetishes, or totems, are important not as units or bearers of meaning but as distributive instruments. That is, they distribute social roles and social space: who can touch the fetish and who cannot; who can carry it and who cannot; who can look at it and who cannot, and so on. In this sense feeding bottles are very much fetishes of modern Western culture, the corollary to earlier rites and rituals surrounding lactation; and their inability to penetrate African culture defines a certain social and spatial contour.

Such a concept of fetish is crucial to Kingsley's incipiently postmodern sense of space, for the cultural practices that she describes clearly do more than distribute themselves across abstract space: like fetishes, what they do, in fact, is distribute space itself. We can see this most clearly when it comes to those fetish practices that revolve around gender relations, that is, around the twin phenomena of kinship relations and the sexual division of labor. Gender, fetish, and space are intimately intertwined in Kingsley's Africa. Her view of this relationship is anything but schematic, however. Though the three concepts are consistently gathered together in some loose fashion, Kingsley conceives of space as something that is gendered and fetishized in a variety of ways, ranging from those that foreground space to those that foreground fetish. At one extreme is the Fan practice of wife-purchase. Males of the Fan tribe must purchase wives with special coins called *bikei,* which in addition fulfill the role of universal currency in an otherwise barter economy.[14] Presumably because of this tradition's special role in regulating kinship (and not simply gender relations), Kingsley identifies it explicitly as fetishistic, leaving implicit the way in which the *bikei* coins implicitly organize social space through the double circulation of women and commodities. At the other extreme is the ceremonial fish-gathering tradition on the island of Corisco. At a duly appointed time each year in August, the various villages on Corisco each appoint one woman to represent them in a communal fish gathering, with the day's catch to be divided equally among the villages (382–396). Though Kingsley details with great precision the various traditional practices of the fish hunt (the parade from villages to the small chain of inland lakes; the "prolonged, intoned" calls for women to join the group; the "specially made baskets"; and the elaborately choreographed techniques for herding and catching the fish), she regards the entire affair as less rigorous and more ad hoc than befitting a properly fetishistic practice: "I cannot find there is any fetish at the bottom of this custom, and think its being restricted to the women is originally founded on the male African's aversion to work; and in the representation of the villages, on the Africans' distrust of each other" (396).[15] Here, the explicit and ceremonial organization of space, labor, commodity, and calendar according to gender is foregrounded but denied the status of fetish. Thus at one extreme the cultural dimension of space is foregrounded, while at the other extreme the spatial dimension of culture is emphasized. Between these two alternatives lies a variety of phenomena: for example, the trader's route marked out with wives or the complex configuration of secret societies, rubber trading, and deforestation. It is this last nexus that I would like to look at in order to highlight Kingsley's distinctively, if incipiently, postmodern sense of the relation between space and culture.

At the foundation of Fan culture lies a highly gendered organization of space, resulting from the sexual division both of labor and of the tribe's secret societies. Essentially, Fan women are concentrated in village huts while the men live in the forests for months at a time, whether rubber hunting (291) or engaged in initiation rituals connected to secret fetish societies (525, 531, 535). But Fan culture shapes the landscape in even more direct and concrete ways as well. In Kingsley's view, Fan rubber farming is something of an ecological disaster. Instead of making small incisions in order to bleed the sap from rubber trees (which can grow only from seed, not from suckers), the Fans harvest entire branches and even plants, producing distinctive patterns of deforestation and, in turn, determining the distinctively nomadic character of Fan society (141–142, 291, 678). At first glance the relationship between "soft" and "hard" transformations of social space (between gender segregation and strip farming) seems casual, even metonymic: "rubber trade and wife palavers sweetly intertwine, for a man on the kill *in re* a wife palaver knows his best chance of getting the life from the village he has a grudge against lies in catching one of that village's men when he may be out alone rubber hunting" (291). But to regard this relationship as metonymic would be precisely to treat the activities that distribute people across landscape and the activities that transform the landscape as related only by contingency. It would be to treat the fetish-driven segregation of the sexes and the fetish-driven patterns of male itinerancy as unrelated to the gender segregation and itinerant quality of rubber farming. Instead, Kingsley is extraordinarily sensitive to the recursive tropes of Fan life. Though she hardly provides a theorization of the complex relationships between secret societies, the gender division of labor, male jealousy, deforestation, and migration—and indeed such a theorization is beyond the scope of this essay—she consistently bundles these issues together in her discussion. Thus it is not only at the coastal margin that the literal making and unmaking of the African landscape is bound up with cultural practices, even if this phenomenon is most dramatic and more perspicuous on the coast. Whether it is the polygamous trader who dots a particular river or trade route with wives to protect himself from local tribes, or the gender division of labor that concentrates women in villages, or the conventions of sexual rivalry that contribute to nomadic patterns of rubber-plant cultivation, space is consistently understood by Kingsley as the "edge of cultural practice." This understanding makes of space something less than the material determinant of culture (for example, malarial swamps inhibiting the development of civilization) but more than the mere allegorical projection of culture (for example, mangroves as an emblem of European colonization).

The Edge of Cultural Practice

The idea of space as the edge of practice suggests the impossibility of a perspective view from which a particular pattern, organization, or system might be grasped in its entirety. Kingsley constantly rejects the possibility of such a view on the African continent. As numerous commentators have noted, hers is not a narrative that privileges the bird's-eye view, or the trope of the "prospect." Yet Kingsley's insistence on the horizontal, ground-level nature of her encounter with Africa does not, alternatively, resolve itself into either a Conradian sublime of indeterminacy or—following Michel de Certeau's schematization of horizontal and vertical perspectives—a confidence in the performative heroism of her own interpretations. The concept of "the edge of cultural practice" compromises between the idea of cultural confrontation as mutual incomprehension, on the one hand (an idea mythologized most famously by Conrad through a thematics of penetration and impenetrability), and, on the other hand, more recent theories in which cultural difference is always already hybridized in performance. The latter model is an apposite one for many colonial and postcolonial spaces, but there is little fascination in Kingsley's text with what is distinctively hybrid about coastal culture. For Kingsley, what is dynamic about African space is less the mutual imbrication or tracing-over of various cultures than the actual malleability of space itself. What fascinates her constantly are those details of language, of trade, of kinship, and of domestic arrangements that suggest just how far—in quite literal, spatial terms— a particular tribe's way of life reaches. Her text is filled with cultural comparisons articulated in the language of geography (to the south they do not sacrifice wives; to the north they do not marry children; on the coast they do not make fetishes). Though space and culture do become mutually imbricated in this scheme, the conflict of cultures—the interlocking and antagonism of economies—is what prevents them from becoming indistinguishable. From each and every point in this bewildering "library" of a continent, Kingsley can make out the edges of an alternative cultural practice; and for her, to make out the edges is to perceive a malleable physical boundary. If she is incapable of apprehending such shifting terrain in anything other than imperialist-evolutionary terms, we should still appreciate the formal distinction between such an account and the more phantasmagorical vision of a "Heart of Darkness."

Notes

1. Mary H. Kingsley, *Travels in West Africa: Congo Français, Corisco, and Cameroons* (New York: Macmillan, 1897), 102.
2. Mary Louise Pratt, *Imperial Eyes: Travel Writing and Transculturation* (New York: Routledge, 1992); Sara Mills, *Discourses of Difference: An Analysis of Women's Travel Writing and Colonialism* (New York: Routledge, 1991); and Anne McClintock, *Imperial Leather: Race, Gender, and Sexuality in the Colonial Contest* (New York: Routledge, 1995).
3. Kingsley, *Travels,* 333.
4. Joseph Conrad, *Heart of Darkness: A Case Study in Contemporary Criticism,* ed. Ross C. Murfin (New York: St. Martin's Press, 1989), 21.
5. Alfred Russel Wallace, *The Geographical Distribution of Animals: With a Study of the Relations of Living and Extinct Faunas as Elucidating the Past Changes of the Earth's Surface* (London: Macmillan, 1876), 262.
6. Kingsley, *Travels,* 502.
7. Paul Gilroy, *The Black Atlantic: Modernity and Double Consciousness* (Cambridge,

Mass.: Harvard University Press, 1993); and Joseph Roach, *Cities of the Dead: Circum-Atlantic Performance* (New York: Columbia University Press, 1996).

8. Kingsley, *Travels,* 103–104.

9. Edward W. Soja, *Postmodern Geographies: The Reassertion of Space in Critical Social Theory* (London: Verso, 1989), 11.

10. Kingsley, *Travels,* 15, 107, 374–381.

11. McClintock, *Imperial Leather,* 31–36, 217–231.

12. Kingsley, *Travels,* 212.

13. Bronislaw Malinowski, "A Scientific Theory of Culture (1941)," in *A Scientific Theory of Culture and Other Essays* (Chapel Hill: University of North Carolina Press, 1944); and A. R. Radcliffe-Brown, "The Sociological Theory of Totemism (1929)," in *Structure and Function in Primitive Society: Essays and Addresses* (Glencoe, Ill.: Free Press, 1952).

14. Kingsley, *Travels,* 320–321.

15. It is worth noting that Kingsley frequently refers to a theory of African laziness in order to explain cultural practices; see, for instance, her explanation for polygamy (*Travels,* 211–212).

Jewish Geography

TROLLOPE AND THE QUESTION OF STYLE

JOSEPH LITVAK

Current Events

I begin with two fables from the recent past—with two articles that just happened to appear on the same day in the *New York Times*.

On 12 December 1997, to be exact—not so long ago, right before the name "Monica Lewinsky" was on everyone's lips—the *Times* ran two stories of Jewish geography, one on the front page, the other further back, in the section of its national edition that it calls the "New York Report."

Entitled "The Diamond District Bands Together to Improve Its Image," the latter article reports on the inauguration of the block of Forty-seventh Street between Fifth and Sixth Avenues, into which "more than 2,500 diamond industry companies are crammed," as the city's newest "business improvement district." Here is how the author describes the street's image problem:

> The president of the newly formed district, Howard Herman, one of the street's most prominent diamond merchants, said the image of 47th Street as a bastion of Jewish control and tradition has become both inaccurate and harmful to business.
>
> "The problem is not who we are," Mr. Herman said. "The street is a very diverse place, with people from India, China, Puerto Rico and other countries [*sic*], but somehow there is a lasting impression among the public that in order to be in the diamond business, you have to wear a long black coat and a beard."

Yet, while the repackaging of the diamond district as "a very diverse place" would allay the anxieties of "the public," the article itself ends on a note that might have the effect not of improving the street's image and business but of harming them further, by insinuating into the minds of that same public—or of those out-of-towners for whom the somewhat bland, attenuated national edition is produced—the suspicion that behind Forty-seventh Street's new, improved, international face, beneath its reassuring United Colors of Benetton mask, a treacherous Jewish plot is nonetheless, and all the more insidiously, thickening. Picking up on the image of more than twenty-five hundred diamond companies crammed into a single block, the article notes that "the 47th Street diamond merchants have not lost their distinction as one of the nation's most highly concentrated industries," and it concludes:

> For those who know the rules of the street, the sheer volume can mean bargains.
>
> "From what I can see, you are paying far less than retail here," said Margaret Ruden, a Manhattan cosmetics industry executive who was shopping for jewelry on 47th Street late yesterday.
>
> "But you have to be an educated buyer," she said. "People can be deterred if they are not familiar with New York City."
>
> And the diamond merchants who have created the new 47th Street district said it is precisely those who are not familiar with the city—or at least the rules of their street—whom they need to attract.[1]

Attracting them may not be any easier, however, once they have read this article. For if no longer exactly a *Judengasse,* Forty-seventh Street seems more than ever a *Rudengasse,* as we might name it after the Manhattan cosmetics executive who knows her way around the block. The diamond district, that is, may want to shed its image as a mean street, but the *Times* reporter, every bit as savvy as Margaret Ruden, has Forty-seventh Street's number: he recognizes, in other words, that underneath its thin veneer of de-Judaized "diversity," the diamond district retains its perilous thickness, its ghettolike, Kafkaesque density and viscosity. Wearing its smiley face, singing "We Are the World," it would indeed assume a diabolically "attractive" disguise, calculated at once to conceal and to promote the business and busyness of a sticky urban spider's web into which innocent Christian tourists, unfamiliar with the rules of the street, can be lured and robbed. The more the street changes, the more it stays the same. Not despite but because of its image-improving plastic surgery, the diamond district—unlike say, Times Square—will always be what it has always been: a highly concentrated, hypercharged, irreducibly Disney- and Giuliani-resistant site, not so much of that "rudeness" for which New Yorkers—why not just say New York Jews?—are famous as of an even more frightening "ruden-ness," by which I mean to designate both the streetwise fluency of the "educated buyer" making her sleek, shrewd way around town and the potentially fatal cramming, jamming, and coagulation of that very fluency into an impenetrable, inescapable mass, into that traumatic experience of the block known, to cardiologists and students of the sublime, as blockage.

This geographical nightmare, of course, expresses the fear, more widespread in today's America than we might have thought, of getting stuck in a Jewish neighborhood. But what of the possibility, at least equally terrifying if less generally acknowledged, of getting a Jewish neighborhood stuck in you? It is this latter possibility that the other, front-page article in the *Times* evokes. This article recounts the mendacious life and career of the late M. Larry Lawrence, the "millionaire businessman" from San Diego whose remains had just been dug up out of Arlington National Cemetery, after it had been learned that he had invented for himself a heroic military record, lying and buying his way first into an ambassadorship to Switzerland and then into the hallowed ground of the heroes' graveyard.[2] As the article makes clear, Lawrence's story boiled down to another Jewish plot—a Jewish plot whose protagonist would have *stayed* in a Jewish plot were it not for the tenacity of Republican investigators on the trail of influence peddling at the Clinton White House. As it also makes clear, Arlington National Cemetery was not the only national ground the lying Lawrence would have desecrated, precisely by lying *in* it. Indeed, as rehearsed in the *Times,* Lawrence's narrative, of upward mobility even in the form of the ultimate *downward* mobility, constituted nothing so much as a grotesquely literalizing parody of Jewish assimilation, whereby the Jew, even this side of the grave, always already embodies something rotten in the state, a corruption of the sacred *social* body into which he would blend himself with a vengeance.

"For years," the article relates, "Mr. Lawrence struggled to fit in in the heavily military, heavily Republican world of San Diego. He loved to tell friends that he had 'three strikes' against him: he was 'a liberal, a Jew, and a Democrat'" (A16). In his assimilationist struggle to fit into that world, however, he himself struck against it, attacking it, one might say, at its very heart: "'We have so many Purple Hearts here and on every corner we have real, real heroes,' said Carol Cahill, a seventy-year-old Coronado resident. 'My next door neighbor was a four-star admiral. Maybe he just figured—merchant marine, you can't trace it, no one will ever know'" (A16). You *can* trace it, of course, as this article demonstrates; and as Carol Cahill ought to have recognized, the *merchant* marine would be just the place for a guy like M. Larry Lawrence. Still, Ms. Cahill usefully articulates the anxiety that Jewish assimilation can induce in the patriotic *imaginaire:* what if the Jewish parvenu, having melted without a trace into the national bloodstream, is thus poised to poison from within the heavily military, heavily Republican system that has been made to absorb him? What if the Jew, not content to stay within the limits of New York City—like the diamond district that it contains, "a bastion of Jewish control" and, therefore, as everyone knows, not really a part of America at all—ends up installing pieces of himself as pieces of New York "on every corner" in America, so that every street in the land might harden at any moment into a diamond district, with the congestive power to stop the heart of the nation as a whole?[3] Against the obscene overflow, or the sclerotic *dis*fluency, that the very name "M. Larry Lawrence" seems to mime, the voice of the mainstream-bloodstream, if not of the heartland, mobilizes the counterredundancy of "real, real heroes." But if M. Larry Lawrence was finally found out and cast out before he could do any more damage, it may require the National Guard to protect America against wilier versions of his kind: racial impersonators who, far from ignoring the rules of the street, violate them by obeying them too well.

Writing Blocks

The fearful (but also excited) fantasy of the Jewish block is no more confined to late-twentieth-century America than it is to New York City. In 1840, for
example, in a letter to John Forster, Charles Dickens tells of how, while walking
in London, he became "mingled up in a kind of social paste with the jews [*sic*] of
Houndsditch."[4] And just as the contemporary image of getting stuck among the
Jews backs up into the perhaps more disturbingly assimilated image of getting a
Jew stuck in you, so the Victorian image of a deindividuating Jewish paste could
travel from the identifiable, circumscribed Jewish "community" to one vile Jewish body in particular. So we might infer, at any rate, from Anthony Trollope's characterization of his Jewish bête noire, the assimilated, indeed converted, Benjamin
Disraeli, in whose novels, and around whose person, "there is a feeling of stage
properties, a smell of hair-oil, an aspect of buhl, a remembrance of tailors, and
that pricking of conscience which must be the accompaniment of paste diamonds."[5]
From the paste district to the paste-diamond district—but how do you tell the difference between a diamond district and a paste-diamond district?—the Victorian
Jew pursues a career whose brilliance comes down to the shininess of so much
hair oil. In a spectacular success story whose title might be *What Makes Benjy
Run?* the Jewish hero starts out as Dickens's Fagin, a "loathsome reptile . . . engendered in . . . slime and darkness,"[6] but transforms himself into a loathsome reptile at the center of English society.

As far as I know, there is no such thing as Jewish pornography, if that means
going to the adult section of the video store and finding, alongside such special
flavors as *Black Stallions, Manhattan Latin,* and *Pacific Rim,* a title like *Ortho-
dox Tastes.* Yet, at least in its Victorian manifestations, the fantasy of the well-oiled,
sticky Jewish body has undeniable pornographic implications, insofar as the word
pornography itself implies an intimate connection between writing and the body.
The mingling of Disraeli's offensive books with his offensive person provides one
Victorian example of a potentially pornographic Jewish body, which is to say, a
Jewish body whose "attractiveness" itself mingles with its "loathsomeness," a Jewish body marked by what Trollope, perceptively describing another such body in
Nina Balatka, calls its "hard, bold, almost repellent beauty."[7]

That this hard Jewish body is not always male is only one of the lessons of
the following passage from *Phineas Finn,* the second of the Palliser novels, in which
Trollope introduces the dangerously alluring Madame Max Goesler:

> The observer who did not observe very closely would perceive that Ma-
> dame Max Goesler's dress was unlike the dress of other women, but see-
> ing that it was unlike in make, unlike in colour, and unlike in material,
> the ordinary observer would not see also that it was unlike in form for
> any other purpose than that of maintaining its general peculiarity of char-
> acter. In colour she was abundant, and yet the fabric of her garment was
> always black. My pen may not dare to describe the traceries of yellow and
> ruby silk which went in and out through the black lace, across her bo-
> som, and round her neck, and over her shoulders, and along her arms, and

down to the very ground at her feet, robbing the black stuff of all its somber solemnity, and producing a brightness in which there was nothing gaudy.[8]

Trollope never explains *why* his "pen may not dare to describe the traceries" of silk that it nevertheless seems to describe, just as he never bothers to specify what "other purpose" an *extra*ordinary observer might discern in the formal unlikeness of Madame Max Goesler's dress. *Not* unlike, say, homosexuality, Jewishness in Trollope, or rather the assimilating Jewishness that most arouses him, is by definition mysterious, essentially a matter of guess- or Goesler-work. As Michael Ragussis has argued, the ominous vagueness that persistently attends Jewish characters in Trollope's novels, while seeming to call their Jewishness into question, in fact *constitutes* it—and constitutes it as a sinuous, sinister aptitude for "infiltration and subversion."[9] To give Madame Max, as she comes to be called, a determinate form or content, to put a finger on her je ne sais quoi of difference that somehow is not different enough, would compromise the work of the passage, which consists, instead, in promoting readerly paranoia vis-à-vis this avatar of Jewishness as a cunningly formalized ensemble of scriptive intricacies and involutions.

Modeling a style whose general rubric might be *l'écriture juive,* Madame Max poses the double threat (the double thread) of a writing at once exotically foreign and uncannily familiar—a writing whose "traceries . . . produc[e] a brightness in which there was nothing gaudy." Leaving no traces of the vulgarity with which other Jews, preeminently Disraeli, would soil Christian space, Madame Max does not necessarily clear her name: on the contrary, it is her very *un*gaudiness that raises the intolerable specter—intolerable even (especially?) for a legendarily prolific author who could boast, "I have known no anxiety as to 'copy'"[10]—of writing block. Trollope's pen dare not describe Madame Max's treacherous traceries, dare not repeat her text, lest it lose all trace of *itself* in a writing too urbanely, too tastefully, too genteelly like its own; what sets the Trollopian pen aquiver is not the possibility of a wounding encounter with the formally unlike but the risk of castration through nondifferentiation.

Did I say *castration?* I meant something much worse, namely, *constipation,* one of the classic Jewish disorders, which, like heartburn, clogged arteries, and acne, testifies to the medical disaster (not to mention the ideological abomination) of a dietary overreliance on animal fat. To be sure, neither Trollope nor even any of his less stylishly anti-Semitic characters ever calls Madame Max a greasy Jew. But she requires a certain ethnic cleansing nonetheless, as suggested when Lady Glencora Palliser, reacting to the prospect of her husband's uncle's marrying Madame Max, reflects: "That such a one should have influence enough to intrude herself into the house of Omnium, and blot the scutcheon, and,—what was worst of all,—probably be the mother of future dukes!" (*PF,* 2:216). As if exuding her pushy inner Disraeli, her bad Jewish blood, in the form, or substance, of greasy ink, as if enacting a parodic, grossly materializing desublimation of her filigreed, darkly bright formalism, as if reducing her gliding metro- and cosmopolitanism to their vulgar racial substratum and condition of possibility, Madame Max converts her assimilative fluency back into the crude oil as which it began.

Admittedly, the blot is in the mind of the lovably flawed Lady Glencora, and even she will *change* her mind—after, of course, Madame Max decides not to marry into the family—and ends up making her one-time bugbear into her best friend and confidante. But the blot is not *just* in Glencora's mind, nor is it confined to the *style indirect libre* by which Trollope might appear to disclaim responsibility for it. The blot or grease stain spreads, in other words, from Madame Max to a series of Jewish male characters. As Bryan Cheyette has shown,[11] the normalization of Madame Max both coincides with and depends upon the emergence, in the next two Palliser novels, of another, more authentically repellent Jewish character, the converted preacher from Prague, "once a Jew-boy in the streets,"[12] whose Christian alias is Joseph Emilius, but whom Trollope delights in unmasking as "Yosef Mealyus," as though the very name bore all the damning self-evidence of a circumcised penis. Like Eastern-European Jewish cooking at its most horrifying, the loathsome Emilius, even more than Fagin, represents a quintessential, emblematic precipitate of the Jew *as* grease, with the result that Trollope seems to need paper towels just to write about him: "The man was a nasty, greasy, lying, squinting Jew preacher" (*ED,* 2:314); "[H]e was a greasy, fawning, pawing, creeping, black-browed rascal" (*ED,* 2:241–242) with "an exuberance of greasy hair" (*ED,* 2:311); "He was a creature to loathe,—because he was greasy, and a liar, and an impostor" (*ED,* 2:314). It is as though Madame Max could be de-Judaized only through a process of siphoning, whereby her nasty racial fluids—blood, ink, oil—are drained out of her and into, or onto, a figure who thus doubles as blot and blotter.

Yet, despite the curious, *Sacred Fount*-like economy that obtains between the de-Judaized woman and the hyper-Judaized man, what is even more curious is how their opposition brings out their kinship. What, after all, is his Disraelite conversion (part of what makes him an "impostor") but the sequel to her assimilation? And although he fails where she succeeds—as we shall see, Trollope likes to plot the failure of a certain Jewish success story—although, well before his downfall in *Phineas Redux,* indeed from the moment he arrives in *The Eustace Diamonds,* which he could turn into *The Eustace Diamond District,* he is doomed to betray the Jewishness that oozes from him accordingly, while she is permitted to secrete hers in the opposite sense, his greasiness is also the secret of the success he does enjoy for a time as a fashionable London preacher. Misrepresenting that greasiness as brilliance, à la Disraeli, Emilius brags to Lizzie Eustace: "I make bold to say of myself that I have, by my own unaided eloquence and intelligence, won for myself a great position in this swarming metropolis. . . . I am known as the greatest preacher of my day, though I preach in a language which is not my own" (*ED,* 2:312–313). Emilius's great position in the swarming metropolis suggests that Lizzie, who "certainly liked the grease and nastiness" (*ED,* 2:314), may not be alone, among the fashionable Victorian audience, in the perversity of her taste. Notwithstanding his boasts of "unaided eloquence and intelligence," however, the "Jew preacher" also owes his bad eminence to grease, understood in less individually charismatic (or chrismatic) and more racially general terms, as the lubricant enabling the smooth international operations of the Jewish mimetic genius. Bad fluid and bad influence, the Jew's greasiness itself mimics his bad fluency, both effect and cause of his notorious age-old skill in mastering foreign languages and manners by seeping into them, as Madame Max's traceries, mimicries again, would meltingly suffuse the Trollopian writing that does not dare to mimic *them*.

The Jewish Hunk

The only thing worse than grease's tendency to flow is its tendency *not* to flow—to stagnate and solidify, cutting off the circulation of whatever system it comes to inhabit. Both Madame Max's fashionable *archi-écriture* and Emilius's fashionable "eloquence" attest to the Jewish talent less for absorbing than for being absorbed into a dominant cultural text that the Jews, once incorporated, implicitly threaten with an oversaturation as alarming as that of a cardiovascular system clogged with cholesterol. But where Fagin's shtick gives Dickens a hard time, what could stop Trollope's pen in its tracks—or what, if it does not exactly impede his famous fluency, keeps him returning to it obsessively, as though he were under a repetition-compulsion that is writing's tribute to blockage—is less the prospect of any theatrical, overtly Faginesque cultural presence than the possibility that an assimilated, dissimulated Jewishness, once having entered the bloodstream of English culture, will begin the murderous, vengeful work of congealing within it, using the fiendish eloquence and intelligence of the race to spread *in*eloquence and *un*intelligence—indeed, an almost bovine stupidity—throughout the English national body, including the body of the English novel.

Nowhere in Trollope—perhaps nowhere in Victorian fiction—does this Jewish revenge plot get elaborated more luridly, or thwarted more violently, than in the penultimate novel of the Palliser series, *The Prime Minister,* whose prime monster is the Jewish villain named Lopez, after the crypto-Jewish doctor—supposedly a model for Shylock as well—who was executed for plotting to poison Queen Elizabeth. Repeatedly referred to as an "incubus,"[13] the new, unimproved Lopez would poison the English social body by penetrating the body of one exemplary Englishwoman; so villainous is he, moreover, that while vaginal intercourse is not his only means of contamination, women are not his only victims. At once picking up where Emilius leaves off, and paving the way for Svengali, Dracula, and Freud—indeed, adumbrating the horror-movie convention of the monster who just won't stay dead—Lopez displays the Jewish villain's particular propensity for language pollution through oral rape, for shoving himself down male as well as female Christian throats.

It is because of this aggression's maleness that Trollope, as Cheyette has pointed out, can clean up Madame Max's act.[14] In deciding not to blot the Palliser family scutcheon, that is, Madame Max renounces the phallic threat implicit in her "androgynous" name, devoting herself thenceforward to the benign narrative project of discrediting Emilius, marrying a nice Irish boy, and becoming, simply, "Mrs. Finn." In the nasty Jewish boy who takes over *The Prime Minister,* however, Trollope conjures up a prince of darkness all the more poisonous for his ability to make gentile mouths water. Obviously a careful reader of *The Eustace Diamonds,* one Christian character in *The Prime Minister* explains to another that "girls . . . like dark, greasy men with slippery voices" (*PM,* 1:138–139). The notion that beautiful young gentile women find Jewish men sexually irresistible is not, it turns out, limited to the films of Woody Allen. But Lopez has this bedeviling effect on male Christian consumers as well, down whose throats he would pour one of the Faginesque commodities (the other being guano) in which he purports to have invested, a "sticky," intoxicating beverage ironically named Bios (*PM,* 2:119, 135). In marrying Emily Wharton, for instance, he violates her father at the same time, managing "to assault and invade the very kernel of another man's heart" (*PM,* 1:29). So widespread and so pernicious is the influence of the sexy Jewish invader that, by the time Lopez finishes his scandalous work in and on the English national libido—with the help of a yellow (if non-Jewish) journalist deliciously named Mr. Slide—the Jew has even screwed with the head of the prime minister himself. When "Mrs. Finn" asks if the depressed Plantaganet Palliser sees his doctor, his wife, Lady Glencora, replies:

> Never. When I asked him to say a word to Sir James Thorax,—for he was getting hoarse, you know,—he only shook his head and turned on his heels. When he was in the other House, and speaking every night, he would see Thorax constantly, and do just what he was told. He used to like opening his mouth and having Sir James to look down it. But now he won't let anyone touch him. (*PM,* 2:219)

Getting hoarse, not from speaking every night but from the general demoralization brought on by Lopez's invasion of his world and assault on his career, the prime minister would seem, in some sense, merely to rejoin his fellow Englishmen in their proper style of tight-lipped virility. But while it sometimes undoubtedly appears that "fair-haired men . . . haven't got a word to say to you" (*PM,* 2:133–134), and while Trollope himself opposes the "uncertain words of an English gentleman" to the "glib tongue of some inferior . . . race" (*PM,* 1:132), the correct gentlemanly style in fact consists in a delicate mixture of diffidence *and* fluency, fluency of the kind that keeps a statesman like Palliser speaking every night, or that keeps a novelist like Trollope fully conversant with the rules of the street, the talk of the town, the way we live now, greasing the wheels of urbanity that enable him to write a sentence beginning "We, who know the feeling of Englishmen generally better than Lopez did" (*PM,* 1:321), or one like "Though [Lopez] had lived nearly all his life in England, he had not quite acquired that knowledge of the way in which things are done which is so general among men of a certain class, and so rare among those beneath them" (*PM,* 2:24). Going so far as to make the prime minister hoarse, then, the Jewish stud almost makes him dumb, which is to say, almost Jewish, like himself.

For though Lopez is duly denounced as "too clever, too cosmopolitan,—a sort of man . . . who wouldn't mind whether he ate horseflesh or beef if horseflesh were as good as beef" (*PM,* 1:152), perhaps the cruelest move in the repertoire of Trollope's genteel anti-Semitism is his gleeful exposure of the stupidity latent in Jewish cleverness, the animal inarticulateness latent in Jewish eloquence—the tendency, in short, of the Jew's cleverness and cosmopolitanism to turn him *into* the horse or ox between which he fails to differentiate, *because* he fails to differentiate. Such indifference, of course, looks like the telltale sign of a bungled assimilation, like the indissoluble residue of the schlemiel, hapless star of a million Jewish jokes, who cannot "quite acquire . . . that knowledge of the way in which things are done" in the host culture. Emilius in disgrace, Lopez humiliated when even Lizzie Eustace spits him out and calls him a "fool" (*PM*, 2:141), Melmotte in *The Way We Live Now*, "kn[owing] nothing of the forms of the House" and therefore making a spectacle of himself in Parliament[15]—all of these would-be assimilators ultimately become M. Larry Lawrence, vomited up by the system they have assaulted and invaded. But while Lopez's indifference obviously betrays a failed assimilation, in-difference—the effacement of difference—just as obviously represents the logical *fulfillment* of assimilation, the consequence not of an assimilation that has failed but rather of an assimilation that has succeeded too well.

Too well, at any rate, from the perspective of a gentlemanly novelistic style impelled by, indeed priding itself on, a cosmopolitan, metropolitan fluency, even if, or rather, precisely because, that fluency needs to be "cut"—as Bios, Lopez admits, needs to be mixed with gin or whisky—by the national stiffness, and not just of upper lip, for which Englishmen "of a certain class" are famous. This Trollopian cocktail, in other words, already too much resembles the grease that hardens the Jew who hardens the gentile. Grease alone, I want to insist, is *not* the word: contrary to a certain critical emphasis on Jewish "slipperiness" and "instability,"[16] on the Jew's heartwarming ability to resist and subvert fixed racial categories, what I am pushing here is a schmaltzy Jewish thematics in which the value of grease lies not in its slipperiness or instability but in its tendency (emblematized by various sleazy Jewish characters in literature and film who are actually named Block or Bloch) to thicken, to produce blockages. As the cosmetics industry executive Margaret Ruden knows, with the intimidating perspicacity of the true city slicker, "slipperiness" and "instability" are merely prerequisites for good citizenship in a culture where the good citizen, or the smart shopper, knows that she'd better keep moving if she knows what's good for her. What's most exciting about the dark, greasy, slippery-voiced Lopez is not that he shakes gentiles up but that he shakes them up by stupefying and animalizing them. Just as he makes the prime minister hoarse, so, by invading the kernel of his wealthy father-in-law's heart, he "take[s] the bull by the horns" and, on the very same page, with typical pornographic perversity, reduces the urbane, novel-loving old barrister to a "milch cow" (*PM*, 1:271). The assimilating Jew is sexy, that is, because he brings out not just the connection *between* fluency and blockage but also fluency's desire *for* blockage, for the *thrill* of blockage; in his supreme act of revenge, he caters to the fantasy of constipation as *jouissance* that underwrites Trollope's own incessantly smooth operations.

Because the Jewish hunk stops traffic, however, because the entire novel could get stuck in his shtick, or vice versa—"The coach has to be driven somehow," as one character says to another; "You mustn't stick in the mud, you know" (*PM*, 1:136)—he himself must be stopped. Hence the novel's climactic scene of counterrevenge, in which Trollope murders Lopez and makes it look like suicide. Or, as Glencora puts it, "I have a sort of feeling . . . that among us we made the train run over him" (*PM,* 2:348). In the name of this "we," the same "we," in any case, "who know the feeling of Englishmen generally better than Lopez did," Trollope allows his prose to assume an uncharacteristically Dickensian thickness, just long enough for one last, ecstatic encounter with the Jew before the novel gets him out of its system. Figuring that system as a vast, intricate railway junction, the novel almost deliriously maps itself as a demonically urbane discursive net-work, presided over by authorities called "pundits," and coterminous not only with greater London but, by extension, with the rest of England as well.[17] The geo-graphical answer to the question posed by Madame Max's inky, slinky dress, the railway junction, "quite unnecessary to describe, as everybody knows it" (*PM,* 2:191), is what Trollope's pen *must* describe, however, if it, and the blood of "ev-erybody," within every *English* body, are to flow freely:

> It is a marvellous place, quite unintelligible to the uninitiated, and yet daily used by thousands who only know that when they get there, they are to do what some one tells them. The space occupied by the convergent rails seems to be sufficient for a large farm. And these rails always run one into another with sloping points, and cross passages, and mysterious me-andering sidings, till it seems to the thoughtful stranger to be impossible that the best trained engine should know its own line. Here and there and around there is ever a wilderness of waggons, some loaded, some empty, some smoking with close-packed oxen, and others furlongs in length black with coals, which look as though they had been stranded there by chance, and were never destined to get again into the right path of traffic. (*PM,* 2:191)

That the adventures of the Jewish adventurer should end in such a place is what everybody calls poetic justice: having slithered down Englishwomen's throats and turned Englishmen into cows, having greased the novel's tracks so profusely, having made it run so smoothly, that he threatens, or promises, to derail it into a blissfully inarticulate, even bestially dumb, state of close-packed noncirculation, Lopez gets what's coming to him:

> There came a shriek louder than all the other shrieks, and the morning express down from Euston to Inverness was seen coming round the curve at a thousand miles an hour. Lopez turned round and looked at it, and again walked towards the edge of the platform. But now it was not exactly the edge that he neared, but a descent to a pathway. . . . With quick, but still with gentle and apparently unhurried steps, he walked down before the flying engine—and in a moment had been knocked into bloody atoms. (*PM,* 2:194)

Unusually dense, Trollope's writing in this chapter avoids stupidity, avoids careering off the tracks into a wagon smoking with "close-packed oxen," only by assuming the annihilating force of an express train traveling at the speed of a thousand miles an hour. Only in this way can the novel's style purge itself of its slick Jewish double. But if Trollope's urbane pen thus removes the obvious threat to its business of making the trains run on time and assures itself that even the smoking, spaced-out oxen will "get again into the right path of traffic," this scene of graphical and geographical cleansing does not necessarily succeed in leaving no traces. The mode of Lopez's murder—he gets "knocked into bloody atoms" by the "flying engine"— clearly expresses its vengeful motive, which is to assimilate the assimilator into nothingness; but the Jew's bloody atomization has the messy effect of spreading him all over the place, including the sacred common-place of the railway junction, the heart that keeps England's economy moving. "The fragments of [Lopez's] body," Trollope tells us, "set identity at defiance, and even his watch had been crumpled up into ashes" (*PM,* 2:195). Yet, unlike Dracula "crumbl[ing] into dust" at the end of Bram Stoker's novel,[18] so that Christian England may breathe free again, the defiantly bloody Lopez, the Jew sent flying into space, can never finally be wiped away. Breaking down Jewish blood only to widen its dissemination, the flying novelistic engine increases, rather than diminishes, the likelihood of its own breakdown. As everybody knows, however, it is just that possibility that really gets it going.

Notes

1. Thomas Lueck, "The Diamond District Bands Together to Improve Its Image," *New York Times,* 12 December 1997, A21, national edition.
2. Don Van Natta Jr. and Elaine Sciolino, "Body, and Tombstone of Lies, Are Removed," *New York Times,* 12 December 1997, national edition.
3. On the oral-alimentary implications of the metaphor of assimilation, see Zygmunt Bauman, *Modernity and Ambivalence* (Ithaca, N.Y.: Cornell University Press, 1991), 103–104.
4. Charles Dickens, *Letters of Charles Dickens,* ed. Madeline House and Graham Storey, associate editors, W. J. Carlton et al., Pilgrim ed. (Oxford: Clarendon Press, 1969), 2:118. I discuss this image in "Bad Scene: *Oliver Twist* and the Pathology of Entertainment," *Dickens Studies Annual: Essays on Victorian Fiction,* ed. Stanley Friedman, Edward Guiliano, and Michael Timko (New York: AMS Press, 1998), 26:40–41.
5. Anthony Trollope, *An Autobiography* (London: Oxford University Press, 1974), 223.
6. Charles Dickens, *Oliver Twist,* ed. Peter Fairclough (Harmondsworth, Eng.: Penguin, 1985), 186.
7. Anthony Trollope, *"Nina Balatka" and "Linda Tressel,"* ed. Robert Tracy (Oxford: Oxford University Press, 1991), 118. Earlier, this same character (tellingly named Rebecca Loth), is described as having "a repellant [*sic*] beauty that seemed to disdain while it courted admiration" (83).
8. Anthony Trollope, *Phineas Finn* (Oxford: Oxford University Press, 1973), 2:26. Hereafter citations will be given in the text with the abbreviation *PF.*
9. Michael Ragussis, *Figures of Conversion: "The Jewish Question" and English National Identity* (Durham, N.C.: Duke University Press, 1995), 242. For a more general discussion of this issue, see 238–260. Two apt formulations: "The most marked characteristic of Jewish identity is its genealogical uncertainty. Jewish identity is represented as a suspicion or rumor that haunts certain characters" (242); "In this group of

novels [the Palliser series and *The Way We Live Now*], unknown ancestry and Jewish ancestry are almost synonymous" (246).

10. Trollope, *An Autobiography,* 104.
11. Bryan Cheyette, *Constructions of "the Jew" in English Literature and Society: Racial Representations, 1875–1945* (Cambridge: Cambridge University Press, 1993), 35–42.
12. Anthony Trollope, *The Eustace Diamonds* (Oxford: Oxford University Press, 1973), 2:128. Hereafter citations will be given in the text with the abbreviation *ED.*
13. Anthony Trollope, *The Prime Minister,* ed. Jennifer Uglow (Oxford: Oxford University Press, 1983), 2:126; 2:175. Hereafter citations will be given in the text with the abbreviation *PM.*
14. Cheyette makes this point more generally: "If Madame Max, as a Jewish woman, could be safely desexualised and deracialised by Trollope, then this form of domestication . . . was not possible for the rampantly sexual, young, upwardly mobile Jewish male. They, after all, were in a position to take advantage of the 'romantic' susceptibilities of Trollope's flawed women" (*Constructions of "the Jew,"* 42).
15. Anthony Trollope, *The Way We Live Now,* ed. John Sunderland (Oxford: Oxford University Press, 1982), 2:179.
16. E.g., Cheyette, *Constructions of "the Jew,"* 38.
17. Robert Tracy (*Trollope's Later Novels* [Berkeley: University of California Press, 1978], 57–59) makes the point that the railway junction serves as a model for both the novel and the social order with which the novel identifies itself.
18. Bram Stoker, *Dracula* (Harmondsworth, Eng.: Penguin, 1986), 447.

✴

Domestic Fronts

When in Rome

Honeymoon Tourism in the "City of Visible History"

Helena Michie

In the introduction to *Walks in Rome,* the two-volume red-and-black guidebook that, from the early 1870s on, was to accompany so many English tourists to the Italian capital, Augustus Hare begins by quoting a euphoric letter from and about the city by Thomas Arnold.

> "Again this date of Rome; the most solemn and interesting that my hand can ever write, and even now more interesting than when I saw it last," wrote Dr. Arnold to his wife in 1840, "and how many thousands before and since, have experienced the same feeling, who have looked forward to a visit to Rome as one of the great events of their lives, as the realization of the dreams and longings of many years."[1]

Hare's appropriation of Arnold's words places them squarely in what we might call an idiom of completion. Rome marks the endpoint of a personal and cultural fantasy, the final destination on a journey undertaken by "thousands," imagined perhaps by many thousands more. By moving in the opening paragraph from the words of Dr. Arnold, that touchstone of Victorian experience, to those other travelers for whom Arnold speaks, Hare signals not only the ubiquity (for a certain class of people) of this Roman fantasy but also its privileged cultural status. The journey to Rome takes on the weight and shape of ritual, the teleology of desire finally, even "solemn[ly]," realized.

What Hare ignores in the short quotation from Arnold, however, is a chronological marker that sets up an entirely different choreography of desire and its satisfaction: Arnold is not writing the date, experiencing the solemnity of time and place for the first time; he is writing the date "[a]gain," and remembering his last trip to the same city. This journey is a revisitation, a repetition of the act of consummation: Rome (at last) again. The two temporal narratives—one climactic, the other reiterative—seem at first reading to be strangely at odds with each other.

By the second paragraph of *Walks,* Hare offers us yet another temporal complication, another version of Rome "[a]gain." Rome is different from other cities, he argues, precisely because to see it, *even for the first time,* is an experience of déjà vu:

> An arrival in Rome is very different to that in any other town in Europe. It is coming to a place new and yet most familiar, strange, and yet so well known. When travelers arrive at Verona for instance, or at Arles, they generally go with a curiosity to know what they are like; but when they arrive at Rome and go to the Coliseum, it is to visit an object whose appearance has been familiar to them from childhood, and long ere it is reached, from the heights of the distant Capitol, they can recognize the well-known form: and as regards S. Peter's, who is not familiar with the aspect of the dome, of the wide-spreading piazza, and the foaming fountains, for long years before they come to gaze upon the reality? (1)

We might think of this second paragraph as resolving the contradiction of the first by extending it and by locating it in the very nature of the city. All journeys to Rome are, it seems, in some sense revisitations: the climactic structure of the fantasy narrative is always, perhaps has always been, undermined by the very experiences that have created the desire for realization. Proleptic dreams both interrupt and anticipate completed fantasies.

Both the climactic and reiterative narratives take from and give new meaning to other contradictions at work in many Victorian descriptions of Rome, which is simultaneously figured by a variety of writers as timeless and as the product of a variety of specific cultures, as a city frozen in time and, especially after 1860, as an emerging modern political state. If we think of Rome's contradictions solely in terms of time, however, we lose the all-important sense of them as they registered again and again in the daily pilgrimages of British travelers to the sights of Rome. These "sights" included dramatic visual contradictions: ruins and modern apartment buildings, statues of classical gods and Christian saints in the Parthenon, Christian churches built visibly on top of Roman temples—as in the case of Santa Maria sopra Minerva. If we linger with the sights of Rome we come to a corrective or supplement not only to the idea of the city's timelessness but to the axis of time itself: we come to an idiom of visual and spatial juxtaposition. And it is visual juxtaposition that takes us, finally, to a more truly geographical account of Rome. Arnold's conventional metonymic slip—calling the word "Rome" with which he heads and locates his letter "this *date* of Rome [my emphasis]"—turns time into a spatial matter, into geographical material. Like the contemporary journalistic practice of calling the geographic marker at the top of a newspaper story a "dateline," Arnold's metonymic use of "date" alerts us to how the reiterative narrative of Rome "[a]gain" gets embodied in the geographical "sights" of Rome.

To quote in anticipation George Eliot's description in *Middlemarch* of Dorothea Brooke's honeymoon in Rome, I would ask the reader of this essay to imagine "yet another contrast," yet another temporal narrative that intersects with the climactic and reiterative temporal narratives of tourism and struggles to represent them simultaneously as effects in time and space. This is the personal narrative of the honeymoon, that journey of physical, psychological, and legal transformation, that time away from others, in a foreign or at least separate space. Elsewhere I argue that the geographical displacements of the honeymoon were reflected in—and in turn reflected—the social, psychological, and cultural shifts supposed to take place in newly married women and men.[2] The honeymoon, I argue, was supposed to do the difficult cultural work of what we might call "reorientation," the movement of men and women to a canonized form of identity that can proleptically be called heterosexuality. Embedded in that narrative of reorientation is, to risk easy honeymoon punning, a climactic structure: honeymoons, according to a culture that came increasingly to expect them as part of the landscape of marriage, changed people permanently. Women and men returned from their honeymoons to take up different personae; women, of course, were assumed to be changed not only psychologically but nominally and legally as well. It is, of course, no accident that these changes were usually imagined to have taken place away from familiar landscapes: whether honeymoon journeys took Victorian couples to the English seacoast for a day or to the Continent for six months, the work of reorientation came by midcentury to be highly dependent on geography.

This brings us (once again) to Rome, this time as a site where we might find the reiterative and climactic narratives of tourism and honeymooning coming spectacularly together. Obviously not all honeymoon couples, even those who could afford to do so, went to Rome. In the larger study of which this is a part, I survey a sample of sixty-three couples who took their honeymoons between 1830 and 1890. Of these only six actually made it to the Eternal City (more made plans to go but were forestalled by illness, lack of money, or other forces beyond their control). Rome, as Eliot calls it, the "city of visible history," offers us as an imagined or achieved honeymoon destination the opportunity to speculate on connections between private lives and public histories, stories of personal change and cultural narratives of decline and fall, time and timelessness, climax, return, and repetition.

In trying to sort out some of the narrative possibilities associated with the Roman honeymoon, I am committing two historicist faux pas. First, as I indicate above, I am working with an extremely small sample of cases. Whereas in other parts of the study I try to do empirical work—moving from my sample to cautious conclusions about, for example, how long people typically spent on their honeymoons, where they went, whether they brought along friends or relatives, how often the women got pregnant on the journey, and how they represented their relation to their families of origin—in this essay I move from wider cultural constructions of Rome to two juxtaposed examples. Perhaps even more galling to the historian, one of the two honeymoons in this article is fictional: I will be juxtaposing the wedding journeys of Emily Jowitt and Dearman Birchall in 1873 and the travels of the fictional characters Dorothea Brooke and Edward Casaubon in George Eliot's *Middlemarch,* published in 1872, a novel about a journey that took place more than fifty years earlier. My object here is not to measure Dorothea's honeymoon against those of a real honeymoon experience as it is captured in a diary. I want instead to see both accounts as working with and against cultural expectations of Rome, of the honeymoon, and of heterosexuality as it was beginning to be understood. While the two accounts can be opposed in terms of genre, detail, and—most notably perhaps—the degree of marital happiness they record, both work with two linked tropes that can illuminate the status of the honeymoon: the conflict between climactic and reiterative narratives and the reorientation from a singular to a conjugal gaze. Both categories—narrative and gaze—for understanding and absorbing the changes that were expected to take place on the honeymoon are deeply imbricated *in* place, not only in Rome as a destination but in the specific sights of that city—particularly, in this case, the sight at the center of the English touristical canon: St. Peter's Basilica.

As *Middlemarch*'s scholar and pedant, Edward Casaubon, prepares to return to England after his famously unsuccessful Roman honeymoon, he expresses with characteristic awkwardness to his wife, Dorothea, the wish that the "time" in Rome "has not passed unpleasantly." He continues with what the narrator calls a "little speech," "blinking a little and swaying his head up and down":

> Among the sights of Europe, that of Rome has ever been held one of the most striking and in some respects edifying. I well remember that I considered it an epoch in my life when I visited it for the first time; after the fall of Napoleon, an event which opened the Continent to travellers. Indeed I think it is one of several cities to which an extreme hyperbole has been applied—"See Rome and die": but in your case I would propose an emendation and say See Rome as a bride, and live henceforth as a happy wife.[3]

The "little speech" is, of course, remarkable for its condescension, as Casaubon rehearses for his wife not only the most familiar platitudes about Roman history but clichés about contemporary English history as well. What interests me here is precisely how those linked histories get expressed as a context for a far more personal kind of history. Casaubon is, throughout the novel, extremely uncommunicative about his own past; despite its pedantry, this moment on the honeymoon carries with it the intimacy of personal revelation. If for Casaubon, then, his first visit to Rome represented an "epoch" in his life, what about this revisitation? this honeymoon? By setting the climactic moment of his climactic narrative of tourism back in time to "after the fall of Napoleon," Casaubon is emphasizing the difference between himself and his wife. His personal story, marked and made significant by its associations with public events that took place when Dorothea was a child, cannot be Dorothea's. Her "epoch"-making journey is the honeymoon, the novel's historical present; his Roman narrative of transformation took place before the events of the (already historical) novel. If Casaubon has already been transformed by Rome, there is little or no place on this second journey for a second, conjugal transformation. The conflict here is not just a matter of age difference; it is, essentially, a highly gendered narrative conflict, an illustration of two different relations to the master-narrative of the Roman journey, embodied, for example, in Dr. Arnold's letter and in Casaubon's reference to the proverb "See Rome and die." Casaubon's amendment to the proverb, "See Rome as a bride, and live henceforth as a happy wife," cannot finally free itself from the sinister implications of the original; indeed, it succeeds only in aligning the happy wife with death and bridal transformation with changes more sinister and equally permanent.

The gendered tension between the story of the young woman's encounter with Rome—which is largely the subject of chapter 20 of *Middlemarch*—and the muted but culturally dominant masculine history of a first journey to Rome "after the fall of Napoleon" can be read back against histories that are even more official. Chapter 19 begins with a sentence whose syntax, characteristic of social-problem fiction, joins the personal and the political, the private and the public, in parallel clauses: "When George the Fourth was still reigning over the privacies of Windsor, when the Duke of Wellington was Prime Minister, and Mr. Vincy was mayor of the old corporation in Middlemarch, Mrs. Casaubon, born Dorothea Brooke, had taken her wedding journey to Rome" (154).

In moving from king to prime minister, to mayor, to private citizen, the sentence simultaneously grants importance to Dorothea as a historical personage and brings the weight of history to bear on the honeymoon journey that transforms her from "Dorothea Brooke" to "Mrs. Casaubon." The syntax of the sentence is, on the surface, democratic, its linking and structuring temporal markers ("when . . . when") arguing, it seems, for inclusiveness by chronological coincidence. If, however, we read the sentence not for its "whens" but for its "wheres," the story becomes more complicated: three out of the four personages named in the sentence are attached to a place—the sentence moves from Windsor to Middlemarch, with an inevitability characteristic of Augustus Hare, to Rome. And it is in Rome of, course, that Dorothea is explicitly described as feeling the weight of history:

> To those who have looked at Rome with the quickening power of a knowledge which breathes a glowing soul into all historic shapes and traces out the suppressed transitions which unite all contrasts, Rome may still be the spiritual center and interpreter of the world. But let them conceive one more historical contrast: the gigantic broken revelations of that Imperial and Papal city thrust abruptly on the notions of a girl brought up in English and Swiss puritanism, fed on meagre Protestant histories and art chiefly of the hand-screen sort. . . . Forms both pale and glowing took possession of her young sense, and fixed themselves in her memory even when she was not thinking of them, preparing strange associations which remained through her after-years. . . . In certain states of dull forlornness Dorothea all her life continued to see the vastness of St. Peter's, the huge bronze canopy, the excited intention in the attitudes of the prophets and evangelists in the mosaic above, and the red drapery which was being hung for Christmas spreading itself everywhere like a disease of the retina. (159)

This passage explicitly connects personal history with a gendered historiography: "meagre Protestant histories and art chiefly of the hand-screen sort" have made Dorothea literally incapable of understanding more cosmopolitan narratives. Elsewhere, Eliot refers to Dorothea's education as producing a "toy-box history of the world" (70); as a woman and a Protestant, Dorothea is made to occupy an infantilized position in relation to historical master-narratives.

Eliot's critique of Dorothea's education makes it clear that the problem is not merely with individuals. Like the proverb, Eliot's analysis points beyond a personal to a structural—indeed, a narrative—conflict. The failure of the Casaubon marriage—and of the honeymoon that foreshadows it—is not the result of a clash of personalities. Both fail because of a conflict embedded in what we might call the marriage plot of Victorian culture, a plot in culture and in literature that assumes that desire, action, and representation end with marriage. In *Middlemarch*— one of the first English novels to move beyond the wedding ceremony, through the honeymoon, and, painfully, through the tribulations, compromises, and accommodations of marriage—there is always at work a conflict between marriage as an ending and as a beginning or, particularly in a novel with this name and focus, as a middle.

We can, then, begin to identify in *Middlemarch* the collision of various temporalities, calendars, and narratives: collisions between the narratives of husband and wife and between the calendar of the marriage plot with its obligatory transformations and Dorothea's mournful looking back. The changes associated with and presumed by the master-narrative of the marriage plot are embedded in a linear structure, a structure that is asymmetrical with respect to gender but nonetheless a matter of temporal sequence. The location of this particular honeymoon offers us a way to rethink the temporal in terms of place, in terms of "visible history" that makes the honeymoon a clash not only of stories but of gazes.

Dorothea's experiences of the honeymoon, as the passage quoted earlier indicates, are refracted through the image of St. Peter's, which serves as an emblematic site of what we might think of as her honeymoon tourism. Obviously, Dorothea is not pictured as deploying the commanding or consuming gaze associated with nineteenth-century English tourists; St. Peter's seems to look back, or at least to take on a sinister agency in the visual economy of the scene. I have suggested elsewhere that the afterimage, the repeated nightmares, and the sense of visual assault are similar to those reported by contemporary survivors of rape and sexual abuse; in the context of the Casaubon marriage we can (perhaps unwillingly) imagine the sexual shocks registered in Dorothea's reaction to the more official "sights" of Europe.[4] What is important here, whether we buy this reading or not, is that these shocks are registered as visual events, and that Dorothea's gaze is radically separated not only from a generalized gaze of the educated (male?) tourist who has moved beyond a "toy-box history" of the world but also from the gaze of her new husband. Casaubon is, of course, literally separated from Dorothea during the honeymoon. While Dorothea spends her days sight-seeing, Casaubon spends them in the Vatican archives, in the "dark labyrinths" that produce and indeed necessitate a different, if equally troubling, kind of vision.

Casaubon's brief reference to his own history of sight-seeing should alert us to the fact that his own gaze is not, at the point of his honeymoon, canonically touristical. The gaze of the educated tourist is associated, in the passage about St. Peter's, with a climactic narrative; this gaze is embodied in "those who have looked at Rome with a quickening power of a knowledge which breathes a glowing soul into all historic shapes." Dorothea's gaze is explicitly contrasted with theirs; when we are told that "forms both pale and glowing took possession of her young sense," the repetition of "glowing," with its very different sense of agency, suggests an inversion in which the sights of Rome actively penetrate Dorothea's mind and body. Casaubon, however, is past such visual enthusiasms; his is the gaze of the scholar, which turns away from the visually accessible to material that is hidden, subterranean, obscure.

And yet, Casaubon's personal revelation suggests not simply the scorn of the scholar for the tourist but a touristic déjà vu that is, if we listen to Augustus Hare, part of the touristic experience for all educated travelers to Rome—all those at least whose education has not been of the toy-box sort. Casaubon's sense of "Rome again" is of course different from Dr. Arnold's; Casaubon cannot capture the enthusiasm of previous visits except by placing them obliquely in a past to which *Middlemarch* denies sustained access. Dorothea's gaze, then, is defined against both the tourist's and the scholar's; ironically, the inscription of her gaze within a climactic narrative that forces Dorothea to confront so many things for the first time makes it impossible for Dorothea to transform herself according to the cultural expectations of the honeymoon.

If we force ourselves, as the Eliot narrator occasionally and self-consciously does, out of Dorothea's point of view and into Casaubon's ("but why," the narrator interrupts herself, at the beginning of chapter 29, to say, "always Dorothea? Was her point of view the only possible one with regard to this marriage?"),[5] we can see that the transformative demands of the honeymoon—with its array of visual landmarks—are equally burdensome for both. Casaubon cannot capture the gaze that was his just after the fall of Napoleon; he can have only what we might think of as a professional visual investment in the sights of Rome. And Dorothea can be haunted only by the scopic requirements of a honeymoon in Rome. The gazes of the scholar and the neophyte cannot be fused into tourism or into the conjugality that honeymoon tourism was specifically designed to produce.

If the honeymoon in *Middlemarch* is remarkable for its failure to bring together differently gendered narratives and gazes, real-life honeymoons offer a far greater variety of possibilities for the production of narrative and visual conjugality. Of all my honeymooning couples, no one claimed to be unhappy with his or her spouse. Although some members of honeymooning couples reported depression or sadness, and at least one a serious discomfort with sexuality, the marriage itself was never questioned. What comes across clearly in reading the entire sample is what is missing or submerged in *Middlemarch:* the ideal of the honeymoon and the cultural work the journey was supposed to perform. This is precisely the work of reorientation and of the production of conjugality whose triumph—if it happened— can be read in the coming together of separate narratives and the fusion of gazes.

One type of source from which much of the information for the study comes, the honeymoon diaries written by one member of a given honeymooning couple, would seem on the surface to be the perfect example of individual narrative. Our cultural conflation of honeymooning with the historical emergence of what we think of as private life, as well as with what Eliot might call the "privacies" of sex, would seem to suggest that honeymoon diaries, even more than diaries written under other circumstances, would fulfill the scholarly (and voyeuristic) fantasy of the archive: the document that reveals the secret of past selves. If the diary is an individual form, the *honeymoon* diary might have an especially intensive and dynamic relation to the individual, the couple, and the family.

It is, of course, almost impossible to deduce to what extent the diaries were seen as private to the writer and to what extent they were shared, either with the traveling spouse or with other family members on or before the couple's return. Some diaries include internal evidence of shared reading, the most obvious of which is Mary Monkswell's 1863 diary of her honeymoon trip to Switzerland and Rome: her description of the wedding day is interrupted by a brief passage written by her husband in which he playfully supplements his wife's description of a young guest's poking of Mary's ex-governess with an umbrella. Mary's diary begins the story: "My dear old Miss Pyman I believe wept a little behind her handkerchief. She was much ill-treated by Thomas Usborne before I arrived;—he had been there some time & was getting rather bored, so he took to poking her with his umbrella between the rails!" Mary's bridegroom, Robert Collier, intervenes: "Ribs she means. Bob (that's me) happened to be looking on from his position close to the altar where he had been standing for some time trying to look unconcerned."[6] Bob's use of his nickname, as well as his playful movement from first to third person, sets up, even at this early point in the honeymoon, a sense of intimacy and informality. His intervention in the diary reads less as a correction than as the continuation and deepening of a shared joke. Bob's introduction of the body—his substitution of the irreverent "ribs" for "rails"—might also suggest a bodily intimacy, or at least an intimacy with the body, that makes this diary quite unusual.

Other diaries show more-subtle traces of conjugal discussion or influence: while we do not, unfortunately, have Charley (Mrs.) Campbell-Bannerman's honeymoon diary for 1860—and thus cannot compare it with her husband's—we do have both diaries for their trip to the same location the following year.[7] Charley notes her husband's corrections to her identifications of works of art on several occasions; it seems also that her husband has made some corrections and additions in his own hand. Juliana Stratton Fuller, whose 1866 diary is perhaps the most complete example of reorientation, starting as it does with a picture of the home of her birth family and ending with a photograph of her husband's house and an account of the welcoming ceremony staged by the servants, quotes her husband on the matter of views, importing and repeating what is clearly his term, "picturesque."[8]

Martha Rolls's extensive honeymoon diary of 1840 is full of loving references to her military husband, but she does not seem to have shared it with him.[9] The diary, indeed, is preoccupied with finding out the secret of her husband's recurrent depressions; the diary depends, formally and narratively speaking, on the idea of secrets between spouses. Although Martha does eventually find out what is bothering Edward, she does not confide anything but the fact of her discovery to her diary. Narrative secrecy, then, remains an element of the diary throughout, as Martha withholds not only from her husband but also from the journal in which she records her attempts to uncover his secret.

While the Rolls's diary clearly gestures in many ways to individual privacy, some of the other diaries at the other end of the spectrum—in terms of the recording of affect—rarely mention the name of the traveling spouse, and when they do, they make no special remark about the relationship. In a diary remarkable for the detail of its menus and quotidian sights, Emma Hardy mentions her husband only once, and then to record his absence in midflow of her idiosyncratic prose:

> Garden of the Tuileries. . . . Little birds dusting themselves around my seat. (A semicircular stone one) Borderings of *ivy* The new fresh green young leaves a pretty contrast with the dark old ones—These borderings all around the flower borders. The birds rejoice in them Vases & Statuary everywhere—the palace of the Tuileries faces this seat *greatly* damaged— Workmen are busy making repairs—The sound of their tools mingles with that of the carriages & voices in the streets & the little birds chirping—It rains softly just a little—
>
> T. has left me for 20m. To get his coat—because of his cold—Three hommes.[10]

Richard Cobden's diary, like Emma Hardy's, reads exactly like any other travel diary; the fact that the journey is a honeymoon never makes its way onto the page.[11]

The point here is not to privilege one kind of diary over another or to make assumptions about the level of intimacy between members of a honeymooning couple based on their acknowledgment of their spouse in their diaries. Some writers, like Cobden, were far more forthcoming in their letters. Others, whose letters are not available, might also have found other venues, other ways, to record the trip as a honeymoon. The honeymoon diary, however, because of its equivocal and generically complex status with respect to individuality, is an important location for tracing specifically narrative signs of conjugality.

Honeymoon diaries and other sources suggest, as does *Middlemarch,* that conjugality had an important visual dimension. Emily Birchall's experiences reported in her meticulous honeymoon journal of 1873 provide a spectacular contrast to Dorothea's. She records a five-month wedding trip to France, Switzerland, Italy, Vienna, and Budapest, a journey whose visual center, like that of the *Middlemarch* honeymoon, is St. Peter's. Despite a radical difference between the Eliot narrator's tone and Birchall's, as she describes her loving and companionable relationship with her husband and the joy she takes in visiting touristic sights, Birchall's account shares with Dorothea's experience a profound investment in the visual as it is embodied in St. Peter's and, more generally, in Rome. Birchall's journal entry for the couple's first day in Rome takes for granted the significance of Rome for educated tourists:

> In Rome! the first thought when we awoke this morning, as it was the last when we closed our eyes last night. It seems almost too good to be true that the longing dreams of one's whole life should actually be realized.

> And the morning is so fresh and bright and sunny that Rome appears to
> us first under the most charming circumstances possible. It is far far more
> delightful in every way than either of us expected, high as were our ex-
> pectations. It is such a bright, healthy feeling, *cheerful* city, such a *clean*
> place, so orderly, that even had it no historic associations, no monuments
> of art or antiquity whatever, it would still be, I think, the very pleasantest,
> charmingest spot on the face of the earth.[12]

More telling than the "charmingest" superlatives with which this passage ends,
more telling even than Emily's most unusual use of the words "clean" and "healthy"
to describe a city that other Victorians associated with filth, beggars, and fever,
are the pronouns that authorize Birchall's point of view. By the time the couple
arrives in Rome, the gazes of husband and wife are completely aligned: the two
open and close their eyes together, look at Rome together, and, finally, evaluate
their experiences in the same way. We do not, of course, have the advantage of
Dearman Birchall's direct testimony, but this is beside the point: we can watch as
Emily carefully constructs a joint point of view from which she is able to assert a
"we" and an "us" that come to life in the act of looking. Emily's sole use of "one"
("the dream of one's whole life") does not so much separate her from her husband as suggest a universality to her dream that extends beyond even this couple's
sleeping and waking.

Emily brings the conjugal gaze with its attendant "we" to bear upon what
for her is clearly a synecdoche for the touristic experience of Rome, the canonical
visit to St. Peter's:

> The first place we go to is of course S. Peter's. Who could be twelve hours
> in Rome and keep away from her Cathedral? We walked to it, guided by
> Perrini . . . over the Ponte di S. Angelo where the first glimpse of the Dome
> came to us, and far from feeling the least touch of that disappointment
> which, so many say, accompanies the first sight of it, it filled me with
> wondering imagination. It was grander, huger, vaster than my dreams of
> it. We walked on till we stood in the glorious Piazza, with S. Peter's in
> front of us, its splendid colonnade on either side, and the Vatican to our
> right. We went into the cathedral, and we stood still in sight of its vast-
> ness, breathless with awe and wonder. Dearman said he never felt so
> excited in his life, and my feelings were certainly beyond description. (23)

Like *Middlemarch,* Emily's journal represents sight as a point of entry into the
body; the hugeness of St. Peter's has the power to take one's breath (and speech)
away. Of course, in this case, not one but two bodies are simultaneously and iden-
tically affected. Emily's own "wondering imagination" melds, several lines later,
with a shared experience of "awe and wonder." In the context of a honeymoon—
and particularly of one in which Emily jokes in her journal about Dearman's "nu-
dity" when he takes off his waistcoat to walk along the French coast—the
excitement Emily attributes to Dearman and the more generalized "feelings" she
claims for herself take on an erotic register. The contrast with Dorothea's experience
is stark.

Emily's gaze, like Dorothea's, takes careful account of the déjà vu. Here, her gaze and Dearman's are contrasted with those of two previous narratives, themselves at odds with one another: the presumed hyperbolic narrative of fulfillment and the counternarrative of "disappointment" that "so many say, accompanies the first sight of" St. Peter's. The newly forged conjugal gaze is, in effect, a talisman against disappointment; while one could imagine an agreement on the part of husband and wife that St. Peter's was, in fact, disappointing, the disappointed gaze, however unifying, would not carry the euphoric power of its transportative opposite.

If Emily Birchall's ecstatic account of visual merging makes the conjugal gaze the product of successful tourism, Louise Creighton's retrospective account, in her memoirs of her 1872 honeymoon to Paris, might complicate the picture by establishing conjugality through touristic withdrawal. Her description begins with a story of shared consumption:

> We visited picture galleries & churches, & looked a great deal in shop windows, hunting for various things for our house. We wanted to have as few ordinary things as possible & looked everywhere for some different kinds of dinner knives, & also for brass fire irons. We were troubled because the sets being made for wood fires never had a poker, & we could not discover the french word for poker so as to ask for one. Our chief purchase was a moderator lamp made out of an oriental vase, which was long our chief lamp. It was many years before parafine [*sic*] lamps came in. We went to the theatre several times.[13]

Here the gaze constitutes itself out of consumption related to but in some ways different from the taking in of sights. Obviously more literal, this consuming gaze depends both on looking and on buying, on "visit[ing] . . . churches," on "look[ing] . . . in shop windows," and on making one major purchase. That these activities are recounted metonymically suggests the closeness of their relation; that the sentences along this metonymic chain depend, like so many of Emily Birchall's, on the plural pronoun "we" suggests the closeness of this consuming pair. If we stop at Louise's ellipses we get a feeling at least structurally similar to the one produced by a reading of Emily's diary; although the activities are more quotidian and more domestic, there is the same shared vision.

After the account of shopping, however, Louise abruptly ends the story of the honeymoon on a more negative note. Without explanation, she moves from a mildly successful shopping trip to a more negative evaluation of place: "I don't think we either of us, either there or at any later time, felt any real love for Paris. We never did more in later years than spend a couple of days there on our way elsewhere" (46). There is something different about the tone of this paragraph, registered, perhaps predictably, in its use of pronouns. The possibly divisive "either of us" lingers briefly on the possibility of disagreement, only to have it disappear with the invocation of the "we" of "later years." There is also a parallel attempt to erase temporal difference with the collapsing of the honeymoon and the "later time." Of course, unlike Emily's diary, this is a memoir written after the death of a husband, but this generic fact makes the collapsing of three time periods (the honeymoon, married life, and the time the memoir was written) even more important. Finally, and perhaps most significantly for the purposes of this essay, the paragraph collapses and indeed fuses space and time. Like Dr. Arnold's "date of Rome" the problematic parallel construction "there or at any later time" suggests the importance of place—in this case Paris—*as* time.

Notes

1. Augustus J. C. Hare, *Walks in Rome* (London: Strahan & Co., 1871), 1.
2. Helena Michie, *Sororophobia: Differences among Women in Literature and Culture* (New York: Oxford University Press, 1991).
3. George Eliot, *Middlemarch* (New York: Oxford University Press, 1988), 163.
4. Helena Michie, "Looking at Victorian Honeymoons," *Common Knowledge* 6, no. 1 (spring 1997): 128–130.
5. Eliot, *Middlemarch,* 228–229.
6. (Lady) Mary Monkswell, *A Victorian Diarist: Extracts from the Journals of Mary, Lady Monkswell,* ed. C. F. Collier (London: John Murray, n.d.), 3.
7. Charlotte Campbell-Bannerman, MS diary (1860), Campbell-Bannerman Papers, British Library.
8. Juliana Stratton Fuller, MS diary (1866–67), Wigan Record Office, Leigh, U.K., M993 EHC 195.
9. Martha Rolls, MS diary (1840), Rolls Family Papers, Gwent Records Office, Cwumbran, Gwent, Eng., MS D361 F/P6.
10. Emma Hardy, *Emma Hardy Diaries,* ed. Richard H. Taylor (Manchester, Eng.: New Cumberland Arts Group and Carcanet New Press, 1985), 41.
11. Richard Cobden, MS diary (1840), Richard Cobden Papers, West Sussex Records Office, Cirencester, Eng., 441–469.
12. Emily Birchall, *Wedding Tour, January–June 1873, and Visit to the Vienna Exhibition,* ed. David Verey (New York: St. Martin's Press, 1985), 22.
13. Louise Creighton, *Memoir of a Victorian Woman: Reflections of Louise Creighton, 1850–1936,* ed. James Thayne Covert (Bloomington: Indiana University Press, 1994), 46.

Erotic Geographies

SEX AND THE MANAGING OF COLONIAL SPACE

PHILIPPA LEVINE

*W*here does the brothel belong? Though nineteenth-century writers may not have put the question so bluntly, it was a problem of considerable significance for a society so committed to the ordering of space and of peoples. The brothel was a difficult site for Victorians because its existence, its presence, and its placement blurred so many of the fundamental categories of Victorian order, challenging proprieties and geographies. The brothel was a paradox, a space of private activity made public, a place where the always opposed ideas of work (for women) and pleasure (for men) melded, where the carefully formulated attributes of "home" and "business" were necessarily blurred. The domestic space turned to business use underscored both the cultural illegitimacy of paid sex and the blurred boundaries that held up, for silent but visible scrutiny, the legitimacy of Victorian order itself.

The invocation, so central to nineteenth-century British life, of "the domestic" is as palpable a spatial form as it is a cultural one, and a geographical reading opens up the varied and connected ways in which the brothel as a problematic space undermined domesticity and all it stood for. The brothel was especially challenging in the colonial arena, for there the very idea of the domestic could be read as under attack from those who lacked the privilege of its refinements. Producing and maintaining the domestic space in the colonial context was always an ambivalent project, for it posited the sensibilities of the West in an environment always figured as antithetical. By attending to the cultural geographies at work in colonial spaces, we can track in detail how these binaries operated, meshing the material and the representational, working simultaneously to "tame" and yet to separate the "metropole" from the "periphery," sexually, politically, geographically. The very notion of domestication in the colonial context proffered the frightening possibility that the colonial might become metropolitan, or that the brothel might, in fact, not be immediately distinguishable from the home. These potential instabilities permit a rich geographical reading, playing on the endless contradictions involved in establishing domestic fronts in colonial settings. Although the siting of the brothel was everywhere an issue of anxiety, colonial settings clearly required an even greater effort at boundary keeping than the cities of island Britain.

The policing of the brothel was always about far more than merely the surveillance of nonmarital sex; in colonial settings the necessity to maintain the divisions that segregated the "native" from the colonizer was critically conceived as spatial and as sexual. In contrast to Britain, which was represented as smoothly ordered and temperate, colonial spaces were seen as too crowded, too fecund, too backward, and these critical images played on the distinction between the alleged chaos of the East and the well-regulated harmony of Britain. Not surprisingly, the brothel figured amply in this debate, the perfect marker of colonial inferiority and, simultaneously, a site of deep anxiety about boundary maintenance.

Space and the need to order it were issues of constant concern in colonial settings. Unordered and chaotic space held many potential dangers, ranging from fears of contagion to the threat of unseen dissent; for the purposes of colonization, there were ordered spaces and those requiring ordering. Eastern spaces were invariably represented as in some way disorderly, whether in the moral sense, in the chaos of a town's layout or buildings, or in the lack of attention to safe sanitation. The very locations of the local population were inimical to propriety. Worrying about the effect of Hong Kong on British soldiers posted there, the China Association (based, tellingly, in Britain) informed the colonial secretary in 1899 that "the barracks in Hong Kong form practically part of the city of Victoria [modern-day Hong Kong island], and . . . men need only overstep that line to find themselves in a locality devoid of restraint."[1] Restraint and respectability were twin features of properly constructed locations, military and civil. In these areas, the private did not spill out-of-doors and intrude into the public, commerce did not mingle indiscriminately with residence, and even the layout of the streets followed an orderliness utterly lacking in the lanes and alleyways of Singapore or Calcutta. Pollution was the inevitable result of the mixing of these spaces for, as the association observed, it was the *overstepping* of carefully marked boundaries (between East and West, between clean and dirty, between foreign and domestic, between controlled and chaotic) that colonialists most feared and that, they claimed, resulted in disease and danger.

The nature of sexual commerce—whether conducted on the street or in enclosed environments such as the brothel—was central to these spatial concerns. The prominence of indoor prostitution in colonial settings was such that, almost everywhere, the brothel dominated the sex trade, even before a British presence was discernible.[2] Although houses of prostitution were, in one form or another, not unfamiliar sights before colonialism, only when the British imported European Victorian notions of the distinctiveness of the public and the private did the brothel acquire new meanings. And alongside these meanings came significant ramifications for the placement of the brothel and for the lives of those who worked there.

Streetwalking—public prostitution—was always emblematic of all that was wrong with the sex trade and was consistently regarded as its most offensive segment. Vice-Admiral Sir Henry Kellett, overseeing the registration of sex workers in the Japanese treaty ports, thought "street prostitution . . . totally corruptive of public decency."[3] In the Nusseerabad cantonment in the Bombay Presidency, women who saw public solicitation as better for business bore the brunt of punitive attention. The result of their streetwalking, claimed a local British surgeon, was a rise in venereal disease rates and a loss of control over the public arena.

> The women left their residences in the Bazar and took to practising their trade in the vicinity of the Barracks, on the public roads, and often in empty bungalows. . . . Control became almost impossible and admissions into Hospital immediately increased. The women's conduct became so offensive that several respectable inhabitants complained to the Police Inspector, who treating the matter as a public nuisance, caused the Magistrate to issue a prohibitory order on them, and thus relegated them to their recognized residencies, bringing them under supervision again, the natural result being their attendance at Hospital as usual.[4]

The links the surgeon drew between public prostitution, the loosening of discipline, and rampant disease were located in a tidy geographical space in which the brothel (the "recognized residence"), whatever its problems, was by implication better, safer, and more smoothly controlled than outdoor prostitution.

Brothels, although preferable, were nonetheless worrisome when they cropped up in the wrong location—near "good" areas or schools and colleges, in well-traveled streets, or in residential areas where housing prices were hefty. The inspector-general of police in Colombo (in British Ceylon) told the dramatic tale of "the notorious case" concerning "the Ladies' College, close to a brothel in Union Place, where a party of sailors refused to believe that the Ladies' College was not a brothel till fortunately a resident who lived between the two houses hearing the noise appeared on the scene."[5]

Such stories clearly had something of a folk quality to them, but they successfully conjured the effects of a sexual spillage with all its soiling associations. The Colombo inspector-general of police ended his story with the arrival of a male neighbor, the rescuer of the Ladies' College damsels, and he notes, "[B]ut for his prompt arrival anything might have happened."[6] Controlling brothels, their placement, and their inmates was a necessary corollary to ensuring European safety and peace of mind. The rights of Europeans to live beyond the visible spectacle of sex was paramount, and, as a result, it was "common" women whose rights and movements had to be restricted. At work here was the fear that if a brothel did not *look* like a brothel, mistakes with damaging consequences were inevitable.

It is not, perhaps, unimportant that it was a "resident" (and, of course, a brave and worldly man) who saved the schoolgirls from violation. This detail of identification drove home the inappropriateness of a brothel's intruding into the secure respectability of a residential area close to schools. But this thinking about boundaries also created contradictions. Since the brothel was to be denied a place in the spaces reserved for conformity and yet also belonged elsewhere than on the public thoroughfare, its literal siting was always at issue.

Brothels did not properly belong in the public eye, for the business they conducted was of a private nature, and scores of commentators expressed distaste at nonwhite disregard for these proprieties. Hong Kong policeman Kenneth Andrew remembered the Wanchai brothels as "very public" because the women sat "in full view of the passers-by in the street outside."[7] The brothel had, then, to be concealed from public view, removed from the public sphere. In northern Queensland, for instance, the local police consistently defended the Japanese brothels of Cairns precisely because "a stranger coming to town would have to seek them before he would find them . . . they are not prominent to public view." Farther south, in the town of Childers, brothels came under fire for their proximity to a school and to "private and business houses." In Singapore, "the red-light districts served as a 'natural' form of boundary maintenance," with the regulated brothel areas kept tied to those places not residentially favored by the colonists. In keeping with his hierarchically racial assessment of the big-city sex trade in India, Police Commander E.C.S. Shuttleworth, in a 1917 report to the government of India, praised Calcutta's European brothels for their "well hid" location in the suburbs, while showing a marked distaste for the "rampant" indigenous practice of "advertisement by sitting at doors and windows and balconies and . . . both by voice and gestures."[8] Gender and geography meld almost imperceptibly in Shuttleworth's assessment. The too public locale of the native woman was mirrored by her unwomanly loudness and her shameless touting of her wares; she appreciated neither the feminine space of domesticity nor the feminine virtues of modesty and soft-spokenness. Yet, written into these sentiments was the challenge to the private-public divide, since the brothel and its inmates could be housed neither wholly in the public space for fear of legitimation nor wholly in the private sphere for fear of pollution and defilement.

Out of reach of the law by dint of their invisibility were women whose services were tied to one man. The mistress, known also in colonial parlance as a concubine and in Hong Kong as a "protected" woman,[9] was routinely granted separate classification from the "common" prostitute willy-nilly degraded by attending multiple clients. But what really set mistresses apart from the women of the brothels and of the streets was their hiddenness from the public gaze, despite the blasphemous parody of "proper" marriage their arrangements proffered. In Bombay, the government "instructed the Commissioner of Police to be careful not to interfere with or attempt to register women in the position of mistresses of wealthy persons, or kept women."[10] Power and class issues were at stake here, and neither politicians nor the military wished to alienate the wealthy, of whatever race or nationality. But the policy went beyond merely protecting the privileges of the rich. The "kept woman" was likelier to be the consort of rich men than was the brothel woman, and her arrangements with a single client guaranteed a level of privacy unattainable in the brothel. Officials in the North-West Provinces in India appreciated that distinction; framing rules under the 1864 Cantonment Act, they stressed that "the proposed Rules are only applicable to *public* prostitutes. There are prostitutes to whom the term 'public' is hardly appropriate. The line will not be difficult to draw in practice."[11] The enduring notion of a mistress as a woman kept in luxury and idleness also helped maintain a perceptual distinction between her and the prostitute. While the mistress could, however unrealistically, be represented as living in comfort, the brothel occupied more ambivalent territory. It was, after all, fundamentally a site of male leisure and female work, however disguised that reality might get. Women peopled the brothels specifically to earn a living, whereas men came there seeking sex and the company of women.

The brothel crossed not only the boundary between work and leisure but many others besides. Brothels, after all, were seldom purpose-built institutions; the word "brothel" describes not so much a space as the activities conducted within that space. Arguments about location and suitability are a posteriori positions arising from how prostitution is moralized, and not from any inherent spatial "fact." Most brothels were—and are—ordinary houses, their only substantial requirement being that there should be sufficient accommodation for business to run smoothly. It was this discomfiting proximity of the brothel to the home that sparked much of the debate over siting. And though such arguments were never a colonial monopoly, they took on extra dimensions in settings where the key and oppositional Western sites of home and workplace, of private and public, had no historical meaning. The constant redrawing of those boundaries, literally and metaphorically noted by Rosemary George,[12] lent a piquancy to their assertion as a necessary part of the imperial uniform.

Nicholas Thomas has shown how, in colonial Fiji, regulations redefined the proper home and its functions, separating animals and humans, thinning out the numbers considered appropriate per house, and so on.[13] The constant British complaints about overcrowding and lack of sanitation in Asian cities and villages and about the filth that was seen as endemic in the Aboriginal camps of Queensland, all spoke to this same sentiment, that the failure to observe the proper boundaries and proprieties of the domestic space and the work space, the indiscriminate mixing of the two with its deleterious sanitary effects, was a moral failure of subject peoples. And the brothel—looking like a home, acting like a business—was always at the heart of that debate.

Hong Kong's colonial surgeon, reporting in 1879 on his visits to the colony's brothels, did not hide his horror at the effect of this indiscriminate mixing of the spheres: the kitchen had permeated the bedroom. "I've a pretty good stomach and don't stick at trifles, but I found the inspection of these places acted as a very unpleasant emetic. The girls' rooms, near the kitchen, nearly all had ventilating openings into the kitchen."[14] Brenda Yeoh has documented the fight in Singapore over the contiguity of commerce and residence, with the veranda as the principal site of conflict between residents and the colonial-municipal authority.[15] Yeoh argues that for the local population the veranda was simultaneously a place of business opportunity and a space for socializing, since the two activities were less categorically separated than in the West. In the brothel—and not just on its veranda—these activities were critically linked as the best means to good business. Just as was so in municipal politics, the argument about what constituted a public site was crucial. In his extensive memorandum on the sale of sex in British Indian cities, Shuttleworth traced the changes in policing in Rangoon, which had pushed brothel residents literally indoors: "European and Japanese prostitutes are now kept behind swing doors quite hidden from the users of the streets."[16] In this instance, the veranda as a place of business lost the battle, and sex workers were reminded of the varied modes of surveillance to which they were subject.

Typically, the brothel was both home and workplace for women, a mixing that raised difficult questions in a whole variety of contexts. The idea of the brothel as an actual place of residence, and not just a house distorted to improper use, had implications both for brothel workers and colonial policy. In India, where soldiers' encampments made the physical ordering of the population an easier task than in less militarized sites, the cantonment acts of the 1890s dictated where women could live. Women suspected of harboring disease and refusing ostensibly voluntary examination could be expelled from cantonments, forbidden to live or work within boundaries laid down by the cantonment magistrate. Since the prostitute had both defiled and de-sanitized the home by turning it into a brothel, she could be ejected summarily from the comforts that accrued only to those who maintained the proper boundaries. The 1897 Aboriginal Protection Act—which allowed Queensland officials to monitor both workplace and residence for Aboriginal women, specifically as a check against immorality—worked similarly. In both cases under the guise of protection, a punitive rearguard action to reassert the proper boundaries separating public and private was at work, and for none more so than for those women whose occupation and livelihood defied the proper spatial and moral divides.

It was not only in the promiscuous entanglement of home and work that the brothel symbolized disorder. Though the women's quarters of the East were always an uncomfortable issue for colonists, the brothel was also a strange distortion of the idea of a women's space. The West, after all, had its own versions of sexual segregation in the linking of domesticity and femininity, but the brothel did not fit neatly into either the Western or the Eastern form. A Madrasi health officer, writing to a government official early in the 1870s, saw the brothel as a paradox that defined Indian immorality: "A brothel is not a brothel as in Europe, where girls live as one family, but in this country prostitution is considered of so little importance that chaste women and prostitutes reside in separate rooms adjoining each other in the same house and are even on intimate terms."[17] Had the health officer looked more closely at Britain's cities, he might have discerned a similar pattern of residence.[18] The reality of the metropolis notwithstanding, his comments are revealing. While European women understood the importance of even ersatz family ties and thus successfully re-created the brothel as home, their Indian counterparts failed to appreciate the qualities that inhered in that practice. For them work and home were not distinct, and neither, it followed, were the divisions between the respectable and the outcast.

The specter of the zenana, the separate women's quarters of many Eastern countries and a space off-limits to colonial eyes, was aligned in colonial thought with the brothel, despite the zenana's association with elite groups. In some sense, both were women's spaces, though obviously in the brothel the presence of men was necessary, though never permanent. For Westerners, both spoke to the idea of a secret and sexually mysterious East. The erotics of the harem have been well rehearsed by many a cultural critic, and I will not revisit them again here.[19] But the links between the brothel and the zenana go deeper than merely the whiff of a hitherto untasted sexuality that both seemed to offer the European sensibility.

In India, purdah was most commonly practiced among the wealthier segments of the north, northwestern, and eastern populations.[20] These were also areas where the British were strong and where military cantonments were most commonly situated. Since it was also cantonments to which prostitute women were drawn by the promise of a large client base, the stage was set for comparisons between the brothel and the zenana, as two "versions" of the woman's space. Neither, of course, owed its existence to colonialism. Rosalind O'Hanlon maintains that, on the contrary, the East India Company had consciously sought to masculinize its operations to counter what it saw as the effeminizing tendencies of the zenana in some areas of company control.[21] And although the shape of the sex business certainly altered as a result of colonization, the exchange of sex for money was not introduced to India by the company or by the British. Both the zenana and the brothel, however, were subject to trenchant disapproval by the colonial state, and often for similar reasons.

Zenana inhabitants and brothel residents alike were regarded as idle women, corrupted by their distance from useful and fulfilling occupations.[22] Though women in the brothels were palpably earning a living, their activities could not be admitted as work without tarnishing work as a source of virtue and moral uplift.[23] Prostitute women were likelier to be convicted as idle and disorderly or on vagrancy charges, rather than on specific charges of prostitution, not only because the former were easier convictions to secure but also, and importantly, because such charges were an articulation of the refusal to embrace prostitution as legitimate work. Inderpal Grewal has argued that the zenana offered the British a useful means of representing the distinction between British industriousness and Asian laziness, a torpor invariably tinged with the erotic.[24] The brothel—where male leisure rather than its necessary counterpart, female labor, was always the dominant personification—played a role similar to that of the zenana in evoking the twin evils of sex and idleness and, of course, the connectedness of those evils. Many an army chaplain and doctor exhorted soldiers to occupy their time fully so that the distraction of sex would ebb.

Idleness, lack of employment, and doing nothing were, in metropolitan terms, evidence of degeneration and immorality. The stigma that unemployment accrued in Britain translated into a kind of pathology in the East, where the zenana and the brothel—although at opposite ends of the spectrum of respectability—evidenced spatially a host of improper cultural attitudes. It was not insignificant that one of the most constant criticisms of the zenana was that it was sunless and airless.[25] The brothel was also depicted, thus, as a place of the night, where sunshine and the wholesomeness of fresh air did not belong and could not penetrate. Airlessness in the vocabulary of Victorian health was synonymous with insalubrity, uncleanliness, and, of course, contagion. Colonial systems of hygiene "privileged openness, visibility, ventilation, boundaries, and a particular spatial differentiation of activities." The zenana, by such standards, was an insanitary and pathogenic environment, "a gendered space of confinement and infection."[26] Women's spaces defied both proper hygiene and respect for the diurnal distinctions that regulated good behavior and separated work and leisure, home and workplace, good habits and bad.

It was not only public health reformers and government-appointed doctors who saw the harem as detrimental to health. This was a common literary trope as well, found in Montesquieu, in Mary Wollstonecraft, in the pornography of the period, in Byron, and perhaps most famously in *Jane Eyre*. The high-walled enclosure that is Lowood experiences fatal epidemic disease. Thornfield is, of course, regularly figured by Brontë via the language of Orientalism, and St. John Rivers, intent on leaving Moor House for mission work in Calcutta, is the epitome of the Christian proselytizer for whom the zenana speaks of irreligion and danger. In the course of the novel, then, the critiques of the women's quarters—unhealthy in spiritual, medical, and physical terms—are all articulated in the name of women's freedoms.

Janaki Nair and Sandhya Shetty have both pointed out that the zenana's threat to colonial rule lay in its impenetrability, in its continuation as a determinedly uncolonized space.[27] One of the principal efforts to get inside the zenana rested on the principles of health, on making it accessible to doctors.[28] Similarly, sanitary principles were the means by which the brothel was brought under control and made visible even while it acted to diminish the visibility of other modes of sexual soliciting. A host of rules, rarely successful, were applied to brothels with the explicit intent of rendering them not just medically safe but acceptable. Of course, such regulations served also to underline that the brothel was and had to be distinguishable from the "home." Regulating private houses, after all, was well nigh impossible in this era of strident individualism. Ironically, the brothel was one of the few institutions in which locals lived where an active sanitation program was even attempted. Both Vijay Prasad and Dane Kennedy have pointed out the considerable inequalities in sanitary service to Indian and to British sections of towns, cantonments, and hill stations, while J. B. Harrison points similarly to the marked difference in density between the Indian city and the European civil lines in his case study of sanitation in Allahabad.[29] Such asymmetries were by no means confined to India; nonwhite populations (and, indeed, much of Britain's own proletariat) were invariably regarded as dirty, and colonial medicine concentrated more on protecting European health than on addressing the pressing health and sanitation problems faced by indigenes.[30]

The brothel, bizarrely, was in its liminality a kind of *cordon sanitaire*. As a quasi-public institution, and especially where catering to a European clientele, it was one of the indigenous spaces that could be most readily regulated and sanitized, peopled as it was by women. Throughout the Asian and Pacific colonies that concern us here, medical and military authorities busied themselves with cleansing and ordering the brothel, even while deploring its existence. One medical officer in India recommended purpose-built state brothels over which supervision could be better exercised.[31] And Hong Kong traveled partway down that path in the early 1870s, when the colonial surgeon closed what he thought were the most noxious of the brothels, as likely breeding grounds in a typhoid outbreak. The consequences were, he claimed, grounds for future optimism, for in his 1874 inspection of the regulated establishments, he found "a wonderful change" had been effected; the houses looked "clean, light and airy for the most part."[32] Fyzabad cantonment in India's North-West Provinces boasted, "[I]n every room in the brothels a permanent brickwork stand has been erected, the height of a man's hips, and on this stand the basin is kept ever ready with water. Soap is always present, and a towel is kept on a nail close by."[33] The celebration in these boasts was of the imposition of hygiene, order, and the boundaries of decency.

All these debates, whether about the merits of a zone of prostitution or about the need for sanitary facilities in the brothels, were far more concerned with maintaining distinct territories of morality, geographies bounded by gender and sex, than they were concerned with the impact of policy on the lives and livelihoods of the subject of these debates. At the same time that these controversies were spatially played out, they were also deeply gendered. As Rosemary George has argued, "[W]ith the establishment of the English house outside England, there was a physical repositioning of the hitherto private into what had been considered the most public of realms—the British empire."[34]

Interestingly enough, it was radical missionaries rather than the political colonizers who most startlingly conveyed this message, through some quite controversial protests against what they saw as British moral failure in stemming the sex trade in colonial settings. In the mid–1890s, sixteen men and five women associated with the Bombay Mission organized a series of "Midnight Missions" based loosely on those in London's theater district, a notorious venue for commercial assignations. The London version had offered streetwalking women a place to sit and have a cup of tea during the night hours, and if in exchange they were obliged to keep quiet during a sermon or to hear out a proselytizer, this was no more than many other recipients of charity endured in exchange for food, shelter, or clothing. But the Bombay version of this old stock-in-trade was of an entirely different order. Focusing on the brothels of Cursetji Sukhlaji Street, the "Midnight Missionaries," according to the city's police, "parade themselves at either end of [the street] from 9 P.M. to past midnight, accosting all (chiefly Europeans) men who enter the street either walking or driving, on the assumption that they do so with the intention of visiting a brothel."[35] This avowedly physical intervention, not surprisingly, had equally physical consequences. The outsiders were resented by the workers and the customers alike and had all manner of things thrown at them alongside the obvious verbal abuse. On occasion, women missionaries were approached by more-secular men with suggestions that provoked actual fighting between the male missionaries and the women's would-be suitors.

This dramatic almost vaudevillian form of protest had little lasting effect on the siting or presence of brothels in Bombay or anywhere else. But it is an interesting episode that brings home starkly the physical and geographical aspects of this issue. The activity of the missionaries was quite defiant, not only in its confrontational aspect but also in its deliberate violation of the public and private divide, mocking or echoing the similarly situated space of the brothel and under-lining the hypocrisy of law and practice. For, of course, at the heart of this stri-dent activity was the point of contradiction. The veil of privacy more usually reserved for an act revered as the sacrosanct moment of marital union was here used for the exchange of sex for money. And the point was to protect not delicacy but the double standard that damned the vendor but not the purchaser.

Colonial space was figured always as a potentially sexual space, a code both for the failings of the locals and for the masculine penetrativeness of the coloniz-ing force; geography and gender permit us perhaps to see the workings of colo-nial power through this curiously eroticized cartography in which mapping and controlling the spaces of women took on an urgency and an importance for the maintenance of colonial power. The very geography of colonial relations had a long-lasting impact on the lives of women engaged in prostitution and on the struc-tures of the colonial marketplace of sex. As a woman's space that defiled femi-ninity, prostitution had to be relegated to the physical world of commerce. Its segregation from residential districts was needed not only to sharpen the distinc-tion of respectable and unrespectable but also to separate the business of sex from the place of feminine domesticity. The failure of the colonial state to implement these sharp divides points to the impossibility of the gendered geography necessi-tated by colonial thinking and simultaneously acts as the signification of its very impermanence. Managing sex in colonial sites was forever an unstable and com-plex business, for the imperial project relied simultaneously on sexual availability and an invocation of sexual respectability. The brothel needed thus to be both spec-tacular and secretive, vulgar and discreet, available but controlled, inviting but off-limits. At the same time, because sex was understood as a guarantor of certain kinds of stability, its very existence disrupted some deeply held kindred sensibilities.

Notes

1. China Association to Joseph Chamberlain, Public Record Office (PRO), 23 February 1899, London, Colonial Office Papers (CO), Hong Kong, 129/296 (4718).
2. African historians, it should be noted, have persuasively argued that the brothel was an outgrowth of the changing patterns of settlement and labor brought about by colo-nialism: see Janet M. Bujra, "Women 'Entrepreneurs' of Early Nairobi," *Revue Canadienne des Etudes Africaines/Canadian Journal of African Studies* 9 (1975): 213–234; Luise White, "Domestic Labor in a Colonial City: Prostitution in Nairobi, 1900–1952," in *African Women in the Home and the Workforce,* ed. Sharon B. Stichter and Jane L. Parpart (Boulder, Colo.: Westview Press, 1988); Benedict B. B. Naanen, "'Itin-erant Gold Mines': Prostitution in the Cross River Basin of Nigeria, 1930–1950," *Af-rican Studies Review* 34 (1991): 57–79; and Jay Spaulding and Stephanie Beswick, "Sex, Bondage, and the Market: The Emergence of Prostitution in Northern Sudan, 1750–1950," *Journal of the History of Sexuality* 5 (1995): 512–534.
3. Henry Kellett to Annesley, H.M. Consul, Nagasaki, 22 October 1870, PRO, Admiralty Office Papers (Adm), China Station, 125/16 (804).
4. Oriental and India Office Collections, British Library, London, India Office Records (IOR), *Lock Hospital Reports for the Year 1889, Nusseerabad,* L/MIL/7/13909.
5. PRO, CO, Straits Settlement, 273/457 (41155).

6. Ibid.

7. Kenneth W. Andrew, *Hong Kong Detective* (London: John Long, 1962), 90.

8. Sub-Inspector John Ferguson, Cairns, to Inspector Lamond, Cooktown, 24 November 1897, Queensland State Archives (QSA), Brisbane, Australia, Home Secretary's Department Papers, HOM/A15 (15952); Mr. Tooth to Chief Secretary, Brisbane, 25 May 1899, and Sergeant J. Kelly to Inspector of Police, Maryborough, 25 April 1899, QSA, HOM/A24 (10551); James F. Warren, "Chinese Prostitution in Singapore: Recruitment and Brothel Organization," in *Women and Chinese Patriarchy: Submission, Servitude, and Escape,* ed. Maria Jaschok and Suzanne Miers (London: Zed Books, 1994); IOR, 1917, Memorandum of E.C.S. Shuttleworth, Extent, Distribution, and Regulation of the "Social Evil" in the Cities of Calcutta, Madras, and Bombay and in Rangoon Town, Judicial and Public Collections, L/P&J/6/1448 (2987).

9. Carl T. Smith, "Protected Women in Nineteenth-Century Hong Kong," in *Women and Chinese Patriarchy: Submission, Servitude, and Escape,* ed. Maria Jaschok and Suzanne Miers (London: Zed Books, 1994).

10. IOR, 7 June 1881, *Proceedings of the Government of India,* P/1664.

11. IOR, March 1866, General Department Proceedings, P/438/27, 438, no. 62 (emphasis in original).

12. Rosemary Marangoly George, "Homes in the Empire, Empires in the Home," *Cultural Critique* 26 (1994): 101.

13. Nicholas Thomas, "Sanitation and Seeing: The Creation of State Power in Early Colonial Fiji," *Comparative Studies in Society and History* 32 (1990): 160–161.

14. PRO, CO 131/11, 66^v.

15. Brenda S. A. Yeoh, *Contesting Space: Power Relations and the Urban Built Environment in Colonial Singapore* (Kuala Lumpur: Oxford University Press, 1996), 245–246.

16. IOR, Shuttleworth Memorandum, L/P&J/6/1448.

17. IOR, December 1872, *Proceedings of the Government of India,* P/674.

18. See Judith Walkowitz, *Prostitution and Victorian Society: Women, Class, and the State* (Cambridge: Cambridge University Press, 1980); and Linda Mahood, *The Magdalenes: Prostitution in the Nineteenth Century* (London: Routledge, 1990).

19. See Malek Alloula, *The Colonial Harem* (Minneapolis: University of Minnesota Press, 1986); Inderpal Grewal, *Home and Harem: Nation, Gender, Empire, and the Cultures of Travel* (Durham, N.C.: Duke University Press, 1996); Janaki Nair, "Uncovering the Zenana: Visions of Indian Womanhood in Englishwomen's Writings, 1813–1940," *Journal of Women's History* 2 (1990): 8–34; Hanna Papanek and Gail Minault, *Separate Worlds: Studies of Purdah in South Asia* (Columbia, Mo.: South Asia Books, 1982); and Sandhya Shetty, "(Dis)Locating Gender Space and Medical Discourse in Colonial India," *Genders* 20 (1994): 188–230.

20. Nair, "Uncovering the Zenana," 11.

21. Rosalind O'Hanlon, *A Comparison between Women and Men: Tarabai Shinde and the Critique of Gender Relations in Colonial India* (Madras, India: Oxford University Press, 1994), 50–51.

22. Grewal, *Home and Harem,* 44–45.

23. Philippa Levine, "Consistent Contradictions: Prostitution and Protective Labour Legislation in Nineteenth-Century England, *Social History* 19 (1994): 23.

24. Grewal, *Home and Harem,* 44–45.

25. See Shetty, "(Dis)locating Gender Space," 208; and Antoinette Burton, "Fearful Bodies into Disciplined Subjects: Pleasure, Romance, and the Family Drama of Colonial Reform in Mary Carpenter's *Six Months in India,*" *Signs* 20 (1995): 562.

26. Thomas, "Sanitation and Seeing," 160; Shetty, "(Dis)locating Gender Space," 209.

27. Nair, "Uncovering the Zenana," 22; Shetty, "(Dis)locating Gender Space," 208.

28. Shetty, "(Dis)locating Gender Space," 190; Maneesha Lal, "The Politics of Gender and Medicine in Colonial India: The Countess of Dufferin's Fund, 1885–1888," *Bulletin of the History of Medicine* 68 (1994): 38.

29. Vijay Prasad, "Native Dirt/Imperial Ordure: The Cholera of 1832 and the Morbid Resolutions of Modernity," *Journal of Historical Sociology* 4 (1994): 253; Dane Kennedy, *The Magic Mountains: Hill Stations and the British Raj* (Berkeley: University of California Press, 1996), 191; J. B. Harrison, "Allahabad: A Sanitary History," in *The City in South Asia: Pre-modern and Modern,* ed. Kenneth Ballhatchet and John Harrison (London: Curzon Press; Atlantic Highlands, N.J.: Humanities Press, 1980), 176.

30. Warwick Anderson, "Disease, Race, and Empire," *Bulletin of the History of Medicine* 70 (1996): 62–67; David Arnold, ed., *Imperial Medicine and Indigenous Societies* (Manchester, Eng.: Manchester University Press, 1988).

31. IOR, 1879, *Lock Hospital Report on the Madras Presidency for the Year 1878,* Madras, India: Government Press, Military Collections, V/24/2287, 10.

32. *Annual Report of the Colonial Surgeon, 1874,* PRO, CO 129/189 (13163).

33. IOR, 1878, *Fourth Annual Report on the Working of the Lock Hospitals in the North-West Provinces and Oudh for the Year 1877,* Allahabad: n.p., Military Collections, V/24/2290, 96.

34. George, "Homes in the Empire," 99.

35. IOR, 16 May 1894, Judicial and Public Collections, L/P&J/6/375 (1083/1894).

Confinements and Liberations

INSCRIBING "WOMAN" IN COLONIAL GEOGRAPHIES OF POWER

BETTY JOSEPH

*W*hen James Rennell, the first surveyor general of India, took off on a survey of the great eastern rivers of Bengal and Bihar from 1764 to 1767, he was ordered by his employer, the East India Company, to keep a journal in which he would "note the appearance and produce of the Countries" he passed through.[1] This journal was not meant to be the private expression of a solitary man in the wilderness or the field notes for the administrator's postretirement monograph. Rather, it was to serve as a filler for what Rennell's subsequently produced maps would not be able to convey: the agricultural and natural resources to be tapped once military control was established. Yet, despite the "cognitive mapping" of India as empty but exploitable space lying before the onward sweep of Rennell's cartographic apparatus, the native arrives as a solitary and curious entry within the company's instructions: "Inform yourself from the Countrey [*sic*] people whether they [the rivers] are navigable all the year" (109).[2] Appearing only as a potential source for information, the natives must unwittingly play their part in inscribing the land, clearing themselves from it in the process. These instructions also tell us something else: in the multiple roles assumed by Rennell, whose survey would produce the first British maps of the newly acquired territories in Bengal, we see the enactment of colonial power as a kind of procession. In a single file, its agents follow one another—the merchant, the soldier, the cartographer, and the writer—all clearing the way for a commercial corporation to come forth as the sovereign body of governance.

Yet as the faint outlines of the native informant in the company's instructions tells us, grasping "India" as "British India" at the end of the eighteenth century was not a smooth cartographic transition engineered by the arrival of science in the colony. Instead, as Matthew Edney suggests in his recent study *Mapping an Empire,* the "picturing" was itself a multilayered conflict between the desire and the (in)ability to implement the perfect panopticist survey.[3] One must remember that the genocidal tactics deployed by colonial powers in sparsely populated precapitalist societies in the New World were not possible in densely populated, already imperial formations in the subcontinent.[4] The conquest of India was always uneven in its impact, and this unevenness would require British power to resort to a wide variety of strategies to maintain control over a territory that could be emptied only in the rulers' imagination.

In the following, I trace the relationships between colonial cartography, political power, and the representation of gendered subjects in nineteenth-century British India. More specifically, I look at representations of space within colonial discourse, their contrapuntal relationship with official maps, and their translation into policies that affected people in actual geographical sites. It is my contention that a close reading of official colonial records reveals that metaphoric representations of space (as British or native, male or female, and so on) and their literal delineation through the use of military, economic, and legal maneuvers moved in synchrony as British colonial rule in India was consolidated in the nineteenth century.

The palimpsestic presence of the British in India was soon made visible in a series of maps issued by James Rennell in the 1780s, maps that replaced those used by the precolonial Mughal rulers. The maps had both the *subas,* or the old Mughal divisions, as well as the contemporary political divisions of the British rulers overlaid on each other. In an "unofficial" memoir written during the surveys that generated these maps, Rennell explains the delineation technique thus:

> It must be observed that since the empire has been dismembered, a new division of its provinces has also taken place. . . . These modern divisions are not only distinguished in the map by the names of the present possessors; but the colouring also is entirely employed in facilitating the distinctions between them. So that the modern divisions appear, as it were, in the *fore ground;* and the ancient ones in the *back ground;* one illustrating and explaining the other.[5]

Here, Rennell opens up a reading space for the literary critic. The interplay between foreground and background is not simply the fading of one color into another, or of dotted lines into unbroken ones, but the representation of the uneven nature of control. This map is not like Conrad's in the *Heart of Darkness,* on which, after the Berlin Conference, Africans were shuttled between blank or black or among blue, green, yellow, purple, or orange. The unevenness and points of overlap in Rennell's map constitute potential sites of anxiety where meaning is contested. For the literary critic, the hermeneutic task of reading the hyphen in the newly inscribed "British-India" is also the task of reading the area in between the declining power of the former rulers and the gradual consolidation by the newly arrived. Here, we will see what Homi Bhabha has called a "presencing," or a beyond, where the estrangement and displacement of home and the world produce new realities and subjectivities.[6] These are sites where natives have to be worked over ideologically and discursively to be made alien to their habitats and to themselves.

By the beginning of the nineteenth century, the foreground-background distinction in Rennell's maps had come to represent not only administrative changes brought about by colonial rule—changes in revenue collection, land tenure, legal jurisdiction and so on—but also the acknowledgment that British power would never be able to wipe the slate clean. Precolonial space continued to show through and signify ambiguous cartographies and unsettled temporalities, and any management of this space had to inescapably deal with the difference posed by three interruptive time zones: the cartographic past of the Mughals, the colonial present into which censuses shunted the population, and the indeterminable future of British rule over India. The rest of this essay will look at a nineteenth-century exercise in people management that culminates at a point where we can view the "woman" emerging in her punctuative function to facilitate new discursive strategies that are also new arrangements of space for the colonizer and colonized.

It is common knowledge among colonial historians that calls for regulating the practices of the East India Company reached a new high after the American Revolution. Fears that the English in India might display secessionist tendencies prompted Parliament-led reforms to curtail the power of the company, which by the end of the eighteenth century had sovereign power over vast territories and peoples. In 1813, one of these newly formed organs of control, the Select Committee of the House of Commons on the Affairs of the East India Company, met to consider a number of issues before renewing the company's charter that year. The ritual of renewing the charter at regular intervals, since its original granting in 1600 by Queen Elizabeth I, had always served as an occasion for the British public to have a say in who was to benefit from the Indian trade. By 1813, the discussions had changed significantly—now, questions of trade were inextricably linked to questions of rule over the immense Indian Empire with its extremely heterogeneous racial, ethnic, and religious peoples. The committee's proceedings took place at a time when it was difficult to ignore that the geographical distance between the rulers and the ruled did create invariably unstable cartographies. Partly as a result of such awareness, there were some departures from earlier exercises in mapping the empire: from an earlier dependence on cadastral surveys that organized revenue and tax collection, the company moved on to large-scale exercises in collecting information about people—human geography lessons that would distribute the native people in new space.

In 1813, the committee members sat together to conjure up for themselves an impression of the place that they had to govern but that many of them had never visited. Reading the minutes of evidence taken before the committee, we see that first on the list of testimonies is the one given by the former governor-general, Warren Hastings, at that time living in retirement in England. Expert witnesses summoned to testify before the committee are asked questions that are remarkably similar to one another; however, it is in the repetition rather than in the order or sequence of the questions that the committee reveals its hidden charge. Similarly, although random sequencing of the questions disperses noticeable agendas, the first testimony inevitably marks a point of departure that cannot be displaced easily. Warren Hastings's answer to the committee's first question does in fact set the trend for the imagined India that will dominate in the proceedings. Here is the committee's opening question as it is posed to Hastings: "If Europeans were permitted to sojourn in India, according to their own pleasure, and without any restraint, have you, from your long experience in that country, the means of stating to the Committee what the effect of that would be?"[7] Hastings's answer is direct and marked with an authority untarnished by his ignominious departure from India thirty years previously or by his long trial on corruption charges, which played out for ten years after his return to England:

> Most hurtful and most ruinous, both to the Company's interest, to the Government and to the peace of the country. Nothing could be more opposite than the characters of the Europeans and the natives of India . . . and though the native Indian is weak in body and timid in spirit . . . there are cases in which a provocation of general grievance would excite a whole people and even a detached number of them to ferocities of insurrection. (2)

Hastings's representation of the colony as a space of timidity and servitude needing protection, especially from the "lower orders of Englishmen" (2) prone to despotism, was a more recent coding for India that was put to work *after* the company's successful military campaigns at the end of the eighteenth century. From this point on in the proceedings, "unrestrained sojournment of Europeans" becomes a code phrase signaling the need to keep strict controls over the English traveling to India. Earlier, after the Regulating Act of 1784, such restrictions had allowed the company to maintain its trade monopoly in the subcontinent, but by 1813, the argument had lost force.[8] The company now resorts to new justifications: the protection of native interests and the maintenance of peace.

These twin concerns explain why, by 1813, the confinement of Europeans ("uncovenanted" servants and civilians) to within ten miles of each major city or port in India could be justified using the very language Hastings employed in his testimony—the opposite characters of the Indians and the British.[9] Because settlement or colonization (as the large-scale occupation of space) was always an unaccomplished project, parliamentary committees would debate well into the 1850s about the numbers of British who could coexist with native populations in India without threatening the equilibrium. To maintain the delicate balancing act of empire, proportions of white to black bodies had to be calculated carefully: how many English were needed to control the native population without the danger of developing also into a seditious community? And how were they to be distributed and at what intervals so as to maintain the visibility of British power without having its legitimacy tested by native perceptions that it was a direct threat to their everyday lives?

Anxieties about native insurrection brought about by the interloping of alien law onto native space were no doubt present in the very act of grasping British India as a picture, an activity not unlike the cognitive mapping of the postcartographic empirical subject, who, with the invention of new instruments of navigation, had to coordinate "existential data with unlived, abstract conceptions of the geographic totality."[10] Maps of the British Empire too often obscured the lived reality of colonial rule and the metonymic relationship of colonial power to the total space of the native. If one were to actually map a nineteenth-century British India unmarked by native habitation, British rule would probably show only as a patchy presence: settlements and port cities governed entirely by the British would remain minuscule dots in a vast territory and would be mostly concentrated along the coast or in the hill-stations. This disproportionate presence of the ruler and the ruled did lead to new forms of inscription on the land that were not always in the service of scientific cartography. For instance, in the growing number of literary accompaniments to Rennell's maps, the genre of the travel narrative, we see not only the changing itinerary of adventure in India, as boundaries of the empire are constantly redrawn with military victories, but also the subsequent arrival of law as another inscription on the land—proceeding where it can in fits and starts but, like Rennell, our colonial surveyor, detained where it cannot. In his *Narrative of a Journey through the Upper Provinces of India* (1824), Bishop Reginald Heber, a missionary, is already able to make an explicit connection between geographical boundaries and British law.

> The road which borders Calcutta and Chowringhee is called whimsically enough . . . "the circular road" and runs along nearly the same line which was once occupied by a wide ditch and earthen fortification, raised on occasion of the Mahratta war. This is the boundary of the liberties of Calcutta and of English Law. All offences committed within this line are tried by the Sudder Awalut or the Supreme Court of Justice—those beyond fall in the first instance within the cognizance of the local magistracy and in the case of appeal are determined by the Dewanee or the Court of the People of Chowringhee, whose proceedings are guided by the Koran and the laws of Menu.[11]

How did the Select Committee members negotiate their own moment of governance ten years before the ditch would catch Heber's attention, when the line between British and native law was not very clear cut? Confining British soldiers and employees to their settlements was one stopgap measure adopted by the company in previous years, and the policy was justified as a law-and-order issue. As entered in its minutes, the committee is advised by William Cowper, a former employee from Bengal, to ensure that the geographical distance from the site of British crime to the place of British justice be kept to a minimum. As Cowper himself states it during his testimony, carrying British criminals eight hundred or one thousand miles under armed escort would be "a most fruitful source of degradation to the English character in India."[12] The paradoxical invocation of law is revealed in two ways: one, as the geographical distance of justice from crime begins to erase the very character difference between the native and the Englishman (its justification in the first place) and, two, as the visibility of British law acting upon British crime undermines its power and legitimacy in native eyes.

In the rest of this essay, I will trace how what began originally as a distinction of foreground and background in a colonial map becomes, in the course of the Select Committee's deliberations, spatial demarcations that are both ideological as well as geographical. The proceedings, which were published as a series of testimonies taken from former employees of the East India Company who had served time in India, initially produce a multiply fractured, hybrid India. Put together through the fragmentary accounts of these colonial functionaries, this "India" is also a kind of entity that no map can represent. It is simultaneously inscribed on different registers—the economic, the politico-legal, the sexual, and the religious. We have already seen how discussions of the economic viability of India are not discontinuous from arguments that the maintenance of feudal forms of production is also one way of maintaining peace. Once "feudalism" is the assigned place for the native, lines that appear to separate geographical spaces also become bars placed between Oriental despotism and British rule, between disorder and order, between tradition and modernity, and, as the next section of the essay will show, between private and public space. Thus, restrictions faced by the British in India are read as implicit guarantees of cultural rights to the natives.

After Warren Hastings's initial remark, spatial management and native subjectivity become inextricably linked, and the committee, paying careful consideration to Hastings's initial warning about likely provocations for native insurrections, now charts the interiority of the native. A close reading of the committee's records shows how British India, when imagined for colonial pacification, sharpens the older relativism of native versus English character into investigations of native subjectivity itself. The native is now subject to cultural or religious discourses that are inescapably gendered (and thus also coded as sexual); moreover, all discussions of gender install new boundaries (psychic and geographical) that cannot be transgressed by company employees. Hence, it is not surprising that after the discussion of restraints to be placed on the movements of the English within British India, the subject who is discussed most explicitly in the language of confinement is the Hindu woman, who, as you will see, has to be released continually into the open space of the committee deliberations from her existence in various native spaces, in order to be given over as one of many native "cultural rights."

To track the itinerary of this discursive journey into womanspace by the committee sitting in its chambers in London, we have to begin with the most significant geographical notation that appears throughout the various testimonies: the distinction between the "coast" and the "interior." It first comes into play in a discussion about the danger posed by "European adventurers" who flout company rules and "penetrate" the interior districts from the largely unguarded Indian coastline.[13] A little later, in Thomas Munro's testimony, these terms reappear to support his view that India could never become a market for British goods but should remain an enclave of cheap labor and raw materials for England's industry. The port city of Bombay, Munro argues, while a large market for English consumer goods, is not representative of the whole subcontinent. Munro makes his point by contrasting the inevitable cosmopolitanism of port cities like Bombay with the equally inevitable inaccessibility of the native's interior: "It is said that the Hindu in Bombay likewise conforms to the European modes of life. . . . But I can have no doubt that, after the ceremony of the visit is over, he retires from his lustre-hung hall to his Hindoo family, in their own Hindu house, sprinkled with cow-dung and water" (168). Using another coastal city, Madras, as a further example, Munro now conflates the interiority of the native with the geographical interior of the subcontinent: "If any person leaving Madras goes to the nearest Hindoo vil-lage, not a mile into the country, he is as much removed from European manners and customs as if he were in the center of Hindostan and as if no foot had ever touched the shores of India" (168).

The real territorial control of India becomes phantasmic in this account as traces of British rule are rendered invisible every day by the native retreating from the public into the private. The coast, which appears as a place of obvious famil-iarity with its older company settlements enclosed by British law, is the only geo-graphical category in the various committee testimonies, which otherwise include long lists of militaristic notations: possessions, cantonments, territories, empire, dominions, and so on. But when the distinction between the coast and the interior becomes an epistemological divide maintained by a line drawn between the self-assured power of knowledge and the failure of that knowledge, the interior ceases to be geographically denoted and instead becomes unpenetrated cultural space. The interior in this new formulation cannot be represented on a map but is constituted by negation, as the space outside British law and cultural influence. In the testi-monies to follow, this space is soon, through a series of substitutions in a chain of signification, equated with the private space of the native. Intersecting with this narrative space is another space where the British in India must tread with care: that of the woman.

The Hindu woman appears without warning in the committee proceedings right after a question about the quality of wool in the Northern Provinces. She disappears almost instantly without a trace, quickly supplanted by questions about the feasibility of translating the Scriptures into Hindi and the truth about the vio-lent character of Mohammedans. Some witnesses are never asked about her, but those who are always have something to offer as evidence. She first appears in the following question posed to John Shore, a former official from Bengal: "What is the condition of the female sex among the Gentoos as affected by their religion and prejudices?" Shore replies, "They are so concealed that we know little or noth-ing of them; nor is it usual to talk with the gentoos about their female sex: I be-lieve that their state in general is merely that of slaves to their husbands" (18). A colon appears in this sentence as though to separate the conclusion from the evi-dence and thereby present slavery as an explanation for concealment and vice versa. The Hindu woman is an epistemological blind spot ("we know little or nothing of them"), but her invisibility to the imperial eye is always a sign of her slave status. We also notice that the woman question is repeated to others, in subsequent testi-monies, often following queries about the supposed immutable character of the Hindus and the possibility of offending them if lower-class Englishmen on drunken sprees helped themselves to women in the interior. The question also marks a way to separate the more rebellious Muslim from the servile Hindu. In its reruns, the woman question is modified enough to place the native woman discursively within the "interior," which is now transformed from geographical notation, epistemological barrier, ethnic enclave, and protected native space to private space, male domain, and native male property. In a question to a later witness, John Malcolm, these discursive graftings are already visible: "Have you known any instances in which it has been necessary to give particular precautions with regard to the conduct of persons in the Company's service going into the interior of the country as to the manner of behaving towards the natives, in respect either to their religious prejudices, their establishments, with respect to the female sex or in any other respect?" (55).

The interstitial region between foreground and background, which drives the notions of coast and interior and where the carefully mapped British India fades back into a potentially uncontrollable space, is now also gendered male.[14]

The committee's discussion of the enslaved Hindu woman produces her as an object of confinement only to release her discursively into the interior. The deliberate shaping of this construction is revealed when we see in the proceedings that several witnesses give very different accounts of their encounters with Hindu women. First, we have the testimony of Thomas Munro, who says that with the exception of one-fiftieth of Hindu women, they overall have "as much liberty, and I imagine more, than the women in Europe" (127). Reading "liberty" as the ability to be seen in public, Munro goes on to describe as his evidence the large numbers of "women of all ranks" who bathe everyday at the public waterworks in European garrisons. When asked by a committee member again if the women's state was that "of slaves to their husbands," Munro counters, "Their state is not that of slaves to their husbands, they have as much influence in their families as I imagine, the women have in this country; I often found them when in charge of the ceded districts, very troublesome tenants as farmers; I have frequently known women of respectable families who kept their husbands, and sons grown up, at home and came to the cutcherry to debate about their rents" (127). Here, Munro reverses two interpretations (already in play) of colonial womanspace. He challenges both the fashioning of privacy and interiority of native men and the seclusion or slavery of Hindu women by establishing rural India as a place of women's public economic activity. It is yet another indication of the powerful work of alien discourses writing native women's subjectivity that Munro feels compelled, later on in his testimony, to refer to this statement and say that his speaking of the women's "bathing in public" should not leave "an unfavourable impression of their demeanour" (169).

If Munro's testimony allows us to see that "publicness" can provide the double bind of liberty and loss of virtue to the Hindu woman, that of Thomas Sydenham provides yet another instance of heterogeneity in gendered perceptions. Unlike other witnesses, he has had access to the interiors of households. When asked if the Hindu women are kept in a state of "slavery, degradation or seclusion," Sydenham answers,

> While I was resident at Poonah I had frequently occasion to transact business with the ministers of the Mahratta court in the private apartments of their dwellings; on most of those occasions I have seen their women and

they have sometimes been present at the transaction of ordinary business between the minister and myself. The only degradation amongst the Hindu women with which I am acquainted is the state of the widow after the death of her husband, the widow has the hair of her head shaved and is obliged to do all the menial offices of the family. I have known many Hindoo women however in this state, when her children had not been of age conducting the business of the family, having the management of the estate belonging to the family, employing agents and attorneys at the Mahratta court, and in short, doing everything in the direction and management of the family which women in similar circumstances in Europe do. (321)

Here, a designated space for the native woman has been momentarily transformed from that of "slavery, degradation or seclusion" into a place of economic deliberation, property management, and legal subjectivity. Sydenham's description, however, splits midway and reinserts the dominant reading of this space as one of oppressed widows, but his description also simultaneously reinscribes the space as one of empowerment (of widows liberated by the deaths of their husbands). The heterogeneity of such accounts will finally be subordinated to the simplicity of John Shore's initial testimony—"I haven't seen any; they are all slaves"—because the woman is staged not to establish her own truth but to service another emerging discourse: that of British noninterference in native custom.

In 1818, in a letter from the company to the East India Company's Board of Control, we see the place finally assigned to the woman in these discussions. The letter, a belated report on the conclusions arrived at by the Select Committee, has the following as its first item: "That the natives of India, though generally speaking weak in body and timid in spirit, are very susceptible of resentment and of peculiarly quick sensibility in all that regards their religion and women."[15] The lid on the woman's heterogeneity is closed tight so that she is reduced to one persistent chronotope in this discourse: that of the unchanging cultural secret of the native in his own cultural space. Moreover, as the mediator between the foreground and the background, she also shuttles those spaces between the time zones of tradition and modernity. She repairs the fractured geography of the colonialists and native collaborators alike. If Oriental despotism, the background, cannot be replaced by British despotism, the foreground, the situation arises not because rights available under the former are now available under the latter but because the distinction was never about political rights anyway. The distinction as it emerges in the committee's deliberations is about being able or not being able to retain cultural rights without political oppression. By deploying the female gendered subject in this manner, native men come forward in the proceedings as despots (religious fanatics and sexual tyrants) who can exist with great ease between two time zones: as captive labor in feudalism but also as a free, cultural, right-bearing Hindu or Muslim householder in modernity.

Finally, in her last appearance in the appendix, or the coda to the proceedings, the woman ensures Warren Hastings the last word. The production of a womanspace as a singular achievement of the committee's cartography will serve to hold together both a policy of noninterference and one of total control over colonial space. The "Hindoo" or "Mohammedan" woman will service both protectionist and interventionist discourses, depending on the interests paramount at a particular historical moment. In the early nineteenth century, emerging discourses of cultural relativism and noninterference could not do without religious or gendered discourses. Like the space of native culture, the space of the native woman in British colonialist discourse of this time will become a protected enclave that cannot withstand foreign influence or intervention without dire consequences (native insurrection). The native woman will be represented as impenetrable space—unavailable to Englishmen for their sexual use—but also as native space and property that must be left exclusively for the use of the native men. Thus, in this imagined India, the woman is not only the most remote geographical space (the interior) but also the very interiority of the native man.

The argument for protection and noninterference that installs a culturalist discourse of "woman" as the guarded secret or the final frontier of colonialism will have far-reaching implications for shaping the "woman" question in subsequent colonialist social-reform legislation, nationalist agendas, and postcolonial struggles, but that cannot be part of my discussion here. Suffice it to say that a history that will do justice to this figure whose heterogeneity has been stifled in these proceedings must be put together from the very accounts of people like Thomas Sydenham—by reading over their shoulders and against the grain of their testimonies.

Notes

1. James, *The Journals of Major James Rennell,* ed. T.H.D. La Touche (Calcutta, India: Baptist Mission Press, 1910), 109.
2. The term "cognitive mapping" is Fredric Jameson's (see "Cognitive Mapping," in *Marxism and the Interpretation of Culture,* ed. Cary Nelson and Lawrence Grossberg [Urbana: University of Illinois Press, 1988]) and is used here not as the means of conceiving "the social totality (and the possibility of transforming a whole system)" (347) but as the projected grid of colonial power, where some "older sacred and heterogeneous space" is reorganized into one of "infinite equivalence and extension" (349).
3. Matthew Edney, *Mapping an Empire: The Geographical Construction of British India, 1765–1843* (Chicago: University of Chicago Press), xiii
4. For a discussion of the "failed parallel" between the American and Indian examples, see Gayatri Spivak, "Constitutions and Cultural Studies," *Yale Journal of Law and the Humanities* 2 (1990): 133–147.
5. Quoted in Edney, *Mapping an Empire,* 11 (Rennell's emphasis).
6. Homi Bhabha, *The Location of Culture* (London: Routledge, 1994), 1–2.
7. Great Britain, Parliament, House of Commons, *Minutes of Evidence Taken before the Committee of the Whole House, and the Select Committee on the Affairs of the East India Company* (London: 1813), 2.
8. In 1813, the company was deprived of its trade monopoly in India and the subcontinent was opened to deindustrialization and massive imports of British goods. Company servants' attempts to show native character as resistant to such goods is thus a

strategy to preserve the monopoly. See E. J. Hobsbawm, *Industry and Empire: From 1750 to the Present Day* (1968; reprint, Harmondsworth, Eng.: Penguin, 1983), 49.

9. Commons, *Minutes,* 55.

10. Fredric Jameson, *Postmodernism; or, The Cultural Logic of Late Capitalism* (1991; reprint, Durham, N.C.: Duke University Press, 1995), 52.

11. Reginald Heber, *Narrative of a Journey through the Upper Provinces of India* (London: John Murray, 1821), 1:26.

12. Commons, *Minutes,* 34.

13. Ibid., 75.

14. The committee's overall exercise to calculate the ideal proportion of white men to natives in the colonies through immigration controls hides the agenda of sexuality that is always already coded into all discussions of settlement. Colonial records elsewhere have shown that the encouragement of the traffic of British women to the colonies began not with the fear of miscegenation (a racist discourse) but with the fear of insurrection from the half-caste or mulatto—populations resulting from sexual contact between Englishmen and native women—since they occupy an in-between space of conflicting loyalties. Meanwhile, in the committee deliberations of 1813, there is no explicit discussion yet of the Englishwoman. Representing a "private" rather than trade interest, she still needs special permission from the company to proceed to India.

15. Great Britain, Parliament, House of Commons, *General Appendix to the Report from the Select Committee of the House of Commons on the Affairs of the East India Company, 16th August 1832, and Index* (London: 1832), 351.

Literacy for Empire

THE ABCS OF GEOGRAPHY AND THE RULE OF TERRITORIALITY IN EARLY-NINETEENTH-CENTURY AMERICA

MARTIN BRÜCKNER

*I*n 1821, after the invasion of the Cherokee, Choctaw, and Chickasaw countries, the magazine *Literary and Scientific Repository* ran a series of editorials responding to British accusations of U.S. imperialism. Closing ranks with other popular magazines, the *Repository* attacked the critical study *America and Her Resources,* in which the British author had claimed that "the American rulers . . . endeavour to direct the whole national mind and inclination of the United States towards their aggrandizement by conquest."[1] In the opinion of the American reviewer, this critique was outrageous—not for its allegation of imperial practices, however, but for its gross misattribution of political agency. Far from denying the reports of U.S. violation of western territories, this author contended that it was not the elected politicians but the will of the American people that mandated imperial actions.

More specifically, the realities of conquest were the direct expression of the people's newly developed geographic sensibility. In the reviewer's analysis, territorial violence was thriving in the United States because its citizens engaged with maps "ten times more frequent than is found among any other people" (211). On the one hand, this surplus of geographic textuality underwrote the belief that the individual American had "vastly more *geographical* feeling than the European" (209), while, on the other, it structured the people's collective aesthetic of territorial expansion. "The Americans are far from being pleased with the irregular figure which the Republic exhibits upon the map," the author argued. "This and that corner of the continent must be bought (or conquered if not bought) in order to give a more handsome sweep to their periphery. . . . In fact, their boundary line is never so exactly round to satisfy the nice eye of an ambitious people; the jagged polygon still needs here and there some trimming; but this perfecting of the figure is to be effected always by increments,—never by retrenchments" (209).

Decrying a mass distribution of geographic texts and map-reading habits as distinctly political and conceptually aesthetic experiences, magazine editors and newspaper publishers contrived more than an unrepentant defense of American territorial practices. Rather, they self-consciously defended the early-nineteenth-century realization that, in the words of William Appleman Williams, "empire as a way of life" was rapidly shaping the cultural habits of the American citizenry. For the United States to become an empire, as Williams proposed for the 1950s, American citizens had to participate candidly in a "process of reification—of transforming the realities of expansion, conquest, and intervention into pious rhetoric about virtue, wealth, and democracy."[2] That process consisted of reconciling two definitions according to which it was understood: first, that "the meaning of empire concerns the forcible subjugation of formerly independent peoples by a wholly external power" (6), and, second, that "a way of life is the combination of patterns of thought and action that, as it becomes habitual and institutionalized, defines the thrust and character of a culture and society" (4). I want to suggest that, in their defense of imperialism in 1821, American citizens formally announced that their everyday political attitude toward the formation of an American empire was directly informed and structured by the proliferation of geographic literacy. They acknowledged that the habits of basic geographic education were becoming the cultural basis of both political rhetoric and action.

Recently, scholars have examined literacy in the early United States in the terms of ideological confrontations, pitting a dominant republicanism against an emergent U.S. nationalism. In this context, they have discussed how the logic of consolidated statehood dovetailed with the antebellum promotion of universal literacy, demonstrating how spellers, grammars, and novels aligned the theory of language acquisition with the politics of domesticity and capitalism.[3] However, by the same token that this discussion confirmed the centrality of literacy in early American studies, the discussion has also demonstrated how a predominantly nationalist definition of literacy has prevented us from taking sight of the imperial practices that continually have shaped the country's literary education. To be sure, the evaluation of literacy has undergone significant revisions. Once considered the touchstone of the country's cultural revolution and the source of American exceptionalism, literacy has now become reevaluated as the site of domestic discrimination, racial confrontation, and cultural creolization producing resistant literacies.[4] Yet, these revisions have underestimated the extent to which the conceptions of literacy have always been bound to the nation's variable geopolitical borders and territories. As suggested by its title, *Literacy in the United States,* for example, and works like it inadvertently subordinate the study of signature rates, institutional spelling bees, and personal reading histories to the uneven territory of the nation-state.[5] Such studies continue to elide the series of predominantly geographical, that is, geoliterary, contexts in which the 1820s onslaught of invasion and expansion—from the Monroe Doctrine to the Indian Removal Act—informed at once the formal diffusion of literacy in the newly emerging common schools and mass-produced schoolbooks, not to mention the innocuous rhetoric of social values predicated on virtue and liberty.[6]

"The power of literacy results not directly from learning the ABC's," Dana Nelson has recently reminded us, "but rather from a powerful set of beliefs that accompanies the act of learning to read and write."[7] For early-nineteenth-century American citizens, geoliteracy provided the discursive outlet for a new strategic belief in the agency of human territoriality. Human territories, as sociologists Erving Goffman and Irvin Altman have shown, are generally conceived as spaces controlled by a person or a family or some other face-to-face collectivity. Control is reflected in actual or potential possession rather than in evidence of physical combat or aggression; acts of signification rather than acts of physical violence are the hallmark of human territoriality.[8] In this context, the geographer Robert D. Sack has recently called attention to the complex interrelation of spatial forms of organization and linguistic modes of interaction, arguing that territoriality is the "geographical bonding agent" that structures, authorizes, and is informed by social relations.[9]

In the following, thus, my concern lies with the literary agency of geography and the ways in which geographic modes of signification became the discursive frontier along which antebellum citizens ciphered and legitimated the acts of territorial violence that were occurring on the margins of the new nation's domain. In the effort to historicize the rise of geographic literacy, I will explore the history of the distribution of geographic letters; how geography underpinned the territorial socialization of readers; and how maps inflected the teaching of reading, writing, and literature in early-nineteenth-century schooling. More generally, I seek to understand how Americans transformed the design of textbooks from a tool of national self-fashioning into the communicative interface mediating between signification and violence.

Masses, Classes, and the Genre of Geography

During the early decades of the nineteenth century, the literary function of geography was undergoing a critical metamorphosis from stabilizing the union to mobilizing the desire of empire. After the Revolution, the new citizens had flocked to national atlases and geography. These texts confirmed recent historical events. Both maps and geography books provided the narrative forms by which the citizens anatomized the form and content of the new nation, while enabling Americans to decolonize traditional customs. Thirty years later, however, the function of geography changed from being less a constitutive national form to becoming more a radical cultural literacy. In the decade following the War of 1812, while the producers and the consumers of geographic writings explored the genre's representational depth for didactic-aesthetic possibilities, mapmakers and map readers ostensibly directed the political gaze away from sites of domestic interest toward the country's margins and neighboring territories. Indeed, American citizens became obsessed with the materials of an emerging imperial cartography.[10] They collected maps and statistical accounts of the borderlands and the trans-Mississippi West; they charted battlefields, burial grounds, and trading routes; and they avidly subscribed to geographic guidebooks that explained the country's fledgling infrastructure. The citizens' investment in maps and other geographic writings corresponded with the domestic policy of President Monroe and his stated goal of creating a continental "home market." As the historian Charles Sellers has explained, beginning in 1816 Monroe's politics forcibly opened to white appropriation the vast territory expanding from central Georgia to the Mississippi.[11]

As ordinary citizens became involved in the fantasy and process of western colonization, the various sectors of the country's emerging culture industry propagated the dissemination of geographic literacy. Indeed, several organizations of the social-reform movement took up the cause of promoting the study of geography. In a process that eventually resulted in the democratization and professionalization of public education, a medley of teachers, religious leaders, and middle-class patrons—for reasons ranging from cultural nationalism to labor concerns to missionary agendas—lobbied successfully for the introduction of geography and map-making into the nation's primary and secondary school curricula.[12] This process was reinforced by the trustees of higher education, who between 1810 and 1820 appointed public school teachers as the custodians of geographic knowledge in the United States, thus affirming that geography was to be considered a basic and popular subject rather than the specialized domain of the educated elite.[13] Personal diaries show that students attending common schools and private academies participated in mandatory geography lessons, including the recitation of place names, map drawing, and the use of the Mercator projection.[14] By comparison, the typical job description for public-school teachers emphasized geography over the conventional three Rs of reading, writing, and arithmetic. When the state of Massachusetts advertised new teaching positions, it emphasized that "*a person who is not qualified to teach geography, grammar, and geometry, and not well recommended for his morals, &c. is forbid, under heavy penalties by law, to take charge of a school.*"[15]

Seizing the opportunity to capitalize on this public demand for geoliterary instruction, the American book industry collectively channeled its energies into the supply of school geographies. After 1815, booksellers started marketing under the rubric "geographical" everything from writing manuals and copybooks to flash cards and dictionaries to atlases and pocket globes. Paramount in the publishers' lineup was the geography textbook. Among the textbooks vying for public approbation and adoption were some of the country's pioneer geographies, such as Jedidiah Morse's *Geography Made Easy* (1784) or Joseph Goldsmith's *An Easy Grammar of Geography* (1804). Largely modernized for republication, these didactic classics competed with a new generation of popular textbooks like Daniel Adams's *Geography; or, A Description of the World* (1814) or the best-selling *Rudiments of Geography* (1819) and its republished version, *System of Modern Geography* (1825), both coauthored by William Woodbridge and Emma Willard.

Three Kinds of Geographic Pedagogy

Geographers, like most educators of the time, engaged in a pedagogic debate about proper reading instruction. At stake was the question, How can a people transform the socializing function of literacy into a national program by which the citizens of the Republic would be able to expand its optimal imperial limits without undercutting its self-stipulated morality of being the "empire for liberty"?[16] The debate pitted the residual teaching philosophy of John Locke against the recent methodology devised by the Swiss pedagogue Johann Pestalozzi. The point of contention was whether students ought to learn the ABCs by rote and orality or by written words and visual stimulation.[17] The conflict between these schools of thought was prominent among the competing school geographers. Textbook authors as different as Morse and Willard endorsed the principles of a Lockean psychology, assuming the mind was a blank slate and destined to be inscribed with permanent impressions bearing linguistic or semiotic properties. In this context, both authors agreed upon the necessity of developing a geographic writing system—including abstract symbols, complex maps, and geographic narratives—by which the "real lines of geography" become transposed into "abiding," "durable," and "deep and distinct impressions."[18]

But the authors fundamentally differed about how geography lessons were to construct a durable geographic memory. For Morse, geography was a matter of national discipline and verbal memorization. He considered *Geography Made Easy* "a reading-book, that our youth of both sexes, at the same time that they are learning to read, might imbibe the acquaintance with their country and an attachment to its interests." On a different occasion he wrote, "[I]n the best maps . . . errors are so numerous, that the mind cannot rest with confidence in their testimony. We want the confirmation of the book."[19] By subordinating geography to literacy, Morse takes control of geoliterary exercises in linguistic and disciplinary terms. He demands a sufficient precognition of words before confronting maps and other geography texts; he also presents geography as a disciplinary regimen that allowed teachers to cudgel the young in order to build their national character.

By contrast, Willard and Woodbridge eschewed this punitive model of geographic learning, offering a more inclusive opportunity that included women, slaves, and the deaf-mute in the expanding web of American readers. Their teaching methods revolved around the nonauthoritarian principles of internal perceptivity proposed by Johann Pestalozzi.[20] Expanding from his method of whole-word and object learning, they argued in *Rudiments of Geography* (1822) that in geography lessons the map must come before geographic description, the visual before the verbal: "*No language can impress ideas so deeply on the mind as information addressed to the eye. . . .* A description cannot give so distinct views of the geography of a country as a *map,* and no words can so fully convey the idea of a remarkable custom or curiosity as a *drawing* or *engraving*" (ix, emphasis in original). The authors prescribe a set of exercises that force the students to engage through slate and pencil with their immediate geographic surroundings. This approach initially denies students the written book in favor of the Pestalozzian exercises of drawing objects before assigning them proper names. By mapping out his or her environs, the student is meant to experience the medium of transcription, here the map, the scale, and pictographic symbols, as a form of individualistic instruction and self-improvement.

Caught between the opposing schools of hard verbalism and soft visualism, geography teachers turned to a third kind of pedagogy, the Lancasterian monito-rial system. Beginning in 1805, Joseph Lancaster proposed a schooling system in which a single teacher could operate a school of five hundred pupils. This sys-tem, emphasizing recitation and classroom competition, used student monitors and a standardized routine of quasi-mechanical exercises intended to replace corpo-real punishment. In order to demonstrate the efficiency of the monitorial system, American educators turned to the Lancasterian method of geographic instruction. "Let a school of ninety-six pupils be divided into twelve classes of eight in each," the *Academician* writes, "the eight classes formed into semicircles, round a map, holding books in their hands, and each pupil taught by his class leader who re-quires each pupil to point out the places named, or answer such questions as re-late to the situation, bearing, relative distance of places and their latitudes, and they will acquire a practical facility and precision in geography."[21] Geographic lit-eracy here becomes the subject of mass production. Students study geography in a didactic climate of alienation and detachment; as group recitation replaces pre-vious demands of local or national geographic memory, they learn, under time pres-sure, to reproduce maps and geographies to the most basic levels of geoliterary knowledge.

The Territorial Work of Orality

The antebellum student encountered geography and its territorial practices in oral language instructions. As recent studies have emphasized, the first literacy lessons taught students how to read before or instead of writing.[22] In this context it is important to note that most spellers, like the ubiquitous *American Spelling Book* by Noah Webster (1817), introduced geographic elements into the textbook sections on ciphering and pronunciation. Following a plan intent on establishing a national reading standard, Webster presented students with a geographic reading table:

United States	Chief Towns	People
Mas-sa-chu'-setts	Bos'-ton	Bos-to'-ni-ans
. . .	. . .	. . .
Lou-is-ian'-a	New Or'-leans	Lou-is-ia'-ni-ans.[23]

Webster's speller organized place names according to a geographical grammar, con-jugating a list of locations according to their geographic coordinates (north to south), their relative distances (in miles), and their territorial size (in square miles). By demanding that students recite this geographic vocabulary, his instructions con-stituted a nationalistic exercise in which the vocalization of geographic writings reproduced symphonically the graphic outlines of the territorial map of the United States. However, this recital of geographic referents also created a soundscape of an outwardly colonizing potential. On the one hand, as Webster explained, his ex-ercise intended "in a copious list of names of places, rivers, lakes, mountains . . . to exhibit their just orthography and pronunciation, according to the analogues of our language" (v). On the other hand, however, he conceded that "the orthography ought to be conformed to the practice of speaking. The true pronunciation of the name of a place, is that which prevails *in and near* the place" (v; my emphasis). Webster here offered an early example of a linguistic theory that, anticipating a Saussurian distinction between langue and parole, demonstrated how territorializing action inhered in every oral performance.

First, the emphasis on a stentorian modulation of geographic vocabulary imitated the typographic hyperextension of place names on maps. Just as the written word "Massachusetts" implied a larger territorial occupation by being stretched across the face of a map, once pronounced as "Mas-sa-chu'-setts" the geographic referent lost its written fix. Being sounded out loud, it now lodged additional and more flexible ("in and near") spatial demands by invoking territorial borderlines between readers and listeners. Second, as the chorus audibly rehearsed its lines of geographic location and affiliation, its tonal scale subordinated regional differences to the largest common territorial denominator: for instance, pupils voiced the North American continent before they voiced the Alleghenies or the Chesapeake Bay; the pronunciation of the United States muffled that of the regional states. In the end, the collective mantra of a unified American geography had triangulated the nation's political geography into a harmonic soundscape whose boundaries rose or fell depending on the volume and proper pronunciation of the student's geographic reading. For the duration of oral geography lessons, as place names, coordinates, and territorial sizes wafted through American parlors and classrooms, students not only learned how to noisily claim a common geopolitical identity but also practiced how certain geographical noises extended territorial rights, at once reinforcing and challenging other linguistic markers that had traditionally focused on birthrights (biology) and cultural rituals (history) by which the antebellum culture of letters defined the student's identity.[24]

Print, Textbooks, and the Reification of Territoriality

Early antebellum geography textbooks reified the territorial work of oral reading lessons by transposing geography into the study of literature. Textbook authors of the first generation worked in the tradition of grammarians, using linguistic taxonomies and methodologies to create order among geographic words as well as things. Thus Joseph Goldsmith assured his readership that even though his *Easy Grammar of Geography* (1810) used "[a] Vocabulary of proper names of Places, divided and accented in the way in which they are usually pronounced, . . . the letters are understood to possess the ordinary powers of the English language." Similarly in *Geography; or, A Description of the World* (1818), Daniel Adams divided the study of geography into *geographical orthography* ("the spelling and the pronunciation of the names of the kingdoms, countries, mountains, rivers, seas, lakes, islands, etc."), the *grammar of geography* ("principles of Geography . . . designed to be committed to memory"), and finally *geography* ("designed for reading in private, or by classes in school").[25] Both Goldsmith and Adams conceived geography in purely logocentric terms, collapsing geographical knowledge into the textual categories of the English language and literary entertainment. Following

these textbook protocols, the student readers graduated from being public choris-
ters practicing their geographic accents to becoming private readers who, after
proper instructions, memorized geoliterary objects for personal edification.

The process of reification revolved around literacy lessons that conceived
geography as a literary artifact. Primers introduced students to the material forms
of geographic texts by using pictographic alphabets. For example, *The Child's Museum* (1804) included the image of the globe next to the spelled out words "A
Globe." The reading book *The Mother's Gift* (1809) pictured a map, spelled "A
Map," and defined it: "MAP. As you get forward in the study of geography you
will find great entertainment in maps. How pleasant is it to trace the progress of a
traveller of whom we read; and still more interesting where a friend is on a voyage or journey." Thomas Wells's primer, *Cries of London* (1814), presenting the
ABCs in the shape of the "human alphabet," showed the figure of the letter *Z* carrying different-sized maps, while asking, "Zealand, or England, & a Map of the
World" (Figure 1).[26] Primers, spellers, and reading books familiarized students with
the diversity of geographic reading materials. They situated geography within a
broader discursive continuum comprised of the visual and the verbal. They suggested that geography was a staple component of everyday narratives; its artifac-
tual forms prescribed the readers' horizon of expectation while binding them to a
certain narrative perspective and persona. This bind becomes naturalized by Wells's
use of the human alphabet in his primer. There, geography established a somatic
link between the reader's body and the reading material: as students were first
trained to conceptualize their bodies as letters of the print alphabet, the pedagogic
tool of the human alphabet as such instituted the notion that the human body, as a
letter, was intrinsically part of a larger geographic text.

Both the form and medium of the various literacy books suggested that just
as geography was a print commodity, the practices of reading and writing maps
were predicated upon the rules of print literacy. Indeed, geographers like Daniel
Adams conceptualized geography as a form of literacy controlled by the technology of printed texts. He contended, "The natural and artificial divisions of the earth,
the courses of rivers, and the relative position of cities, are mechanical in their
nature, as much as the letters of the alphabet."[27] By comparing real geographical
phenomena to the technical construction of the letter alphabet, this author compared geographic literacy either to the movable types of the printing process or to
the graphic design of handwritten letters. In the opinion of many first-generation
geographers, map literacy conducted most of its territorial work by being as much
a printed work of art as a mechanical reproduction of the alphabet. The patriarch
of American school geographies, Jedidiah Morse, took the print analogy even further by introducing the territorial work of map literacy in a set of experimental
"word maps" (Figure 2).[28] In four continental maps, published in the first editions
of *Geography Made Easy,* Morse reduced the conventional map image—consisting of the geodetic grid, contour lines, words, and pictographic symbols—to a blank
rectangular frame into which he spelled the names of existing nation states, arranging the printed letters so as to imitate their respective geographic dimension
and relation. As a result, each map approximated the visual components of the

FIGURE 1. "Zealand, or England, & a Map of the World." (*Thomas Wells,* Cries of London as They Are Exhibited Every Day *[Boston, 1814]. Courtesy of the American Antiquarian Society, Worcester, Mass.)*

world's political geography, using verbal signs and the printed page as equivalent substitutes for physical and human spaces. As Morse subordinated geography to the machine-made appearance of the printed text, his verbal typography evened out the usually uneven representation of continents and geopolitical territories. For readers, the word maps established a spatial order by which, as Walter Ong has observed, "typographic control typically impresses more by its tidiness and inevitability: the lines perfectly regular, justified on the right side, everything coming

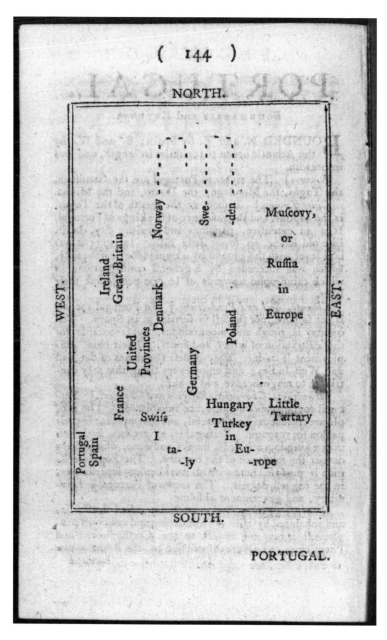

FIGURE 2. One of Jedidiah Morse's word maps. ("*A Newly Constructed Map*," *Geography Made Easy* [*Boston, 1784*], *144. Courtesy of the Houghton Library, Harvard University.*)

out even visually."[29] In this context, the territorial work of geographic print literacy consisted of reducing the world first into a flat horizontal text, constituting an environment that people control—that they not only read but also create and manipulate. The map reader operated under the illusion that the literary acts of spatial invasion and occupation resembled at once mechanical and ludic activities. Just as the printer's apprentice shuffled the typeface inside the letterbox, Morse's students learned how to scramble playfully the verbal markers of social and political territories.

The Practice of Everyday Geographic Writing

For the empire to become the dominant literary environment, textbook authors had to reconcile a vexing strain that lies at the heart of all geographic representation: their lessons had to negotiate the reader's imaginary freedom of movement offered by the map and textbook while aligning this mobility with the pedagogic (and political) authority by which the geographer and mapmaker shaped the signification and interpretation of American geography. The pedagogic use of the verbal and the visual in textbooks exacerbated this tension because it enabled novice geographers to perform what Michel de Certeau has called a "Cartesian move." This move consisted of placing the reader in the scopic position not only of the geographer but also of the politician, frontier officer, or urban planner—or, in de Certeau's words "into the position of having to manage a space that is his own and distinct from all others and in which he can exercise his own will."[30] By offering the map as a quasi-blank, verbally elastic, and colorful page, the apprentice mapmaker was able to create a space of his or her own. Indeed, geography lessons generated the notion that geoliteracy propagated creative activities. Just as new and beautiful maps created a place of imaginative production for the reading subject, they became a self-reflexive place in which the map reader assumed control not only over the geographic text as a mediative form but also over its immediate and seemingly unauthorized memory. Antebellum geographies working within the parameters of the verbal and visual sought to curb and redirect this mapping impulse. Having fused geography to literacy, authors introduced as controlled writing lessons the grid and isomorphic mapping.

Above and beyond the verbal and visual, the image of empire took hold of early-nineteenth-century geography textbooks when they subordinated lessons of map reading to the written (and printed) lines of the global geodetic grid. Like most contemporary textbook authors, Joseph Worcester included in his *Elements of Geography* (1819) a chapter on the construction of maps, in which his instructions depended on the cartographic use of nonglottic writings: mathematics and graphics (Figure 3).[31] By promoting the skeletal outline of the Mercator map and its distorted representation of the Northern Hemisphere, Worcester encouraged his students first to establish geographic coordinates by drawing the grid as a network of uniformly spaced parallel lines intersecting at right angles. Unlike the associative word and literary maps, the Mercator map served as an abstract storage system, urging the geographic reader to compartmentalize the world before writing

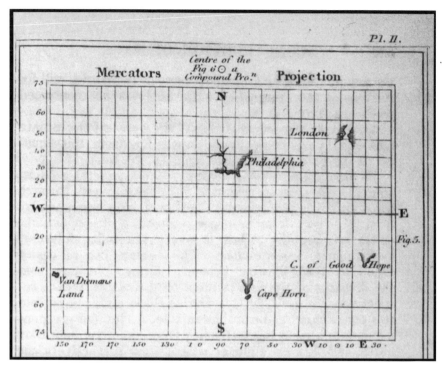

FIGURE 3. Worcester's Map of Mercator's Projection. (*"Construction of Maps,"* Elements of Geography *[Boston, 1819], 320. Courtesy of the American Antiquarian Society, Worcester, Mass.)*

it. Only after the grid had taken control over the blank page did the textbook allow the novice mapmaker to inscribe other carto-literary elements, such as contour lines, pictographic signs, and place symbols. Thus, on the one hand, the completed Mercator map contained all geographic spaces by collapsing spatial differences inside the totalizing framework of geometrical lines. On the other hand, Worcester's mapmaking lesson anticipated a new hemispheric world order. Inside the training grid, this author sketched out the territorial markers of the competing empires of the United States and the United Kingdom. Following the grid's north-south lines, the map presented Philadelphia and Cape Horn as the western endpoints of a trans-American empire, whereas London and the Cape of Good Hope demarcated the maximum eastern extension of the British Empire. Antebellum students here practiced what in a few years would become the basic tenet of U.S. imperial policy when, in 1823, President Monroe laid claim to all of the Western Hemisphere in the name of the American people's interest.

Compared to Worcester and his global writing lesson of the grid, Emma Willard and William Woodbridge encouraged a more localized system of demarcating territories. In their *System of Modern Geography* (1825), they domesticated the overtly abstract and defamiliarizing appearance of the grid map by urging that the student "draw simple maps, beginning with a plan of his table, or the room in which he is, proceeding to delineate successively a plan of the house, garden, neighborhood, and town, until he has represented with tolerable correctness, the relative situation and outlines of the principal objects within his view."[32] Willard and Woodbridge designed lessons in "home geography" that not only foregrounded the visual perception of the world but also commuted the reading subject into a local object. This geographic writing process now redefined the individual reader in relational terms, using spatial proportions and material similarities as territorial markers.

Ideally, this mapping exercise shaped the students' horizon in ways similar to those by which Renaissance scholasticism navigated knowledge—by creating a memory house in which the map is the conceptual base and the grid the textual superstructure containing the writer's cognitive and moral faculties. In practice, however, this geographic memory was contingent upon a set of isomorphic maps, a cartographic method of representation introduced by Alexander von Humboldt, in which the map traced out lines of equal value or subject, such as height and distance, temperature, and vegetation.[33] Following Humboldt, Woodbridge supplemented his companion atlas with two thematic maps: the "Isothermal Chart, or View of Climates & Production" and the "Moral and Political Chart of the Inhabited World" (Figures 4 and 5). Already predicated upon the grid, these new maps transformed the world from a comparative field into a competitive one, while the textbook called for a process of group selection rather than self-abstraction. Woodbridge instructed American students, first, to trace out the isodynamic properties differentiating climate zones *and* the civilized world (defined by religion and government, with Christianity and democracy on top, paganism and tribal culture at the bottom); and, second, to write up the value of territorial dimension measured by a color scheme in which areas deemed civilized were colored light and barbarian areas dark. In the process of this writing exercise, geography effectively fused the printed world inside the book to the emerging realities of the imperial state outside: while the map trained the new geographic writer to stake out his or her collective claim to a latitudinal slice of global territory, the textbook authors Woodbridge and Willard fostered the internalization of global land claims by having the student write them out in "Geographical Copy Books."

Ultimately, between the grid and comparative mappings the most aggressive form of geoliteracy was the "geographical running hand." Because early antebellum education introduced writing after reading, for most students the study of writing was considered to be part of their practical business rather than their cognitive training.[34] In this context antebellum scriveners encountered the chirographic form of the "geographical running hand," a round-hand style used for writing out place names in their relative *geographic* context. Manuals like *Milns's Geographical Running Hand Copies* (1822) specialized in exercises like this: "Amsterdam, Holland, Netherlands. Europe. 52° North."[35] Similarly, Abner Reed's *A New Plan and Easy Set of Geographical Running Hand Copies* (1801) included warm-up exercises in which the student copied out first the alphabet ("a, b, c, etc.; A, B, C, etc.") before writing out the actual geographical exercise ("Boston, the

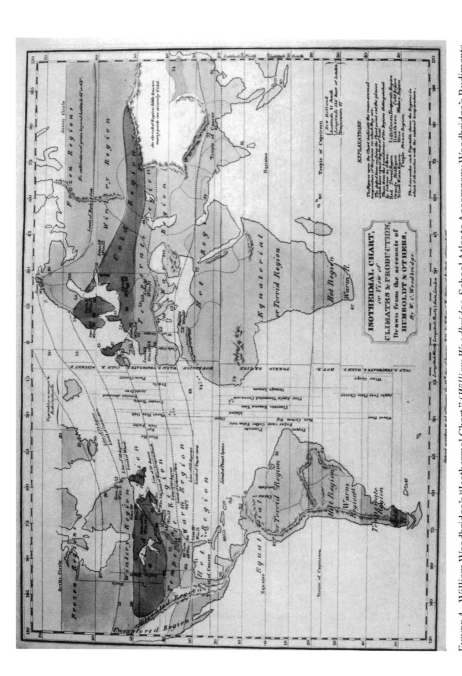

FIGURE 4. William Woodbridge's "Isothermal Chart." (*William Woodbridge, School Atlas to Accompany Woodbridge's Rudiments of Geography [Hartford, 1823]. Courtesy of the American Antiquarian Society; Worcester, Mass.*)

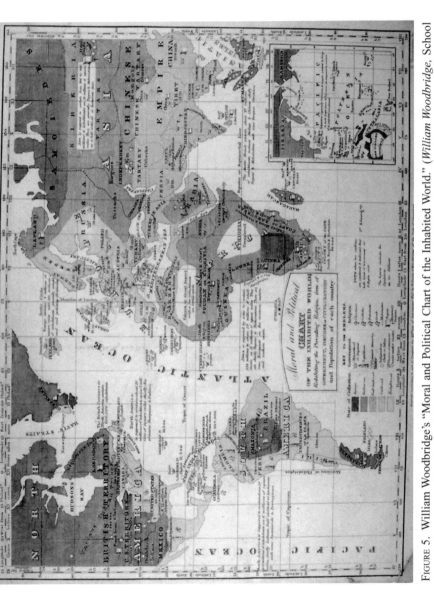

FIGURE 5. William Woodbridge's "Moral and Political Chart of the Inhabited World." (*William Woodbridge, School Atlas to Accompany Woodbridge's Rudiments of Geography [Hartford, 1823]. Courtesy of the American Antiquarian Society, Worcester, Mass.*)

capital of Massachusetts and of New England").[36] What comes as a shock, however, is that in between these exercises, American writing masters, like Reed, had students copy out a self-consciously designed list of words: "and, band, bind, mind, hill, kill, brood, blood" (1). Buried between the letter alphabet and the geographic longhand was a writing exercise that by a process of free association and phonemic minimal pairing connected the manual task of writing with the rhetoric of territorial aggression.

Geographic handwriting here generated the fantasy of direct chirographic control over the spaces touched by the pen; the act of putting a pen to paper was apparently considered to be equal to the task of taking possession of territories. In contrast to the geographic aesthetics borne out by orality and print, the geographical hand moreover suggested a new faith in personalized manual labor. Because the images of the running hands defined the process of writing in terms very similar to print technology, the previously spoken words of geography here discretely slipped into the discursive realm of standardized mechanical reproduction. Yet, in the context of the geographical writing manuals, these exercises acknowledged the individual author as the source of spatial agency; just as writing allowed individuals to inscribe themselves directly on the map, geographic writing, along with the map, the grid, or other narrative devices, became the territorial agent through which antebellum citizens practiced their hands at empire building.

Conclusion

In the end, becoming geoliterate in the antebellum United States meant studying the art of territorial aggression. Literary exercises using picture alphabets, printed books, and handwriting suggest that traditional literacy and territorial violence were considered as correlative activities, which geography teachers, with an eye on coy propriety, instilled according to the levels of age and literary competence. The discursive scrambling of letters, maps, and narratives instructed students to practice—at least in spirit if not in person—how to subvert, inscribe, and brutalize any given form of geographical space. In this application, geographies fulfilled the logic of modern literacy, namely, that reading and writing enacted representations of limited and detached agency. Just as the printed word disguised the identity of the author, geographic writing protected a whole population from accusations of actual participation in the contemporaneous acts of territorial violence. Given the instructions in geographic writings—moving from the grid to the "geographical running hand"—the letter alphabet and its practical application had become aligned with a general will that increasingly was claiming land and territory by manual, mechanical assertion. As geographies lowered the threshold at which people began to fear the risk of individual accountability—of exposing the identity of the territorial aggressor—this general will could no longer be exclusively aligned with a republican print ideology derived from a Habermasian public sphere or a nationalistic ideology as defined by Benedict Anderson. Rather, this will was the product of a process of self-education and pedagogic principles by which a people had collectively internalized the dictates of imperial authority.

Notes

1. Cited in "Art. X.—*America and Her Resources* . . . By John Bristed [London, 1818]," *Literary and Scientific Repository* 2, no. 3 (1821): 209.

2. William Appleman Williams, *Empire as a Way of Life* (New York: Oxford University Press, 1980), ix.

3. Ronald J. Zboray, *A Fictive People* (New York: Oxford University Press, 1993); and William J. Gilmore, *Reading Becomes a Necessity of Life* (Knoxville: University of Tennessee 1989).

4. I am summarizing more complex arguments, such as those in Lawrence Cremin, *American Education: The National Experience, 1783–1876* (New York: Harper and Row, 1980); Lee Soltow and Edward Stevens, *The Rise of Literacy and the Common School Movement in the United States* (Chicago: University of Chicago Press, 1981); Harvey Graff, *The Legacies of Literacy* (Bloomington: Indiana University Press, 1987); Carl Kaestle, Helen Damon-Moore, Lawrence C. Stedman, Katherine Tinsley, and William Vance Trollinger Jr., *Literacy in the United States: Readers and Reading since 1880* (New Haven, Conn.: Yale University Press, 1991); and Grey Gundaker, *Signs of Diaspora, Diaspora of Signs: Literacies, Creolization, and Vernacular Practices in African America* (New York: Oxford University Press, 1998).

5. The function of empire is missing in Kaestle et al., *Literacy*. More strikingly, the imperial dimension escapes Malcolm P. Douglass in his study titled *The History, Psychology, and Pedagogy of Geographic Literacy* (Westport, Conn.: Praeger, 1998).

6. Here I expand from Amy Kaplan's observation that the concept (and habit structure) of national discourse in the United States was always already contingent upon the spatial formation of U.S. borders. See her introduction to *Cultures of United States Imperialism,* ed. Amy Kaplan and Donald Pease (Durham, N.C.: Duke University Press, 1993), 14–17.

7. See Dana Nelson, "The Word in Black and White," in *Reading in America: Literature and Social History,* ed. Cathy N. Davidson (Baltimore: Johns Hopkins University Press, 1989), 141–142.

8. The essay follows what I understand to be the constructionist rather than behaviorist definition of "territoriality," as explored in Erving Goffman, *Asylums* (Garden City, N.Y.: Anchor Books, 1961); Irvin Altman, *The Environment and Social Behavior* (Monterey, Calif.: Brooks/Cole Pub. Co., 1975); Stanford Lyman and Marvin B. Scott, "Territoriality," *Social Problems* 15, no. 2 (fall 1967): 236–249; and Robert Sommer, *Personal Space: The Behavioral Basis of Design* (Englewood Cliffs, N.J.: Prentice-Hall, 1969).

9. See Robert D. Sack, *Human Territoriality: Its Theory and History* (Cambridge: Cambridge University Press, 1986), 26.

10. On the proliferation of cartographic materials in U.S. print culture, see Walter W. Ristow, *American Maps and Map-Makers* (Detroit, Mich.: Wayne State University Press, 1985); and Seymour T. Schwartz and Ralph Ehrenberg, *The Mapping of America* (New York: Abrams, 1980).

11. See Charles Sellers, *The Market Revolution* (New York: Oxford University Press, 1991), 79–93. For a survey of Monroe's territorial policies, see Jack Ericson Eblen, *The First and Second United States Empires* (Pittsburgh, Penn.: University of Pittsburgh Press, 1968), 138–170; for a general introduction to territorial policies, see Reginald C. Stuart, *United States Expansionism and British North America* (Chapel Hill: University of North Carolina Press, 1988).

12. See the comprehensive study by Daniel H. Calhoun, "Eyes for the Jacksonian World: William C. Woodbridge and Emma Willard," *Journal of the Early Republic* 4 (spring 1984): 1–26.

13. William Warntz, *Geography Now and Then* (New York: American Geographic Society, 1964); John A. Nietz, *Old Textbooks* (Pittsburgh: University of Pennsylvania, 1961).

14. In 1810 Catharine Beecher studied geography and map drawing; see Nancy Cott, *Bonds of Womanhood* (New Haven, Conn.: Yale University Press, 1977), 115. Beecher's pupil, Fanny Fern, who attended the Hartford Seminary in 1828, satirizes antebellum geography lessons in the novel *Ruth Hall* (*Ruth Hall and Other Writings*, ed. Joyce W. Warren [New Brunswick, N.J.: Rutgers University Press, 1986], 101–103).

15. Cited in "Massachusetts' Schools," *Niles' Weekly Register* 20, no. 7 (1821): 108.

16. This phrase echoes Wai Chee Dimock, *Empire for Liberty: Melville and the Poetics of Individualism* (Princeton, N.J.: Princeton University Press, 1989).

17. Soltow and Stevens, *Rise of Literacy,* 96–102.

18. Jedidiah Morse, *Geography Made Easy* (Boston, 1820), iii; Jedidiah Morse, *Modern Atlas Adapted to Morse's New School Geography* (Boston, 1822), i; and William Woodbridge [and Emma Willard], *Rudiments of Geography* (Hartford, Conn., 1822), viii. I attribute the authorship of *Rudiments* in equal parts to both Willard and Woodbridge. In her part of the introduction to *System of Modern Geography,* Willard explains her share of authorial labor: "The arrangement entered into between Mr. Woodbridge and myself, was predicated solely on my having compiled and taught a system of modern geography similar to his: whereas my writing the ancient was merely an accidental consequence of my becoming a partner in the concern" (Woodbridge and Willard, *System of Modern Geography* [Hartford, Conn., 1824], xx).

19. Jedidiah Morse, *Geography Made Easy* (Boston, 1818), iv; Morse, *Geography Made Easy* (1820), v.

20. Several of the second-generation geographers affected the school curriculum as teachers of "special education." For example, textbook author Samuel Howe taught blind students in Massachusetts; William Woodbridge and Thomas Gallaudet both taught at schools for the deaf (Calhoun, "Eyes," 11–13). The impact of the Pestalozzian pedagogy was noted by magazines such as *Niles' Weekly Register,* suppl. to vol. 15, "Education" (1819), 52, and by one of the nation's first teaching journals, the *Academician* 1, no. 16 (February 1819): 245–246, 264. For the role of Pestalozzi in American education, see Carl F. Kaestle, *Pillars of the Republic* (New York: Hill and Wang, 1983), 67; Soltow and Stevens, *Rise of Literacy,* 96–97; and Cremin, *American Education,* 77–79.

21. Cited in "The New School; or, Lancaster System," *Academician* 1, no. 5 (May 1818): 100–101. Here I would like to thank Patricia Crain for her enthusiastic comments and for sharing her forthcoming work on Lancaster.

22. See Gilmore, *Reading;* and Cathy N. Davidson, *Revolution and the Word* (New York: Oxford University Press, 1986).

23. Noah Webster, *American Spelling Book* (Hartford, Conn., 1817), 120.

24. Here I depart from John Brian Harley's contention that maps are inherently silent signifiers. See his "Maps, Knowledge, Power," in *The Iconography of Landscape,* ed. Denis Cosgrove and Stephen Daniels (Cambridge: Cambridge University Press, 1988), 290. Also of interest are his essays "Deconstructing the Map," *Cartographica* 26, no. 2 (summer 1989): 1–20; and "Cartography, Ethics, and Social Theory," *Cartographica* 27, no. 2 (summer 1990): 1–23.

25. J. Goldsmith, *An Easy Grammar of Geography* (Philadelphia, 1810), 109; Daniel Adams, *Geography; or, A Description of the World* (Boston, 1818), 3–4.

26. *The Child's Museum, Containing a Description of One Hundred and Eight Interesting Subjects* (Philadelphia, 1804), table 3, figure 13; *The Mother's Gift; or, Remarks on a Set of Cuts for Children* (Philadelphia, 1809), 14; Thomas Wells, *Cries of London as They Are Exhibited Every Day* (Boston, 1814).

27. Adams, *Geography,* 3.

28. Jedidiah Morse, *Geography Made Easy* (Boston, 1784), 144. I discuss this further in "Lessons in Geography: Maps, Spellers, and Other Grammars of Nationalism in the Early Republic," *American Quarterly* 51, no. 2 (June 1999): 311–343.

29. Walter J. Ong, *Orality and Literacy* (New York: Methuen, 1982), 122.

30. Michel de Certeau, *The Practice of Everyday Life,* trans. Steven Rendall (Berkeley: University of California Press, 1984), 134.

31. Joseph Emerson Worcester, *Elements of Geography* (Boston, 1819), 320. On the range of signification engendered by nonverbal writings and discussions on cartography, see Roy Harris, *Signs of Writing* (London: Routledge, 1995); and Edward R. Tufte, *Envisioning Information* (Cheshire, Conn.: Graphics Press, 1990).

32. Woodbridge and Willard, *System,* xvi.

33. On the impact of isodynamic maps, see Margarita Bowen, *Empiricism and Geographical Thought: From Francis Bacon to Alexander von Humboldt* (Cambridge: Cambridge University Press, 1981), 222–259.

34. See Tamara Plakins Thornton, *Handwriting in America* (New Haven, Conn.: Yale University Press, 1996).

35. William Milns, *Milns's Geographical Running Hand Copies for the Use of the City Commercial School* (New York, 1822), 3.

36. Abner Reed, *A New Plan and Easy Set of Geographical Running Hand Copies* (East Windsor, 1801), 1–2.

From House to Square to Street

Narrative Traversals

Robert L. Patten

*Geography is not an inert container, is not a box where cultural history
"happens," but an active force, that pervades the literary field and
shapes it in depth.*
Franco Moretti, *Atlas of the European Novel*

*F*ranco Moretti's *Atlas of the European
Novel* deploys maps to highlight the "place-bound nature of literary forms: each
of them with its peculiar geometry, its boundaries, its spatial taboos and favorite
routes." His maps also "bring to light the *internal* logic of narrative: the semiotic
domain around which a plot coalesces and self-organizes."[1] Moretti's study ranges
widely and offers a succession of dazzling empirical and theoretical formulations.
This essay has more modest aims and evidence. It seeks to practice a hermeneu-
tics of literary space within a particular culture and time period. Are there some
ways that British nineteenth-century literature organizes the geographic places of
its narratives? At different times during the century, do different kinds of spaces
seem to predominate in literary representations? Do these different spaces incarnate particular sets of values and "belong" to particular kinds of people? Does
movement to, into, and away from these spaces signify differently not just in different works but more generally in different periods or sites? And might there be
suggestive connections between particular spaces within texts and the formats of
those texts, between particular kinds of travel to and from those spaces and specific modes of publication? Reading, this essay suggests, may be not only about
the spatial situatedness of literature and "the *internal* logic of narrative" but also
about the ways places and journeys to and from those sites trope the processes of
reading texts and environments, the methods of making sense of sensory data.

    One advantage of looking more closely, over a period of time within a national culture of spaces and their traversal, is that doing so allows us to contemplate character, setting, and art freshly. Thinking about the spaces constructed in and by texts foregrounds construction: the made environments (houses, parks, streets) within the text, the made environment of the text (principally, formats), and the made environment of reading (the spaces, times, and social constructions of readerly activity). A hermeneutics of literary geography helps to recover the aesthetic, in an enlarged sense of the word: those designs within the text, of the text, and outside the text, which embody ways a culture organizes itself. But each text, each format, even, to a large extent, each reading constructs in specific ways, according to a unique though culturally shared logic of the aesthetic object. A hermeneutics of literary geography allows us to ask why a place is made, why it is a place to start from or come to, and why, because of those starting and ending points, the spaces and journeys signify in particular ways. In other words, although language, texts, spaces, houses, travel, and reading are continua, the deliberate fashioning of particular configurations of word, text, house, and reading demarcates a particular organization of experience. To differences in those particular organizations of experience we now turn.

We will begin by considering country houses as destinations for the deserving in early-nineteenth-century literature. The three-decker novel and the long poem seem to be formats homologous to such narrative traversals. Then we will journey in time to the decades around midcentury and in space to the city, where journeys and arrivals are more uncertain and the formats more evanescent, as in periodical publications. Maps, I will argue, seek to reassure readers and reestablish purposeful travel; in some ways they enable shorter and more completed narratives. But by the end of the century, it is hard to map travel through the labyrinthine or merely extensive streets of any urban space, or to predetermine an appropriate format for the narrative of living.

> *To Pemberley, therefore, they were to go.*
> Jane Austen, *Pride and Prejudice*

Taking a synoptic view of late-eighteenth-century and early-nineteenth-century fiction, the settings, even the novels themselves, were often country houses: *Castle Rackrent, Northanger Abbey, Mansfield Park,* and *Kenilworth* are instances; and *Waverley* is the title of the book, patronym of the protagonist, and name of his estate. This last is not a surprising observation; many subjects of fiction are middle class and inhabit houses and books named for them. And certainly the country house had become in its own right the subject of a genre of poetry.[2] What may take us a bit further along, however, is an allied observation, that the characters are identified with their spaces: identified not only as to fortune, a ruined house and a penniless character being almost interchangeable tropes, as in *Castle Rackrent,* but also as to religion—overt or secret Catholicism and priests' holes or other hiding places are virtually synonymous in works by Sir Walter Scott, Harrison Ainsworth, Charlotte Brontë, and others.[3] Spaces also identify character—with furnishings often serving as an index to morals. And in some cases, quite elaborate equivalencies are discursively established between the spaces one inhabits and the complex economic system of, say, a country house or other economic center.

Take, for example Fanny Price's rooms at Mansfield Park. They provide her with privacy and a retreat from the grandeur and the "too large" spaces of the house while defining her dependency and subsidiary-ness. As Fanny grows in "consequence," she inhabits upstairs and downstairs more comfortably.[4] The meanness and remoteness of her chambers eventually critique Sir Thomas Bertram for his neglect of his "blooming" niece as well as for his suddenly awakened interest in her after his return from the West Indies: "How comes this about [that you have no fire?]," Sir Thomas demands of Fanny; "[H]ere must be some mistake. I understood that you had the use of this room by way of making you perfectly comfortable.—In your bed-chamber I know you *cannot* have a fire. Here is some great misapprehension which must be rectified. It is highly unfit for you to sit—be it only half an hour a day, without a fire."[5]

Rooms are often crucial indices of their inhabitant's fate. No one would ever suppose, from Pip's first meeting with Miss Havisham at Satis House, that the old lady might meet a kindly, rich gentleman who would take her away from all that decay and make her giddy with happiness, true love, and the delayed fulfillment of desire and the marriage plot.[6] The room is against it; it is no room with a view to the future. In *Wuthering Heights,* Emily Brontë mismatches room with temperament: Cathy Earnshaw betrays herself in aspiring to the Lintons' drawing room at Thrushcross Grange, "a splendid place carpeted with crimson, and crimson-covered chairs and tables, and a pure white ceiling bordered by gold, a shower of glass-drops hanging in silver chains from the centre, and shimmering with little soft tapers."[7] By contrast, the effete visitor Mr. Lockwood and the feeble boy Linton Heathcliff cannot abide the rude strength of Wuthering Heights, with its white stone floors, primitive furniture, and half-tame animals.

Country-house novels are not just about marriage; they are also about getting the right inhabitants into possession of the right rooms. In *Pride and Prejudice,* Mr. Bingley's temporary residence at Netherfields is only a beginning for Mrs. Bennet and Miss Austen; what is at issue is how the Bennet daughters will be permanently housed and whether Mr. Collins will usurp their own home. Charlotte Lucas's trajectory is not so much a marriage plot as a residence plot; she gets a house to keep and a function to fill within the social economy of Lady Catherine's estates. Time and again in these early-nineteenth-century novels the resolution involves revitalizing domestic architecture: the restoration of a home (for instance, the Baron of Bradwardine's manor house, Tully Veolan, in *Waverley*), or the restitution of a home to its rightful owner, or the institution of an appropriate person into his or her appropriate room, role, and status in the economic and emotional center of the text.

And thus many of these texts figure travel not only as a necessary, realistic aspect of living within country spaces but also as a symbolic, metanarrative event that tropes the characters' own psychological, social, and moral journeys. *Waverley* might be said to be a narrative about a young, unformed man's maturation, figured as a succession of sometimes voluntary, sometimes involuntary trips from the homes of his contesting parents (father and uncle) to destinations that realize the fantasies awakened by his early reading.[8] Traveling under different assumed or imposed identities, book and eponymous hero visit homes and situations that allow the text to explore his options: loyalist or rebel, country squire or courtier, Saxon or Celt, dead legend or live librarian.

What happens along the way in the early-nineteenth-century novel is thus not only the "stuff" of the narrative but also in some ways the structure, the design, of the narrative. When, in Jane Austen's *Emma,* Miss Woodhouse gets into the coach with Mr. Collins as they are returning from Randalls to Hartfield on Christmas Eve, everything about that journey has been set up to reinforce the falsity of the marriage proposal that will be offered therein.[9] *Emma* is in many senses a novel about the dangers of being on the road, between residences: Jane Fairfax's experiences in Weymouth and at home, Harriet Smith's rescue from the gypsies when she goes out walking, Emma's pretext of a faulty shoelace that allows entry into the vicarage so, she hopes, Mr. Elton can propose that Harriet become its mistress, Frank Churchill's journeys between Yorkshire and Highbury and the miscarried messages between him and Jane, Mr. Woodhouse's fear of traveling outside his narrow circle of rooms and company. Or, to cite another "road" novel, *Frankenstein* might be said to be about finding the right inhabitant, both of the bonehouse that is the body and of a home; a homeland (for the exiled DeLaceys and Safie); and a resting place for the fevered imaginations of Robert Walton, who thinks the Arctic is "the region of beauty and delight," and for the homeless, mateless monster.[10]

So getting "home" becomes an alternative to the marriage plot, a narrative line as old as the Hebrew Scriptures' narrative of Eden, exile, and arrival at the promised land. Home figures centrally in Tennyson's *In Memoriam.*[11] Hallam's corpse, grave, and dark house in Cambridge are counterpointed to the observances of Christmas and Hallam's death at Somersby Rectory: the homeless narrator seeks some place in the universe for his friend and his friendship, a place eventually found within a new rectory, a revitalized Cambridge house, a marriage displaced from the protagonist and his deceased mate, and a fruitful womb that will bring forth a higher type of life. Home becomes, in many ways, the key term in Browning, whether it be Casa Guidi incarnating Barrett Browning's transcendent love, England ("Home Thoughts, from Abroad"), the heart or soul, or the diurnal, what Abt Vogler calls the "C Major of this life."[12] Getting to that home becomes the burden of many nineteenth-century texts—fictional, poetic, and dramatic. Whether the story is Gothic or picaresque, historical or romantic, both the journey and the arrival matter. Reading a novel in particular becomes troped as the designedly middle-class reader's taking a journey in some ways parallel to, in some ways oblique to, those undertaken by the characters.[13]

Some fundamental epistemological assumptions underlie this homology: life is a journey, and one's journey is supposed to develop and reveal character and to define one's relation to others—worldly and otherworldly. The three-decker novel, introduced by Scott with *Waverley* and flourishing until the 1890s, seems to incarnate that notion of life as journey to one's proper home; three-deckers are named after the great three-deck men-of-war ships that won Britain's naval battles during the Napoleonic Wars. There are countless versions of the journey-home plot: *Rookwood* and *Oliver Twist* are obsessed with the topic, as are *Wuthering Heights, Jane Eyre, The Mill on the Floss, The Egoist,* and *Tess of the d'Urbervilles,* to name a few. Many of Anthony Trollope's novels center on the question of who will occupy or inherit the family home; *Can You Forgive Her?* which narrates the struggles of Squire Vavasor to find an appropriate heir to Vavasor Hall, and *Orley Farm,* in which the estate passes through a forged will to the wrong heir, are just two of the many examples that might be adduced.[14]

> *Fantastic idea: the city*—the generalized spatial proximity unique to
> the city—*as a genuine enigma: a "mosaic of worlds."*
> Franco Moretti, *Atlas of the European Novel*

Toward midcentury the homologies between country house and journey, three-decker and life, end of story and finding a proper home, and middle-class protagonist and middle-class reader get disturbed. To begin with, the setting of fiction, even poetry, moves to urban environments; homes are replaced by squares and streets and unmarked houses.[15] This is not a trivial change, even at the level of finding one's way around. In the earliest of Dickens's *Sketches by Boz,* a middle-aged, timid Somerset House clerk, Augustus Minns, who hates children and dogs, is given directions to the home of his cousin, Octavius Budden, a retired corn-chandler residing on the edge of the country who is hopeful that the bachelor Minns will leave his godson, Budden's boy, his fortune of ten thousand pounds. "'Now mind the direction,'" Budden tells his city cousin Minns,

> "the coach goes from the Flowerpot, in Bishopsgate-street, every half hour. When the coach stops at the Swan, you'll see, immediately opposite you a white house."
>
> "Which is your house—I understand," said Minns, wishing to cut short the visit and the story at the same time.
>
> "No, no, that's not mine; that's Grogus's, the great iron-monger's. I was going to say—you turn down by the side of the white house till you can't go another step further—mind that—and then you turn to your right, by some stables—well; close to you, you'll see a wall with 'Beware of the Dog' written upon it in large letters—(Minns shuddered)—go along by the side of that wall for about a quarter of a mile—and any body will show you which is my place."[16]

This story tells several things about the consequences of moving from a country house to the city and traveling in its environs. First, there is more need for varied transport in Dickens's *Sketches* (hackney cabs and omnibuses and stage-coaches and steamboats, rather than simply the carriages and horses used in, say, *Mansfield Park*); next, it takes more complex directions to get somewhere, with decreased likeliness of arriving ("any body will show you which is my place") and increased danger along the way ("Beware of the Dog"); third, homes become identified, not by their name, identical to the name of their owner, but by less visible indicators (the white house is not Budden's, but "Grogus's, the great ironmonger's"—Dickens will get many stories and several farces out of urbanites entering at the wrong door); and finally, the analogy between journeying and story ("wishing to cut short the visit and the story at the same time") in such a confusing geography means that neither trip nor tale can be cut short or certainly sketched.

Maintaining oneself in the city, at a good address on or adjacent to a square or other notable architectural or geographical feature, is a precarious business. In the city, one inhabits houses but seldom owns them. Think, in *Vanity Fair,* of Amelia's precarious tenure at a home in Russell Square or how badly Mr. Sedley's ex-clerk and Becky's landlord are treated by their respective tenants; think of the uncertainty at the parsonage about who will inherit Queen's Crawley.[17] Or recall Dickens's exactly contemporary novel, *Dombey and Son,* and the instability of homes there, from Miss Tox's "squeezed" house in the "anxious and haggard" No Thoroughfare of Princess's Place to Mr. Dombey's "great house in the long dull street" and Captain Cuttle's room at Mrs. MacStinger's.[18] In these midcentury British novels, the issue no longer seems to be who will be worthy of the inherited place but, rather, who can get, or keep, a place at all? Just how precarious one's lodging might be is well brought out by J. Mordant Crook. Even the richest of the industrial age plutocrats found it hard to preserve their fortunes much beyond their own generation (if they were lucky enough not to go bankrupt in their lifetime), and their grand houses were transferred into the hands of other plutocrats, or divided into flats, or demolished, in less than a century.[19]

And so *Vanity Fair* figures Becky's roulette wheel as a major trope for all the circular motions in the story, the buyings and sellings of places, persons, horses, medals, ancestors (Becky's Montmorency forebears), and descendants (Amelia's son Georgy); and in its narrative structures, it returns again and again, redoubles, stops and goes backwards, utilizes its serial structure to propel a deceptive forward motion that leads to death on the battlefield or to regressivity (Miss Crawley enters a second childhood) or, in the last chapter, to a return to one's beginnings.[20] Dickens does a similar thing with *David Copperfield:* a serial that begins by declaring unresolved the question whether the protagonist will turn out to be the hero of his own story has no destination in mind, only a series of "new beginnings," which is what several of the chapter titles name and what much of the journeying in that novel accomplishes.[21]

If lives, now, are often more circular than telic, and if houses are temporarily inhabited decentered spaces rather than centers of economies and texts, then journeys are more difficult to narrate because they are so uncertain. To begin with, roaming the city becomes a different kind of experience, in life and in texts, for women and for men, as Deborah Epstein Nord has shown: "If the rambler or flaneur required anonymity and the camouflage of the crowd to move with impunity and to exercise the privilege of the gaze, the too-noticeable female stroller could never enjoy that position."[22] Thus, to narrate a bigendered perambulation leads to the kinds of divisions that Charles Dickens structures overtly in *Bleak House* and that Wilkie Collins does more covertly in, say, *The Woman in White.*[23]

But many mid-Victorian journeys are unnarratable. While the coach trip may figure, from Dr. Johnson forward, as the apex of human felicity—and all the more obnoxious, therefore, is Mr. Elton's proposal in a coach—and while a regular coachman is, for Tony Weller in Dickens's *Pickwick Papers,* "a sort o' con-nectin' link betwixt singleness and matrimony," a train trip is another matter. Trains annul narrative. In *Dombey and Son,* the railway mows down Staggs's Gardens—despite its name, "a little row of houses" and "miserable waste ground." Railway development has eradicated location ("no such place as Staggs's Gardens"), transformed the neighborhood, eliminated addresses, destroyed houses, dislocated families, and mangled their stories. In their places have sprung up railway accessories, "villas, gardens, churches, healthy public walks," all growing "at steam's own speed," until the "very houses seemed disposed to pack up and take trips." Now people do not travel to houses; houses travel on their own. And when, for instance, Mr. Dombey takes a train ride, the landscape rushes past so fast that he cannot order his thoughts in any sequence or make any sense of what he sees. In a grim parody of troping life as journey, Mr. Dombey, on his train trip to Leamington Spa for rejuvenation, experiences an iron power "defiant of all paths and roads, piercing through the heart of every obstacle, and dragging living creatures of all classes, ages, and degrees behind it"; the train and its passenger are types "of the triumphant monster, Death."[24] Trains obliterate and silence: James Carker, Anna Karenina. A fatal railway accident at Staplehurst nearly killed Dickens in June 1865; five years to the day later, Dickens did die. No wonder Agatha Christie and Patricia Highsmith set death on trains.[25]

The city, too, frustrates narrative. One cannot find one's way around. The scene in which Bill Sikes drags Oliver Twist through Smithfield Market on their way to the burglary at the Maylies has been analyzed as a paradigm of "a world in chaotic action."[26] The city's assault on the body and senses produces an overwhelming confusion in an innocent observer: "[T]he crowding, pushing, driving, beating, whooping, and yelling; the hideous and discordant din that resounded from every corner of the market; and the unwashed, unshaven, squalid, and dirty figures constantly running to and fro, and bursting in and out of the throng; rendered it a stunning and bewildering scene, which quite confounded the senses."[27] The novel depicts this London as a Manichean place: at the Angel pub in Islington, where Oliver enters the city, there are two routes forward: one leads southwest into the West End, to salvation and reclaimed identity at the Brownlow house and eventually farther west in the country with the Maylies; the other plunges directly south into the heart of the Seven Dials and the filthy rookeries where thieves had lived for hundreds of years.[28] There are those—Bill Sikes, Fagin, Nancy, and Monks—who can make their way between the two geographies. But for the most part this city and its streets trope a moral typology, a geography of the soul that cannot be traversed with impunity. And these journeys cannot, in any meaningful sense, be narrated: Oliver can never say how he gets from one side of the Manichean divide to the other, and the novel's narrative logic breaks down whenever anyone tries to put Oliver's history of traversals into an orderly, causal sequence. The broken-backed, inconsistent, changes-of-mind-and-character narrative discloses, at the outset of the great midcentury city texts, how threatening to fictional and narrative coherence is the anarchy of streets and squares, markets and trains, lodgings and hideouts.[29] Significantly, Jo the crossings-sweeper, existing at the very ground zero of the city's grid in Dickens's midcentury novel *Bleak House,* "don't know nothink about nothink at all."[30]

There are those who can make their way through the city—Fagin and Sikes, as well as Inspector Bucket, Inspector Field, Sherlock Holmes. The labyrinths, cloacal streets, no thoroughfares, and secrets of the city can be understood by three classes: criminals, detectives, and readers.[31] (Victor Hugo's *Les Misérables* is a classic instance of this configuration.) The criminalization of the reader, who becomes detective and villain, and whose reading is increasingly policed by media fearful that readers will learn too much or learn the wrong things, is a central feature of later Victorian literature, strikingly prevalent in the lurid sensationalist journalism that allows stay-at-homes imaginative participation in the crimes of Jack the Ripper or the deflowering of virgins or the luxurious excesses of decadence.[32]

> *I have the English [translation of the story] in my pocket-book, and a*
> *fac-simile of the map, if it can be called a map.*
> Allan Quatermain, in H. Rider Haggard, *King Solomon's Mines*

To rescue the novel from structural chaos and indeterminacy, to redeem knowledge from moral ambiguity, some writers constructed maps leading to treasures, power, and reclaimed identity. Robert Louis Stevenson's *Treasure Island* is a paradigmatic instance, in its own right and as it inspired successors such as Rider Haggard's *King Solomon's Mines.*[33] Stevenson and his stepson Lloyd Osbourne concocted a map of an island holding buried treasure while on holiday at Braemar in Scotland in the late summer of 1881. The book was written from the map. The harbors of the imagined island, Stevenson says, "pleased me like sonnets," and as he and Lloyd "elaborat[ed] the map," "the future characters of the book began to appear there visibly among imaginary woods; and their brown faces and bright weapons peeped out upon me from . . . those few square inches of a flat projection."[34] Moreover, it was not a book shaped by adult-fiction conventions; it was a tale published in seventeen weekly installments in a boy's adventure magazine, *Young Folks,* from 1 October 1881 to 28 January 1882. Stevenson got twelve shillings and six pence per column, thirty pounds for the whole, from this publication, and another one hundred pounds when it was published as a single volume, the greatest amount he had yet earned from writing.

The novel retains its connection to what I have been describing as the nexus of home spaces and journey structure in nineteenth-century fiction, but as usual for Stevenson's slyly subversive imagination, the novel inverts and adapts these protocols. *Treasure Island* is the story of Jim Hawkins's leaving home. His father, innkeeper of the Admiral Benbow Inn on the coast somewhere near Bristol, dies at the same time that their mysterious lodger dies, leaving a sea-chest containing coins and a map. Villains wreck the inn just moments after Jim and his mother escape with the map; significantly, they hide out under a nearby bridge until res-cuers on horseback thunder by, trampling to death one of the former pirates who have spread out to search for Jim. His home—a lodging for travelers—despoiled, Jim never resumes life at the inn. He lives with the squire's gamekeeper while a ship is fitted out and packed, and on the one occasion when he returns to visit his mother at the inn, he makes the life of the boy apprentice who is to stay behind in Jim's "place" a dog's life. Here are all the elements we have been tracking: home, now figured as temporary lodging; expulsion from home, in this case with a map ambivalently pointing to a new patrimony; terrors on and even, in this case, under the road; and no place to which Jim might return—so he must voyage.

One might wish to speculate further about the relation between the buddy plot "on the road" and the discovery of buried treasure that restores virility. In *Treasure Island* the silver bars and the armor, the kinds of things that might sig-nify potent and defended manhood, remain buried on the island. Only those sym-bolic tokens of power exchange, the coins, are brought back home. And they are obtained for Jim and his fellow mariners by a dead man—a character Stevenson based on Robinson Crusoe, who, thought dead, is brought back to life. The imbrication of stories with lives and money and books is resonantly layered.[35]

Treasure Island thus maps, as its central embedded text and as its total text, a journey from the loss of family and home to the establishment of a social com-munity of adult males within which Jim can take his place. He earns that place by saving his multitude of foster fathers, lawful and lawless; he does so by stealing their mobile home, the ship, another of those houses that, as *Dombey and Son* puts it, seems "disposed to pack up and take trips." And Jim himself becomes another mapmaker when, at the request of his foster fathers, the squire and the doctor, he writes "down the whole particulars about Treasure Island, from the beginning to the end."[36]

It takes only a single volume, however, to narrate and enact this voyage, be-cause the map is the sign and story of a voyage already completed: the voyage that took pirates to the island where they buried their treasure and were, with one exception, then killed. That voyage had a beginning, a middle, and an end, and the map is both a guarantee that the voyage may be repeated and the prompt for Stevenson and Jim to write that journey.

Maps constitute narratives in graphic form, perhaps the last instance in nine-teenth-century literature of the attempt of the visual to assert some kind of equal-ity with, if not primacy over, the increasing domination of the verbal narrative.[37] In an age when journeys anywhere, from prison to home or from childhood to adulthood, become increasingly problematic; when homes are only spaces and ad-dresses, squares and streets, and flats and inns; when getting home or earning a permanent space is difficult and often involves apparently illegal usurpations or collusion with criminals (for example, Dickens's *Little Dorrit* and *Great Expecta-tions*), then the very structures of narrative, like the labyrinths of the city, have a devil of a time articulating a structure and physical embodiment of "progress." Although the character Little Dorrit has her birthday recorded in the first volume of the church register, her childhood in the second, and her marriage in the third, the book of her life, *Little Dorrit,* was issued in twenty monthly numbers as nineteen (the final being a double number) and therein divided into two "books," "Poverty" and "Riches." Dickens initially designed the story of Pip's life, *Great Expectations,* as a three-decker. Although he had to rush it into the pages of his weekly, *All the Year Round,* he continued to plan in monthly installments and to structure the whole as a three-stage narrative homologous to the three volumes in which the novel was issued at the conclusion of the serial run. The struggles of the three-decker amid the competing forms of fiction in the second half of the century and the eclipse of the serial by magazine novels, short stories, "tit-bits," and slices of life are not purely brought about by economic Darwinianism and Gresham's Law. The contents of literature, I am insisting, are inextricably bound up with the symbolism of their structures; and the homologies between life and story, home and book, and journey and narrative are in a process of disruption.

> London is the possibility of the unexpected inscription.
> Julian Wolfreys, *Writing London*

If urbanization disturbs the assumptions about identity, progress, and narrative that shaped the bourgeois text, then those who can read the city and reconstruct coherent stories are the new heroes.[38] And those are precisely the criminal, a bad plotter; the detective, a good plotter; and the reader, who shares attributes with both. Dickens required some three hundred thousand words to explore the incoherences of a city troped as a "Bleak House," whose central figure "don't know nothink" and whose key connecting links, Lady Dedlock and Esther, barely speak to each other. Four decades later, Arthur Conan Doyle, a doctor, reconceptualized the mysteries of the city as, on the one hand, a labyrinth through which the detective Theseus threads his way, and, on the other, a kind of diseased body amenable to physiological diagnosis. Exit from the maze and prescription for the ailing body politic can be performed in a few pages and a few hours of the doctor's (Doyle's, Holmes's, Watson's) and reader's time. Everything in Sherlock Holmes's London is legible; but one must be a master reader to articulate the narratives of calluses, cigar ashes, footprints, and silent hounds. (Doyle bet his wife a shilling that she could not guess the solution to his next story until she finished the chapter.)[39] Turning cities and their inhabitants into puzzles, like mapping journeys, restores coherence, guarantees endings, and reduces texts from volumes to pages.

Paradoxically, journeys that can be reliably mapped and repeated, such as those provided by London Transport, may be represented both as coherent and as indeterminate. One who composed both representations in a single story was a writer named Arthur Morrison, the son of an engine mechanic for the London Transport system. (In his case, the train did yield narrative, but only of kinds that were in their own way destructive of the world the train destroyed.) Morrison wrote a series of street sketches, the East End 1890s equivalent of Dickens's *Sketches by Boz.* But whereas Dickens's sketches produced modes of knowing the city, penetrating its byways, and going behind the exteriors of buildings and facades to narrativize interior spaces and states—even to the point of narrativizing the empty condemned cell in Newgate Prison—Morrison's stories of the streets refuse any coherent narrative of the higgledy-piggledy life lived therein.

In one of his most startling early achievements, Morrison tells of the "eleven-five" nighttime tramcar running from Stratford to Bow Bridge.[40] The story is all middle, with virtually no beginning or end, just as the Continental realist writers were, in their various ways, prescribing.[41] The story, in good realist fashion, tells a slice of life. But this tram ride does not go anywhere in particular—not from somebody's house to somebody else's house, not from square to square—and the famous locations, Stratford and Bow Bridge, are drained of any historicity; they are simply tram stops. The tram passes through a dim, humming, dark city, picking up and dropping off passengers after the pubs close at 11 P.M. Snatches of conversation are overheard, preludes to the voices in *The Waste Land:* "So I ses to 'er, I ses . . . I'm a respectable married woman, I ses. More'n you can say you barefaced hussy, I ses."[42] People get on and get off, and the final paragraph of the story leaves the tram lurching on its way. This streetcar, *not* named desire, has simply passed through the streets, and the story records its passing, without implying by structure, imagery, characters, events, setting, tone, or rhetorical tricks that there is any beginning or end; all is middle, muddle.

Or at least that is the way the lives of the characters are portrayed. As H. L. Mencken points out, Morrison's short stories collected under the title *Mean Streets* "got a kind of double fame, as a work of art and as social document."[43] The social document conforms to the aesthetics and structures of realism. The "work of art" inheres in the function of the narrator, who does "begin" the story by boarding the tram at the beginning of its journey, and who records his observations and sensations along the way, until only he and a quiet mechanic are left. Then the text annuls even their presence, moving to an almost observerless closure: "the tram-car, quiet and vacant, bumped on westward."[44]

It is possible to make short stories, even a book, out of slices of life, as Harold Biffin tries so effortlessly to do in writing a one-volume novel about a single day in the life of "Mr. Bailey, Grocer," in George Gissing's *New Grub Street,* a novel contemporaneous with Morrison's *Mean Streets*.[45] But the shortness of Morrison's stories and Biffin's unsuccessful novel do not imply that a pattern has been discovered and reimposed or that the chaos and secrets of London's *Bleak House* have been elaborated and exposed.

When Conan Doyle tired of Sherlock Holmes, wanting to return to more "respectable" and traditional historical fiction, he sent his detective on a final voyage, over many landscapes, to perish in an environment that expresses the antithesis of static, ordered, mapped, material coherence: the violent, unending, indiscriminate liquid of the Reichenbach Falls. Into that publishing breach rushed Morrison, hoping to capture Conan Doyle's audience. He invented a detective, "Martin Hewitt, Investigator," who—like Holmes, and unlike the utterly unconstructive narrator of *Mean Streets*—could make sense of the strangest events: in one case, the criminal turns out to be avian, not human, a mischievous thieving parrot.[46] Animal as criminal is a kind of logical devolution from tropes of criminal as animal and surely owes much to Edgar Allan Poe's "Murders in the Rue Morgue"; but Morrison's parrot might be more closely connected to Captain Flint, the original pirate who buried the treasure in *Treasure Island,* and subsequently to the name of the hundred-year-old parrot that sits on Long John Silver's shoulder and squawks the last words of Stevenson's commodified text: "Pieces of eight! Pieces of eight!" The world has truly turned unnarratable when the only ones who can draw maps and secure treasure are thieving birds whose speech and acts relate only to money.

Morrison's detective did not usurp Conan Doyle's territory in any sense of the word—space in the *Graphic* or in the public mind or in the imaginative space of London and Holmes. Doyle, by far the better artist at writing detective fiction, usurped his own ending, bringing Holmes back to life for further adventures. For representing the indeterminacy of life in Bethnal Green, Morrison had an undoubted gift; at inventing a readable city or a reader of the city he proved incompetent. So although the city enables the telling of a million stories, Morrison ceased writing fiction and instead took up collecting Oriental art. By World War I, he was the leading authority in the West on Japanese painting. Morrison found his buried treasure in the veiled and decentered graphic narratives of the East, patterns steeped in tradition and resonance that contrasted to the incoherences and disconnections of the tram rides through the East End streets.

> *The city represents one of those nodes or points of confluence where*
> *the status of textuality is today being articulated and tested.*
> Alexander Gelley, *City Texts*[47]

If this essay were to go on like a three-decker or a serialized narrative or a traveling house or a tram, it would speculate on the restoration of domestic architecture in Edwardian literature. In John Galsworthy's *Forsyte Saga,* the house at Robin Hill that Soames Forsyte commissions from Philip Bossiney, the architect who cuckolds him, is a structure charged with significations: a new style of architecture, declaring its difference from the stuffy West End Forsyte mansions, it also declares its difference from Victorian inheritances of all sorts—moral, behavioral, stylistic, and material.[48] Getting to, maintaining, and leaving home are major themes in E. M. Forster's novels; the Marabar Caves in *A Passage to India* might be thought of as the incomprehensible Other to Western civilization, a twentieth-century imperial analogue to Dickens's city. Voyages to, from, and between homes and spaces are essentially the structure and narrative of *Ulysses* and *Mrs. Dalloway* and *The Great Gatsby.* Much of T. S. Eliot's poetry consists of journeys, at first from ruined spaces and later to restored ones. For Hollywood in the 1930s, the house was a star: more than one home was, as Alfred Hitchcock said of Manderley when filming *Rebecca,* "one of the key characters."[49] And the essay might press on to recent fiction, touching on the tense and gendered relations between domestic interiors—sites of marriage plots and Egyptian bourgeois security—and streets—as deadly as any in *Treasure Island,* releasing nationalist and revolutionary forces—in Najib Mahfuz's great Cairo trilogy.

Notes

1. Franco Moretti, *Atlas of the European Novel, 1800–1900* (London: Verso, 1998), 5.
2. I have been influenced in many ways by Raymond Williams's *The Country and the City* (New York: Oxford University Press, 1973). Although I do not share his political conclusion, that overcoming the division of labor is the solution to society's ills, I agree that images of the city and the country, changing over time and in relationship to one another, mark alterations in consciousness, perception, and relationship (297).
3. Examples of architectural spaces that conceal hiding places include the library of Osbaldistone Hall in Scott's *Rob Roy,* Rookwood in Ainsworth's novel of that name, and the attic of Madame Beck's school in the Rue Fossette in Charlotte Brontë's *Villette.*
4. Jane Austen, *Mansfield Park,* ed. R. W. Chapman, Oxford Illustrated Jane Austen, 3d ed. (Oxford: Oxford University Press, 1934), bk. 1, chap. 2: "The rooms were too large for [Fanny Price] to move in with ease" (14–15); and ibid., bk. 2, chap. 4, p. 205.
5. Ibid., bk. 3, chap. 1, p. 312. Sir Thomas "knows" that Fanny cannot be heated in her bedroom; this insight may not only indicate his knowledge of the structure and incendiary dangers of Mansfield Park but also intimate his awakening to her sexual maturation. His insistence on "here," Fanny's chambers, as the place where she should have fire and warmth may also signify his desire to keep her from circulating too freely and warmly within the house. See John Wiltshire's extensive remarks about the functions of setting in the novel in his essay *"Mansfield Park, Emma, Persuasion,"* in *The Cambridge Companion to Jane Austen,* ed. Edward Copeland and Juliet McMaster (Cambridge: Cambridge University Press, 1997), 58–83: "As the figures move, disperse, and reassemble within the various venues Sotherton and Mansfield and Portsmouth offer them, one is made vividly conscious not only of the opportunities and inhibitions of these spaces, but of their being at issue—contested over, claimed, and owned" (65).
6. Charles Dickens, *Great Expectations,* ed. Edgar Rosenberg, Norton Critical Edition (New York: W. W. Norton, 1999).
7. Emily Brontë, *Wuthering Heights,* ed. William M. Sale Jr., Norton Critical Edition (New York: W. W. Norton, 1963), 47.
8. Sir Walter Scott, *Waverley; or, 'Tis Sixty Years Since,* ed. Claire Lamont, World's Classics (Oxford: Oxford University Press, 1986).
9. Jane Austen, *Emma,* ed. R. W. Chapman, Oxford Illustrated Jane Austen, 3d ed. (Oxford: Oxford University Press, 1933).
10. Mary Shelley, *Frankenstein; or, The Modern Prometheus,* ed. M. K. Joseph, Oxford English Novels (London: Oxford University Press, 1969).
11. Alfred, Lord Tennyson, *Poems,* ed. Christopher Ricks, Longmans' Annotated English Poets (London: Longmans, Green and Co., 1969).
12. Robert Browning, *Poems,* ed. John Pettigrew and Thomas J. Collins, 2 vols., Penguin English Poets (Harmondsworth, Eng.: Penguin Books, 1981).
13. I say "designedly middle-class" because the whole economic structure of commodity publishing functioned to sustain a middle-class and patriarchal environment of class labor and gendered leisure: see N. N. Feltes, *Modes of Production of Victorian Novels* (Chicago: University of Chicago Press, 1986); and Feltes, *Literary Capital and the Late Victorian Novel* (Madison: University of Wisconsin Press, 1993).
14. These did not all appear in three volumes; some (e.g., *Oliver Twist, Tess*) were serialized first. *Wuthering Heights* and *Agnes Grey* were combined together to make up three volumes by the publisher, Thomas Newby; and the two Trollope novels were issued in two volumes each.

15. In *The Presence of the Present: Topics of the Day in the Victorian Novel* (Columbus: Ohio State University Press, 1991), Richard D. Altick notes in his chapter "The Sense of Place" that "time stood still" "inside the great country houses" of Victorian fiction—one might add that it moves rather slowly in the great city mansions of Mr. Dombey and Mr. Osborne as well—whereas modes of transport were among "the most visible signs of change" in cities and, by later in the century, the country as well (339–381, 361, 339). And it is change above all, with its accompanying hurry and accelerating transportation, that characterizes urban life and fiction.

16. Charles Dickens, "Mr. Minns and His Cousin," in *Sketches by Boz,* new ed. (London: Chapman and Hall, 1839), 335–345, 338–339.

17. William Makepeace Thackeray, *Vanity Fair: A Novel without a Hero,* ed. John Sutherland, World's Classics (Oxford: Oxford University Press, 1983).

18. Charles Dickens, *Dombey and Son,* ed. Alan Horsman, Clarendon Dickens (Oxford: Clarendon Press, 1974), quotations from chap. 7, p. 85, and chap. 59, p. 786.

19. J. Mordaunt Crook, *The Rise of the* Nouveaux Riches*: Style and Status in Victorian and Edwardian Architecture* (London: John Murray, 1999). Similar points are made by Peter Thorold in *The London Rich: The Creation of a Great City, from 1666 to the Present* (London: Viking, 1999). For instance, Brooke House, Park Lane, designed by T. H. Wyatt for Sir Dudley Coutts Marjoribanks, later Lord Tweedsmuir, was built in 1870 and demolished in 1933; Rothchilds Row (141–148, Piccadilly) lasted barely one hundred years, from 1862 to 1961.

20. On Thackeray's peculiar style, see John A. Lester Jr., "Thackeray's Narrative Technique," *PMLA* 69 (1954): 372–409; and John Loofbourow, *Thackeray and the Form of Fiction* (Princeton, N.J.: Princeton University Press, 1964). Many more-recent studies examine closely the details of *Vanity Fair*'s verbal and visual structures.

21. See Robert L. Patten, "Autobiography into Autobiography: The Evolution of *David Copperfield,*" in *Approaches to Victorian Autobiography,* ed. George P. Landow (Athens: Ohio University Press, 1979), 269–291.

22. Deborah Epstein Nord, *Walking the Victorian Streets: Women, Representation, and the City* (Ithaca, N.Y.: Cornell University Press, 1995), 4.

23. Charles Dickens, *Bleak House* (London: Chapman and Hall, 1852–1853); Wilkie Collins, *The Woman in White,* ed. Julian Symons, Penguin English Library (Harmondsworth, Eng.: Penguin Books, 1974).

24. Charles Dickens, *Pickwick Papers,* ed. James Kinsley, Clarendon Dickens (Oxford: Clarendon Press, 1986), pt. 18, chap. 52, p. 808; Dickens, *Dombey and Son,* chap. 6, p. 66; chap. 15, pp. 217–218; chap. 15, p. 218; and chap. 20, p. 275.

25. For instance, Agatha Christie, *Murder on the Orient Express, The 4:50 from Paddington, The Blue Train;* Patricia Highsmith, *Strangers on a Train.*

26. J. Hillis Miller, *Charles Dickens: The World of His Novels* (Cambridge: Harvard University Press, 1959), 60. Sergei Eisenstein maintains that D. W. Griffith got his film technique of "montage" from Dickens's practice of intercutting between scenes; Eisenstein provides a close analysis of the filmic qualities of the description of Oliver's passage through Smithfield in his 1944 essay "Dickens, Griffith, and the Film Today," in *Film Form: Essays in Film Theory,* ed. and trans. Jay Leyda (New York: Harcourt, Brace and Co., 1949), 195–255. The *"montage exposition"* (217) that Eisenstein identifies in this scene (214–217) produces for Oliver and his readers "a stunning and bewildering scene, which quite confounded the senses." But for Griffith and Eisenstein the cinematic density—visual play of light, movement, panorama, close-up, and intercut—and the aural structure of individual sounds, cacophany, and crescendo from

rumble to roar provide an instance of what the cinema by the Second World War was aiming toward: a synthesis of unity and diversity yielding "*a unity of the whole screen image*" (255). Thus Dickens's verbal representation of Smithfield chaos is converted by early filmmakers into the visual and aural synthesis of a unified screen image.

27. Charles Dickens, *Oliver Twist,* ed. Kathleen Tillotson, Clarendon Dickens (Oxford: Clarendon Press, 1966), chap. 21, p. 136.

28. Miller speaks of London as being represented in this novel "at the deepest imaginative level," not as "a realistic description of the unsanitary London of the [eighteen] thirties" but as "the dream or poetic symbol of an infernal labyrinth, inhabited by the devil himself" (Miller, *Charles Dickens,* 58). For a description of the housing in these criminal rookeries and how they were so interconnected over the centuries and so cleverly designed for escapes that they seemed constructed expressly for the purpose of evading police searches, see Donald S. Thomas, *The Victorian Underworld* (London: John Murray, 1998). Moretti points out that for the most part the two halves of the novel and its geography cannot see one another (*Atlas of the European Novel,* 86).

29. Burton M. Wheeler, "The Text and Plan of *Oliver Twist,*" *Dickens Studies Annual* 12 (1983): 41–61.

30. Dickens, *Bleak House,* pt. 5, chap. 16, p. 158.

31. Moretti would add lawyers, at least in Charles Dickens's last completed novel, *Our Mutual Friend* (1864–1865): when Mortimer Lightwood and Eugene Wrayburn journey from the Veneerings' dinner in the West End to the other half of London, Docklands, Dickens makes a "pathbreaking discovery: once the two halves are joined, the result is *more* than the sum of its parts. London becomes not only a larger city (obviously enough) but a more *complex* one; allowing for richer, more unpredictable interactions" (*Atlas of the European Novel,* 86).

32. See Judith R. Walkowitz, *City of Dreadful Night: Narratives of Sexual Danger in Late-Victorian London* (Chicago: University of Chicago Press, 1992).

33. Robert Louis Stevenson, *Treasure Island,* ed. Wendy R. Katz, Centenary Edition of the Collected Works (Edinburgh, U.K.: University of Edinburgh Press, 1998); H. Rider Haggard, *King Solomon's Mines* (London: Cassell and Co., 1887). Haggard wrote his boy's adventure tale after seeing how successful Stevenson had been the year before in turning a map into a best-seller. Maps have accompanied literary journeys for centuries; more recently, the Library of Congress has produced maps to imaginary places, depicting ideas as much as locations: Oz as well as Mark Twain's "St. Petersburg," modeled on Hannibal, Missouri; see Martha Hopkins and Michael Buscher, *Language of the Land: The Library of Congress Book of Literary Maps* (Washington, D.C.: Library of Congress, 1999).

34. Quoted in Roger G. Swearingen, *The Prose Writings of Robert Louis Stevenson: A Guide* (Hamden, Conn.: Archon Books, 1980), from Stevenson's own account, "My First Book: *Treasure Island,*" *Tusatala* 3: xxiii–xxxi.

35. There are several good readings of the father-son dyad in this novel; one that both comprehends earlier interpretations and pushes beyond them to speculate about the voice of the parrot and the authorization of adulthood is Alan Sandison's chapter, "*Treasure Island:* The Parrot's Tale," in his *Robert Louis Stevenson and the Appearance of Modernism* (London: Macmillan, 1996), 48–81. Sandison points out that Jim's last glimpse of the island's central protuberance, the hill known as the "Spy-glass" first on Stevenson's map and then in the story generated by that map, "suggests that Jim's anxiety has everything to do with his psychosexual development and bespeaks a troubled awareness of a highly vulnerable masculine identity." The island becomes the symbol and site of Jim's fight for identity (72).

36. Stevenson, *Treasure Island,* 11.

37. For a complex generation of narrative from map, see my chapter on *The Tower of Lon-don,* an illustrated novel that Harrison Ainsworth and George Cruikshank produced as a twelve-part serial in 1840, in *George Cruikshank's Life, Times, and Art* (New Brunswick, N.J.: Rutgers University Press, 1992, 1996), 2:129–152.

38. See Julian Wolfreys, *Writing London: The Trace of the Urban Text from Blake to Dickens* (London: Macmillan Press; New York: St. Martin's Press, 1998). Wolfreys dis-cusses some of the ways authors have written London—not just written about it, but written it out, made it provisionally and continuously: "[T]he London being read here is that which *takes place* in the texts in question" (4).

39. Harry How, "A Day with Dr Conan Doyle," *Strand Magazine,* 4 (August 1892): 182–188, reprinted in *Sir Arthur Conan Doyle: Interviews and Recollections,* ed. Harold Orel (New York: St. Martin's Press, 1991), 62–68.

40. Arthur Morrison, "To Bow Bridge," in *Tales of Mean Streets,* preface by H. L. Mencken (New York: Modern Library, n.d.; preface dated "Baltimore, 1918"), 51–59.

41. See George J. Becker, ed., *Documents of Modern Literary Realism* (Princeton, N.J.: Princeton University Press, 1963).

42. Morrison, "To Bow Bridge," 55

43. Morrison, *Mean Streets,* x.

44. Morrison, "To Bow Bridge," 59.

45. George Gissing, *New Grub Street,* ed. Bernard Bergonzi, Penguin Classics (Harmonds-worth, Eng.: Penguin Books, 1968).

46. Arthur Morrison, "The Lenten Croft Robberies," in *Martin Hewitt Investigator* (New York: P. F. Collier and Son, n.d.).

47. Alexander Gelley, "City Texts: Representation, Semiology, Urbanism," in *Politics, Theory, and Contemporary Culture,* ed. Mark Poster (New York: Columbia University Press, 1993), 237–260. Gelley opens with a discussion of Jean-Luc Godard's *Deux ou trois choses que je sais d'elle,* a film and script (Paris: Seuil, 1971). Godard's film is in a sense the endpoint of Nord's meditation on the woman of the streets as prostitute: his protagonist, Juliette, and the city she perambulates are both known and unknown, the old semantic structures of communication having broken down: "To paraphrase a remark by Walter Benjamin that I shall return to later," Gelley comments, "Juliette 'is not a flaneur,' instead 'one may rather see in [her] what would become of the flaneur when the context in which [she] belonged would be taken away'" (240).

48. John Galsworthy, *The Man of Property,* vol. 1 of *The Forsyte Saga* (New York: Charles Scribner's Sons, 1926).

49. Gavin Lambert, "Origins of the *Sunset Boulevard* Mansion," *Architectural Digest,* April 1998, 70–86, quote on p. 70. J. Paul Getty's 1924 Italianate residence provided the exteriors for Billy Wilder's *Sunset Boulevard* (1950). Films of the 1940s featuring houses include *Rebecca* (1940), *The Magnificent Ambersons* (1942), *Gaslight* (1944), and *The Spiral Staircase* (1945).

✷

Orientations

Those "Gorgeous Incongruities"

POLITE POLITICS AND PUBLIC SPACE ON THE STREETS OF NINETEENTH-CENTURY NEW YORK

MONA DOMOSH

$\mathcal{E}$llen Olenska, the heroine of Edith Wharton's novel *The Age of Innocence,* is immediately marked as an outsider to New York society when she returns from Europe and strolls on Fifth Avenue with Julius Beaufort, a man of questionable virtues. As Mrs. Welland, a prominent social figure in the city, thinks to herself, "It's a mistake for Ellen to be seen, the very day after her arrival, parading up Fifth Avenue at the crowded hour with Julius Beaufort."[1] This first social faux pas defines her character irrevocably, since Ellen Olenska could have declared her impropriety no more extensively than by her transgression on Fifth Avenue, the most public thoroughfare in middle-class New York, on the day after her arrival in the city. Similarly, Lily Bart, the tragic heroine of Wharton's *The House of Mirth,* positions herself outside the bounds of decorum when she is seen on the wrong street in New York, at the wrong time of day, and is forced to lie about her destination, a lie that ultimately leads to her destruction.[2] Edith Wharton saw the streets of New York as a public stage where the intricate scripts of bourgeois behavior were played out each and every day. And on this public stage, the scripts were monitored closely.

Wharton's images of the streets of nineteenth-century New York seem to bear little resemblance to the images of streets created by recent scholarship on modernizing cities. Scholars lamenting the loss of public space in the postmodern city depict the streets of the nineteenth century as the preeminent sites of "democracy and pleasure."[3] Michael Sorkin, for example, speaks of the nineteenth-century city as a "more authentic urbanity," comprised of "streets and squares, courtyards and parks." He counterposes this "authentic urbanity" of the past with the cities, or "theme parks," of the present, places that have lost their traditional moorings in space and time (xv). In drawing these conclusions, Sorkin is pulling together different threads of recent cultural criticism and political theory that posit connections between the decline of the democratic, public sphere and the disappearance of public spaces.[4] Other urban scholars, such as Mike Davis and Edward Soja, suggest similar scenarios, particularly as they describe a Los Angeles that has lost any connection to real communities, and whose public spaces have become "militarized"—that is, fenced in and controlled by private interests.[5]

Yet analyses of behavior in the public spaces of nineteenth-century American cities suggest that these spaces too were often controlled by private interests and were not necessarily any more democratic in the sense of tolerating deviant behavior than are our postmodern "theme parks."[6] Edith Wharton's characters were not free in their behavior on the streets of New York; they were intensely guarded in their displays, aware all the time of how their public behavior communicated their identities. Wharton's, then, is a different sense of "public," where "public space" refers to places under public scrutiny, removed from the privacy of the domestic.[7] In these public spaces, a governing set of social norms controlled behavior, and therefore it is difficult to suggest that these spaces contributed to a completely democratic public sphere, where people were free to express themselves.

Through an analysis of three select images of street life in mid-nineteenth-century New York City, I provide case studies of how social norms were embodied in the everyday, public actions of people on the streets. I also suggest that those social controls were never completely hegemonic. I argue that socially controlled street spaces could serve as sites of political and social transgressions, but in ways different from those suggested by Sorkin and others. It is only by looking carefully at the often hidden codes of social performance that such slight transgressions can be made apparent. Our recent conceptual frameworks for analyzing the nature of public space seem to direct our attention elsewhere. By providing this analysis of the streets of a nineteenth-century city, I hope to show that the democratic potential of public spaces may still be possible, even in our contemporary "theme parks," if we direct careful attention toward slight, everyday transgressions, or what Patricia Mann calls acts of a "micro-politics."[8]

Conceptualizing Public Space

The ideas expressed by Sorkin and others about the nature of public space draw on what Don Mitchell argues are the two "predominant ways of seeing public space in contemporary cities," ways that correspond to a dichotomy set up by Henri Lefebvre.[9] On the one side, space serves as the arena for human action, and public spaces are the sites of human action unconstrained by political or social laws or mores. Public spaces are the places where people are free to voice opinions and to actively engage in behaviors that run counter to dominant societal norms. Mitchell counterposes this view of public space with that referred to by Lefebvre as "representations of space." In this view, space is purposely representational of certain societal ideals, and therefore the holders of these ideals attempt to control its use. Public space in this sense is a "controlled and orderly *retreat* where a properly behaved public might experience the spectacle of the city."[10] According to Mitchell, the publicness of spaces is constantly being negotiated through a dialectical relationship between these two visions: "Whatever the origins of any public space, its status as 'public' is created and maintained through the ongoing opposition of visions that have been held, on the one hand, by those who seek order and control and, on the other, by those who seek places for oppositional political activity and unmediated interaction" (115). This schema for thinking about the publicness of spaces is powerful and useful. It helps us understand the passionate fights over the uses and control of public parks, shopping malls, and urban plazas. It also helps to explain how scholars have interpreted the public spaces of nineteenth-century cities, particularly how they contrast what they believe were the unconstrained public spaces of the past with the controlled spaces of the present. Yet, as a schema, it outlines only the broad contours of the arguments about what constitutes urban public space. What I suggest here is that on close inspection, and in particular cases, these opposing visions as outlined by Mitchell do not provide a sufficient descriptive framework.

First of all, by creating a framework of opposing positions, theorists set up two extreme positions that cannot adequately describe the complexities of real life. Second, by setting these two views in opposition to each other, the schema serves to focus attention on the times and places where these views actually come into head-to-head contact—that is, when there are violent struggles between the people who represent these two views. In this way, the schema limits definitions of a counterpolitics to those highly noticeable events that arise when, for example, police face rioters in People's Park in Berkeley or when homeless people are forcibly removed from urban streets, where "the stakes are high and the struggles over them might very well be bloody" (127). This means that the schema does not help us understand situations in which transgressions are being made within, and in some instances because of, heavily surveyed and controlled public spaces.

Here I emphasize that there are "polite" political events that happen within, and because of, the highly surveyed and scripted public arena of everyday life in the streets. Through an analysis of specific minor events on the streets of nineteenth-century New York, I suggest that theorists adopt a more nuanced approach to thinking about public space, an approach that sees public space as it is interpreted and confronted daily, and not as an ongoing opposition.

The Streets of New York

New York City's population at the close of the Civil War was a little less than a million people, of whom 85 percent lived less than two miles from the city's population center, Union Square, where Broadway crosses Fourteenth Street. That density of population reflected the economic growth of a city that would become the capital of capitalism in its next quarter century and, of course, the relative lack of intraurban transportation systems. All movement of people, goods, and animals took place on streets designed, as the commissioner of public works said in the 1870s, to "impede rather than to facilitate travel."[11] On those streets all types of people could be found, although not in the same proportion or at the same time or in the same manner. New York in the 1860s was a city characterized by extremes in wealth and poverty, by ethnic and racial diversity, by economic elites competing for political power, and by an unstable social-class system. As public spaces, then, the streets provided not only transportation corridors but also sites for the displays of social class and political power.

The three images analyzed in this essay depict scenes on Broadway and Fifth Avenue. By the mid–nineteenth century, these two streets had become important icons for portraying the city and, with the addition of Wall Street, constituted the range of symbolic streetscapes that were usually highlighted in contemporary accounts.[12] Already by 1860, Wall Street had come to symbolize the economic power of the city. The commercial dominance that New York had attained over the country by the 1840s was translated into financial dominance on the eve of the Civil War.[13] And that control over the nation's capital was forcibly expressed in the tight clusters of banks, insurance offices, and financial traders that surrounded the exchange buildings on Wall Street, "the great financial centre of America."[14] In the small, often cramped offices of financial institutions along Wall Street, the business of America was conducted. The symbolism of Wall Street as the capital of finance was so powerful that as early as 1850 it was known simply as "the street"— home to the "favored and powerful individuals who [exerted] this immense control over society and the world."[15]

Broadway was the grand boulevard of display, extending the whole length of the island and, therefore, carrying along its edges an incredibly diverse array of people and activities. Below Wall Street, Broadway was home mainly to business offices, particularly shipping, and farther north were the offices of real estate agents, insurance companies, and bankers. Beyond City Hall, on Broadway at Chambers Street, were the beginnings of the retail district, centering in 1860 on Stewart's Store, just north of City Hall. This retail area, surrounded on side streets by wholesalers, extended almost to Fourteenth Street and Union Square, where businesses were taking over what had been a residential area. The built-up area of Broadway extended to about Twenty-third Street, at its intersection with Fifth Avenue. Along this path, and particularly on the stretch of businesses south of Fourteenth Street, throngs of New Yorkers passed daily. And those New Yorkers were a diverse lot: "Every class and shade of nationality and character is represented here. America, Europe, Asia, Africa, and even Oceania, has each its representatives here. High and low, rich and poor, pass along these side-walks. . . . Fine gentlemen in broadcloth, ladies in silks and jewels, and beggars in squalidness and rags, are mingled here in true Republican confusion. . . . From early morning till near midnight this scene goes on."[16] This account may not have been an exaggeration. In 1850, almost 60 percent of New Yorkers were foreign born. The largest group of the foreign-born were the Irish, who constituted about 30 percent of the city's population in the 1850s.[17] In distinction, the proportion of people with known African heritage was relatively low—1.6 percent of the total population in 1860.[18] Their population was centered in the lower West Side, particularly along the narrow streets of Greenwich Village.[19] Although there was a small "social aristocracy" within the black community, most African Americans were poor. According to a state census of 1855, the unemployment rate among blacks was almost 60 percent, and those who were employed worked largely in services, with domestic service as the primary occupation.[20] Historian Rhoda Freeman argues that the small numbers and relatively low social status of the pre–Civil War black community prevented it from wielding any form of political or economic power within the city (xx). Yet all these groups frequented the sidewalks and main thoroughfare of Broadway. It connected their tenements in the lower East Side and West Side with the factories, offices, warehouses, and homes of the wealthy farther north, where they worked.

Fifth Avenue was the center of the upper-class residential district of New York in the 1860s, and Wall Street was "constantly sending fresh 'stars' to blaze on Fifth Avenue."[21] The upper classes of the city had been on a northward march throughout the nineteenth century, seeking refuge from the expanding commercial areas below Fourteenth Street and, particularly, from the immigrant and working-class neighborhoods of the lower East and West Sides.[22] By the 1850s, after several prominent New Yorkers built brownstone mansions there, Fifth Avenue became the new fashionable area. It was lined with costly private residences, private clubs, and churches, the magnificence of which increased as one moved farther north. When Alexander Stewart built his mansion on the corner of Fifth Avenue and Thirty-fourth Street in 1864, he hastened the movement north to this newest of fashionable areas in the city. The street, with its displays of wealth, was the symbolic center of "society"—the space where people could exhibit their good taste, both in fashion and culture. It was the preeminent site of promenades, rivaling the retail areas of Broadway for ladies parading their new fashions:

> Nowhere else in America are there such fine opportunities for the display of dress as in New York. Where else in the broad world can there be found such a magnificent week-day promenade as Broadway, or such a Sunday morning strolling-place as Fifth Avenue? . . . The spacious sidewalks [of Fifth Avenue], bowered in the most luxurious of foliage, make it a tempting place to walk in the fashionable season, especially on a bright and sunny Sunday morning.[23]

As public spaces, therefore, these two streets were highly scripted arenas for social display.

Images of the Streets

All three of the images examined here are from the *New-York Illustrated News,* a weekly newspaper modeled after *Harper's Weekly.*[24] The first image appeared in January 1860, showing the crowds on Broadway at different times of day (Figure 1). The captions are telling: "At 7 a.m.—laborers, shop boys, and factory girls, begin the moving panorama of the day"; "At 9 a.m.—merchants and clerks hurrying to their place of business"; "From 12 to 3 p.m.—beauty and fashion on the promenade"; "At 6 p.m.—general rush for home"; "Midnight revelry in Broadway." In the first three images, an almost total segregation of classes is evident—including the working classes, the middle and aspiring middle classes, and the leisured class, particularly the bourgeois women who are allowed to parade in their fashions while window-shopping at the new dry-goods stores along Broadway.

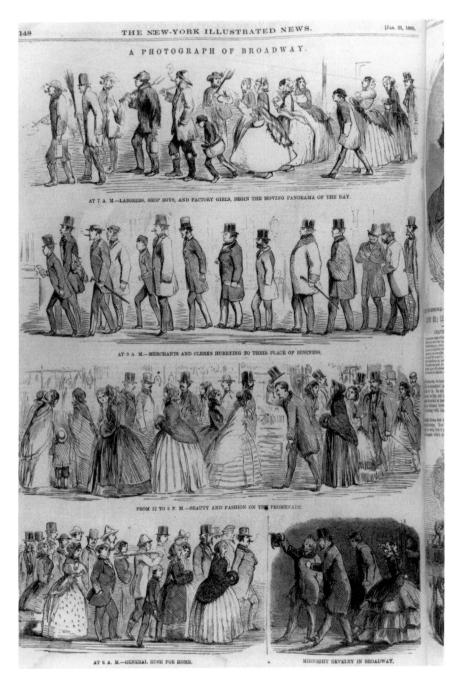

FIGURE 1. "A Photograph of Broadway." (New-York Illustrated News, *21 January 1860.* ©
Collection of the New-York Historical Society.)

Yet in the fourth and fifth images, both working and middle classes are commingled—merchants and laborers, factory girls and fashionable women, prostitutes and male consumers. Look carefully at the faces in the fourth image. The very depiction of certain facial features was, at least in the nineteenth century, enough to indicate to a general audience the particular class and type of a person. According to Mary Cowling, this systematic connecting of physical and mental attributes was part of a belief in physiognomy—the "science" of classifying people according to physical characteristic—that was widespread in nineteenth-century England and America.[25] Physiognomy formed an important part of nineteenth-century anthropology, borrowing methods of classification developed by natural historians. In this system, such features as a jaw and face that are large in proportion to the forehead and head (where the so-called higher faculties reside) indicated people of a lowly, possibly criminal sort. Other lowly signs were a convex chin, "a long, flat upper lip and coarse formless mouth" (297). We can see in this image, then, attempts to depict members of what was considered the lowest class in New York, most likely the new Irish immigrants, mingling in the crowds with members of several other classes. Indeed, all five of the images can be read as excursions into an anthropology of the modern city, depicting for those at home the various specimens of human life. According to Cowling, this was a fairly common form of imagining the nineteenth-century city, creating and then satisfying the curiosity of the middle classes about how their new industrial cities looked and functioned (297).

But, like Wharton's Lily Bart, fashionable women found on Broadway at the wrong time of day were in danger of losing their bourgeois status. Even more so, of course, if they were seen on the street at midnight, when different classes and sexes mixed in the revelry of "dark" Broadway. Most accounts of the lives of middle-class New York women in the 1860s indicate that it is indeed true that they rarely ventured out alone to walk after four in the afternoon.[26]

And yet, in this image there are fashionable, middle-class women to be found in the crowds of late afternoon. They were, in Tim Cresswell's terms, out of place.[27] At certain times, then, Broadway brought together different classes of people. And that diversity was frightening—it presented a challenge to nineteenth-century bourgeois life, in which each group was meant to inhabit its own place.[28] As historian David Scobey says of the promenade, it was "the bourgeois woman who figuratively condensed the class requirements and sexual risks of polite sociability. Like the proverbial canary in the coal mine, her presence marked what had to be protected in and from public exposure."[29] To see a bourgeois woman on Broadway beyond what he calls the "canonical" hours (11–3 P.M.) was a breach of "respectability" (215). But such breaches occurred often.

In fact, it was the very "publicness" of Broadway that allowed such behavior. Because it was so open to public scrutiny, any threat of potential evil behavior could be assuaged. Yet, under the surveillance of the bright lights of public scrutiny, immoral behavior was less likely. The bourgeois codes of the street could be violated, at times, if those violations occurred within the purview of the public.

FIGURE 2. "Our Best Society." *(*New-York Illustrated News, *31 January 1863. © Collection of the New-York Historical Society.)*

The second image, from 31 January 1863 (Figure 2), was accompanied by a caption and an explanation that followed in the text:

> Our best society—A scene on Fifth Avenue, the fashionable promenade on Sunday afternoons. From a sketch taken opposite the [blank] club, 15th Street and Fifth Avenue. Our city readers will not fail to recognize the faithfulness of the picture on page 196, having probably experienced the difficulty that attends a stroll through this fine avenue any pleasant Sunday afternoon.
>
> Our influential Colored Citizens have recently taken this magnificent promenade under their supervision, turning out on Sundays and holidays, with a degree of splendor and enthusiasm quite startling to a reflective mind. The gorgeous incongruities of costume, and the highly intellectual countenances (as seen in the illustration) which proudly sail by the humble white pedestrians, are enough to make a sorrowful man laugh.
>
> The air of satisfaction and nonchalance that characterizes our friends on these occasions, is irresistible. The way they ignore the privileges of their white brethren is not, however, so agreeable.
>
> The scene of our sketch lies in the vicinity of one of the fashionable club-houses. If the reader imagines, for a moment, that our artist has yielded to his satirical propensities, we beg that skeptic to make a pilgrimage through Fifth Avenue the next unclouded Sunday afternoon.

Certainly this image is in some senses a satire of proper, white society and its discomfort with those who are "in" yet "out of place" in their space. But I believe that it also tells us about a world in which such displacements were possible and probable. To situate this point, recall that Fifth Avenue was the most prestigious residential address in 1863, and that the area between Union and Madison Squares was home to several upper-class men's clubs and fashionable churches. The Sunday morning fashion promenade had become a standard activity for middle-class New Yorkers—after Sunday morning services at the Presbyterian or Episcopal church on Fifth Avenue, families paraded in their finest up and down the avenue. As a contemporary commentator noted: "There is the Sunday stroll, with pensive face and prayer-book in hand, on Fifth avenue. . . . The time will be immediately subsequent to morning service. The scene may be scarcely appropriate, following so soon upon the religious exercises that have preceded it, but it is very fascinating in its freaks of worldly frivolity. . . . all the extremes of the latest fashions mingle in one vast stream of wealth and luxury."[30]

This image certainly represents a freak of worldly frivolity. A relatively large group of African Americans are walking up Fifth Avenue on their Sunday promenade. At first glance, they seem appropriately middle class and fitting to the scene. The central figures form a traditional family unit, and they are dressed in what seem to be the latest fashions. Their grouping fits the norms of the Victorian family—the woman's arm is resting on her husband's, she is leaning toward him in a diminutive manner, and their daughter walks along the mother's side.

Yet middle-class and fashionable "Colored Citizens" upset and shocked their white counterparts. First of all, there were, of course, no African American churches on Fifth Avenue, and no blacks lived there. They have come to Fifth Avenue purposely and solely to promenade—to show off. According to Freeman, there was indeed a "social aristocracy" of blacks in New York at this time,[31] as concerned about their dances, parties, social visits, clothes, and promenades as their white "brethren." Her reading of African American newspapers suggests that "ladies of the Negro community were as concerned with fashion and elegant attire as were their white counterparts" (318). In terms of their ability to parade in their finery on New York's display avenues, members of this socially elite class were on a par with other upper-class residents of the city. As Scobey says, this was indeed the point of the promenade—to disengage from any sort of personal or concrete relationships in order to engage in a ritualized behavior whose raison d'être was the performance itself: "Not only private sentiments, but also social affiliations, material interests, indeed all concrete grounds of relationship were to be disengaged from the performance of respectability itself. As one expert put it, the 'passers in the street know no difference in individuals.'"[32] The family may be black, but they are completing the performance. Yet even the tightness of behavioral codes that governed the promenade cannot prevent some obvious disruptions, such as staring. Look particularly at the white women's faces. Clearly, a violation of sorts is taking place here. The most obvious violation is that the African Americans have taken over the sidewalk and are pushing their "white brethren" onto the street. Notice the positioning here, as a black woman walks ahead while the white couple bend to keep their balance on the sidewalk—obviously white "privileges" are being ignored. The literal space of the promenade, then, is being appropriated by African Americans.

There are other, less obvious violations here. The clothing of the African Americans is telling—most are depicted with clothes a bit over-the-top, some quite literally, with hats that are more ornate and unusual than those of their white counterparts. The couple in the center are certainly dressed in their finest and have outdone the whites. The woman's skirt is heavily flounced, her cape is edged in fur, her bonnet is topped with flowers, and she completes her outfit with a parasol, quite an unnecessary item in January in New York. The man's fully displayed white vest is topped with an elaborate cravat, and his hat is decorated with a wide band. He carries his walking stick out from his body, resting its end against his face. Both are wearing white gloves. Fancy clothes, white gloves, parasol, walking stick—all items completely dedicated to fashion, to leisure, without function on a cold winter's day. Scholars of African American culture have documented the importance of clothing to both slaves and freed blacks in distinguishing the "hours of work from the hours of leisure and, in the case of those still enslaved, the master's time from the slave's."[33] As signifiers of status, then, clothes were extremely important to freed blacks. And such status threatened the white order. To see African Americans in clothes clearly unsuited to work must have seemed particularly threatening to whites, who had difficulty fathoming a leisured black class.

But the "gorgeous incongruities" alluded to in the text get more to the heart of the issue. It is the juxtaposition and contrast between, on the one hand, black skin (with all the racist meanings this carried to white culture) and, on the other, top-of-the-line fashions and middle-class family structure that apparently shocked not only the viewers depicted here but also the lithographer of this scene. This juxtaposition represents an inversion of the "natural order," and when the world is thrown upside down in this way, dire consequences are sure to follow. Such dire consequences seemed just around the corner in January of 1863—this image appeared in print four weeks after the Emancipation Proclamation had been signed, abolishing slavery in America. Most "emancipation" images that appeared in New York newspapers carried a much less threatening message—one of freed blacks as good laborers for the American industrial powers, often depicted as laborers barely above animals in the evolutionary chain. But in this image, the fear of freed, leisured blacks marching north to New York (as they are here walking north up the avenue), promenading in white space, pushing whites out on the street, is given form and voice. Six months later, in July of 1863, New York City experienced the most violent civil disorder in nineteenth-century America, when protests against the conscription act for the Civil War turned into riots. The major targets were black laborers, many of whom were hung on the streets; and a large percentage of the rioters were Irish workers who felt threatened by black labor and feared the consequences of a large labor supply if freed slaves moved north.[34] That the draft riots, as they were called, differed from other instances of racial violence in New York City, suggests, in the words of historian Iver Bernstein, a "citywide campaign to erase the post-emancipation presence of the black community."[35] But on Fifth

FIGURE 3. "Club House, Fifth Avenue and Fifteenth Street, New York." *(New-York* Illustrated *News, 7 February 1863. © Collection of the New-York Historical Society.)*

Avenue, polite society stopped, stared, and, at least for the moment, allowed the parade to continue. To do otherwise would have upset the script far too much.

A week later, on 7 February, the image shown in Figure 3 appeared with the short caption "Club House, Fifth Avenue and Fifteenth Street, New York"—a scene just across the avenue from the previous one. The men of a social club are staring intently at the fashionable crowd passing in front of their plate-glass window. We could argue that the image illustrates the powers of the bourgeois male flaneur of the modern city, surveying the scene, choosing which delights he will indulge in.

An inversion of the expected order, however, is apparent in the text that accompanies the illustration:

> [T]he reader is now called upon to respectfully admire, at a distance, some of the approved types of "our best society" as they languidly lounge at the Club House window, ogling the pretty women, who, we are bound to say, do not always seem sufficiently impressed with the honor done them.
>
> Mesdames! take your revenge, and look them out of countenance—in our picture! Without a blush, Miss Crosspatch, without so much as a drop of your eyelash, behold Tittlebat Tittlmouse in his element! behold the elegant Adolphus; and the famous Fitz-Clarence, (as carefully gotten up as a venerable ballet-dancer), and, close behind him, observe Sir Loin Beef, the young English baronet, who carnt se, for the life of him, why we 'aven't such fine women in this blasted country as he has been in the 'abit of meeting at 'ome. Behold them all—those pretty hot-house plants, native

and exotic, as they faintly bud and bloom, and languish in their plate-glass conservatory.

The caption suggests how the image in some ways subverts social and spatial norms. First, several of the women seem in no way "impressed" by the men in the window; in other words, they are not participating in the expected rituals of the fashion parade. They are simply getting on with their business. Second, the men are inside the house; the women, outside on the street. This spatial reversal is echoed in the reversal of gender roles. As occupiers of the interiors, the men here are emasculated. Think of the words and images used to describe them: "languidly," "pretty," "ballet-dancer," "hot-house plants," "exotic," "bud and bloom"—these are undoubtedly feminine descriptors. These clubmen are decorative objects, as delicate as hothouse flowers in a conservatory, as useless and silly as titled English aristocrats. After all, real (bourgeois, American) men work.

Yet it is difficult to read the reversal of roles and spaces in the image as subversive in and of itself, for as a signifier of the feminine, the relative subordinate positioning of the domestic in society is simply reinforced—even when occupied by men, it serves a denigrating function. So the men are made fun of precisely because they occupy women's subject position. In this sense, the image does not suggest resistance to the status quo but supports traditional beliefs by using the idea of the feminine to denigrate a certain group of men. Yet smaller, tactical transgressions are also evident. Women in the image are both watched and watching, but so are the men. The caption is directed at the women readers of the newspaper, who are given the final authority as observers. They are being invited to view these men at a distance and to see them, with a long and accurate gaze, for what they really are—mere fops who would wilt in the cold February air. The women are conducting the important business of the streets while the men attend to decorative matters. Again, this presents an inversion of the natural order. In the public space of Fifth Avenue, women were, in some senses, in control of the business of life, as they ventured out daily to participate in the commercial city, paying bills, visiting stores, eating at the new restaurants set up for them. In fact, just a block south of the site of this image was Delmonico's, the most fashionable, although exclusively male, restaurant in town. When the first women's club was organized in New York in 1868, it held its meeting at Delmonico's, in direct and conscious challenge to the status quo.[36] The club's membership was limited to professional, middle-class women, but these women used their access to these new spaces of the city to renegotiate their identities. And the women viewers of this image, who are directly addressed in the caption, are invited to take "revenge," to stare back, to become the "looker"—in other words, to take advantage of what the modern city allows, to switch identities, however intermittently or ambivalently. So the image suggests the possibilities of transgressions at the same time that it supports existing power relationships. Some women could stare back and gain power. Although for some, their relegation to the domestic sphere meant that their power was annihilated, they nonetheless could, on the streets, at least for a brief time, "take revenge."

Conclusion

These last two images are particularly ironic because they consciously juxtapose bourgeois norms of behavior with depictions of actual behavior that runs counter to those norms. Whether that irony is apparent only to us or was intended by the artist is impossible to determine. But what we can say is that these images show that the public streets of nineteenth-century New York were not the "democratic" spaces of an authentic urbanity, nor were they completely manipulated and exclusionary. A more nuanced analysis suggests that the metaphor of theater might be more appropriate, but a theater where scripts could be manipulated. Even in the heart of middle-class space, on Broadway and Fifth Avenue, classes mingled, different "races" fashionably paraded, and gender roles could be reversed.

A polite politics was possible on these publicly guarded streets of 1860s New York, but not one immediately apparent from our historical record. Nor was it a politics that corresponded to notions of an "authentic urbanity" of democratic possibilities. It was a politics made possible by the conditions of social surveillance, not surveillance by the state or institutions using technological means, but one constituted of minute activities of seeing and being seen. Because bourgeois norms of behavior encouraged people to parade in their finery up Fifth Avenue, and because that space was heavily surveyed, African Americans too were allowed to engage in the promenade. Their behavior simultaneously disrupted and supported bourgeois standards. Because Broadway was the most public thoroughfare of the city, the mingling of different classes was tolerated when it took place in the "light" and out of the "shadows." And middle-class women could be as much the subject of the gaze as the object because social norms positioned some men as displays in the windows of Fifth Avenue, similar to the frocks and corsets seen in other plate-glass windows.

Edith Wharton's Lily Bart was eventually destroyed by her transgressions, not by violent struggles where the "stakes [were] high" but by the accumulative effects of a social system that tolerated certain "polite" forms of transgressive behavior and punished those who pushed the borders of politeness too far. Her fatal error, if we can call it that, was to refuse to marry the "appropriate" man (the wealthy and well-positioned Percy Gryce) and to seek instead her own individual fulfillment. Her "micro-politics," then, was transgressive to the established norms but was evident only in the smallest of ways, and only to those who understood the complex and contextual script of polite performance.

If we know how and where to look, it seems we will find similar "polite" politics being enacted every day in the "theme parks" that we now call our cities. Broadening our definitions of politics to include a micropolitics of complex and contextual agency should direct our attention to the "tactics" that many of us, who cannot afford the emotional and spatial distance required of an oppositional politics, embody in our everyday transgressions.

Notes

This paper has benefited greatly from the comments and suggestions of colleagues, particularly Stephen Daniels, Richard Dennis, Peter Goheen, J. P. Jones, Don Mitchell,

Joan Schwartz, and several anonymous referees. I am grateful to Nick Fyfe for organizing a session on "the street" at the Glasgow Institute of British Geographers meeting, where I presented a paper that contains the kernel of the ideas expressed here, and to the participants' comments at seminars in geography departments at Queens University in Canada, Royal Holloway College of the University of London, the University of Vermont, and Middlebury College. The research for the paper was funded by the National Science Foundation, Anthropological and Geographic Sciences, #9422051.

1. Edith Wharton, *The Age of Innocence* (New York: Charles Scribner's Sons, 1968), 32.

2. Edith Wharton, *The House of Mirth* (New York: Bantam Books, 1984).

3. Michael Sorkin, ed., *Variations on a Theme Park: The New American City and the Decline of Public Space* (New York: Hill and Wang, 1992), xv.

4. Sociologist Bruce Robbins aligns this scenario of decline with a "myth of general decline" of the role of the academy, and he argues that, like all myths, it is a "defense of a very particular group—in this case, perhaps, white, male, native-born intellectuals who once had something of a monopoly of American 'public' discourse but since the 60s, when the universities in fact became more 'public' by letting some new people in, no longer do" (Robbins, "Intellectuals in Decline?" *Social Text* 25/26 [1990]: 258).

5. See Mike Davis, *City of Quartz: Excavating the Future in Los Angeles* (New York: Verso, 1991); and Edward Soja, *Postmodern Geographies: The Reassertion of Space in Critical Social Theory* (London: Verso, 1989).

6. See Susan G. Davis, *Parades and Power: Street Theater in Nineteenth-Century Philadelphia* (Berkeley: University of California Press, 1986); Mona Domosh, *Invented Cities: The Creation of Landscape in Nineteenth-Century New York and Boston* (New Haven, Conn.: Yale University Press, 1996); and Elaine S. Abelson, *When Ladies Go A-Thieving: Middle-Class Shoplifters in the Victorian Department Store* (New York: Oxford University Press, 1989).

7. For a very interesting and useful assessment of the importance of private space as a political possibility, see Judith Squires, "Private Lives, Secluded Places: Privacy as Political Possibility," *Environment and Planning D: Society and Space* 12, no. 4 (1994): 387–402.

8. Patricia S. Mann, *Micro-Politics: Agency in a Postfeminist Era* (Minneapolis: University of Minnesota Press, 1994).

9. Here I am using not Henri Lefebvre's work (*The Production of Space* [Cambridge, Mass.: Blackwell Press, 1991]) but rather Don Mitchell's "take" on it ("The End of Public Space? People's Park, Definitions of the Public, and Democracy," *Annals of the Association of American Geographers* 85, no. 1 [1995]: 115). Lefebvre's analysis of public space exceeds the bounds presented by Mitchell.

10. Mitchell, "End of Public Space?" 115.

11. Quoted in Seymour J. Mandelbaum, *Boss Tweed's New York* (New York: John Wiley & Sons, 1965), 12.

12. See Edward Spann, *The New Metropolis: New York City, 1840–1857* (New York: Columbia University Press, 1981); Domosh, *Invented Cities;* and Roy Rosenzweig and Elizabeth Blackmar, *The Park and the People* (New York: Henry Holt and Co., 1992).

13. See David Hammack, *Power and Society: Greater New York at the Turn of the Century* (New York: Columbia University Press, 1987).

14. Edward W. Martin, *The Secrets of the Great City* (Philadelphia: Jones Brothers & Co., 1868), 141.

15. George G. Foster, *New York by Gas-Light and Other Urban Sketches,* ed. and with

an introduction by Stuart M. Blumin (Berkeley: University of California Press, 1990), 224.

16. Martin, *Secrets,* 46.

17. Spann, *New Metropolis,* 24.

18. Seth Scheiner, *Negro Mecca: A History of the Negro in New York City, 1865–1920* (New York: New York University Press, 1965), 6.

19. Iver Bernstein, *The New York City Draft Riots: Their Significance for American Society and Politics in the Age of the Civil War* (New York: Oxford University Press, 1990), 287.

20. Rhoda Golden Freeman, *The Free Negro in New York City in the Era before the Civil War* (New York: Garland Publishing, 1994).

21. Martin, *Secrets,* 80.

22. See Kenneth Scherzer, *The Unbounded Community: Neighborhood Life and Social Structure in New York City, 1830–1875* (Chapel Hill, N.C.: Duke University Press, 1992); and Charles Lockwood, *Manhattan Moves Uptown* (New York: Barnes and Noble, 1976).

23. George Ellington [pseud.], *The Women of New York; or, The Under-World of the Great City* (New York: New York Book Co., 1869), 34–35.

24. The *New-York Illustrated News,* which tried unsuccessfully to compete with *Harper's Weekly,* was published only for four years. It would seem from an analysis of its articles and images that the newspaper was Republican in leaning. See Frank Luther Mott, *A History of American Magazines* (Cambridge, Mass.: Harvard University Press, 1957).

    My method is informed here both by Robert Darnton's incredibly rich book *The Great Cat Massacre and Other Episodes in French Cultural History* (New York: Basic Books, 1984), in which Darnton explored singular depictions of what seemed to him as "odd" behavior in seventeenth-century France in all their complexities in order to gain insight into the social history of the past, and by Mary Poovey's book *Uneven Developments: The Ideological Work of Gender in Mid-Victorian England* (Chicago: University of Chicago Press, 1988), in which she claims the richest of insights come from exploring "border cases"—those that just do not quite fit into what we expect of the past—and following the path where it takes us. The three images discussed in the text struck me as the most interesting I had seen in several weeks of examining mid-nineteenth-century illustrated newspapers, and I knew that there were insights to be gained by examining them closely.

25. Mary Cowling, *The Artist as Anthropologist* (Cambridge: Cambridge University Press, 1989).

26. This generalization is based on my reading of the diaries of six New York women (written between 1854 and 1898) and from secondary accounts in Abelson's *When Ladies Go;* Christine M. Boyer's *Manhattan Manners* (New York: Rizzoli Press, 1986); and Carroll Smith-Rosenberg's *Disorderly Conduct: Visions of Gender in Victorian America* (New York: Oxford University Press, 1985).

27. Tim Cresswell, *In Place/Out of Place: Geography, Ideology, and Transgression* (Minneapolis, Minn.: University of Minnesota Press, 1996).

28. For analyses of the class dimensions of street culture in New York City, see Boyer, *Manhattan Manners;* Scherzer, *Unbounded Community;* and Rosenzweig and Blackmar, *Park and the People.*

29. David Scobey, "Anatomy of the Promenade: The Politics of Bourgeois Sociability in Nineteenth-Century New York," *Social History* 17 (1992): 214–215.

30. Ellington, *Women of New York,* 35.

31. Freeman, *Free Negro*, 317.

32. Scobey, "Anatomy of the Promenade," 217.

33. Shane White, *Somewhat More Independent: The End of Slavery in New York City, 1770–1810* (Athens: University of Georgia Press, 1991), 195.

34. Noel Ignatiev situates New York's draft riots within the context of how the Irish "took up arms for the White Republic" and became "white," whether they were fighting in the army, or in the streets of New York. Their desire to become "white" was rooted in the very American context of a racism born of miserable economic conditions (Noel Ignatiev, *How the Irish Became White* [New York: Routledge, 1995], 89).

35. Bernstein, *New York City Draft Riots,* 5.

36. Karen Blair, *The Clubwoman as Feminist: True Womanhood Redefined, 1868–1914* (New York: Holmes & Meier Publishers, 1980).

Dickensian Dislocations

TRAUMA, MEMORY, AND RAILWAY DISASTER

JILL L. MATUS

Mid-Victorian Concepts of Psychic Shock

"Between the outer and the inner ring, between our conscious and our unconscious existence," wrote E. S. Dallas in *The Gay Science,* "there is a free and a constant but unobserved traffic forever carried on. Trains of thought are continually passing to and fro, from the light into the dark, and back from the dark into the light." Mental physiologist William Carpenter also relies on images of trains and railways in his discussion of memory and unconscious association: "[O]ur ideas are . . . linked in 'trains' or 'series' which further inosculate with each other like the branch lines of a railway." And as Dickens has Mr. Toodle in *Dombey and Son* sagely remark, "What a Junction a man's thoughts is . . . to be sure!"[1] By mid-century the railway was far more than a common means of travel; it had radically altered perceptions of space, communication, and connection. We should not therefore be surprised that railway tracks, networks, trains of thought, and lines of communication should come to the aid of those explaining the activity of invisible modes of thought and, indeed, influence the very way in which the mind's operations could be visualized.

As a technology of traffic and speed that sometimes went horribly wrong, the railway was also the locus of Victorian anxieties relating to large-scale, public disaster. Around the mid–1860s, the concept of psychic trauma began to percolate in Victorian Britain, and what brought it to consciousness was, arguably, the railway. To place the railway more squarely within the history of trauma studies, we may say that the railway accident was to Victorian psychology what World War I and shell shock were to Freudian. The railway accident was the exemplary instance that propelled the prevailing pathological bias in relation to injury in the direction of a psychic interpretation of injury.

Industrialization and the rise of the insurance company were the twin economic factors in the development of medical interest in psychic injury. As Henri Ellenberger notes in his magisterial history of the discovery of the unconscious,

"The development of industry and the multiplication of industrial accidents on the one side, and the development of insurance companies on the other" meant that "more and more 'official medicine' was on the search for new theories and new therapeutic methods for these neuroses."[2] Similarly, Wolfgang Schivelbusch's 1977 study of the railway journey, which laid the tracks for all future studies in this line, points out that in England by 1864, railroad companies had become legally liable for their passengers' safety and health;[3] because only "pathologically demonstrable damage qualified victims for compensation," those victims who suffered damages "without a demonstrable cause" created a legal and medical problem whose solution depended on the medical profession (135).[4] Through a focus on railway shock, Schivelbusch's work goes a long way to illuminate Victorian thinking about the unconscious. Although I am indebted to his excavation of the medical literature in this regard, and to his discussion of pre-Freudian traumatic neurosis, my emphasis, shaped by trauma theory of the last decade, is more specifically on the dysfunctions of memory in the registration of trauma.[5] This essay, to pursue the spatial metaphors introduced at its outset, explores how the concept of the mind's well-regulated traffic, the relations of "conscious to unconscious cerebration," could accommodate the event of a traumatic experience. Theories about the relations of conscious and unconscious thought and memory and the traumatic effects of railway accidents intersect in my focus on Dickens's experience of a disastrous train accident near Staplehurst in 1865 and on the story, "The Signalman," that he wrote in its aftermath.[6]

In the 1860s the "phenomenon of accident shock," the traumatization of a victim without discernible physical injury, became the object of systematic investigation by the medical profession.[7] Thomas Buzzard, for example, a doctor whose series of articles appeared in the *Lancet* in 1867 was very interested in cases in which external injuries were negligible but effects on the nervous system were severe. In one case, he noted, the shock changed the very national constitution of an individual "from that of the most thorough Englishman in all his habits to the manner of the most coxcombical Frenchman."[8] Herbert Page, whose work of the 1880s and 1890s is influential, was interested primarily in fright or shock, but he paid attention largely to its effects on the nervous system—hysterical fits, spasms, vomiting, pulse rate, and so on. And though he noted shock's effect on memory, he merely says that shock affected energy and concentration rather than recall of events and incidents of past life.[9] He recorded that patients suffering from traumatic hysteria sometimes had a "great dread of impending evil."[10] They usually slept badly and were constantly troubled by distressing dreams: "Depend upon it that the man who can sleep naturally and well after a railway collision has not met with any serious shock to his nervous system" (158). He noted too the element of delay or belatedness that would become so important in the Freudian conceptualization of trauma: "Warded off in the first place by the excitement of the scene, the shock is gathering, in the very delay itself, new force from the fact that the sources of alarm are continuous, and for the time all prevalent in the patient's mind" (148). The emphasis on a "continuous" and "prevalent" source of alarm suggests the possession of the patient by the shocking event. William James explained delay by means of the following example in his 1894 review of the French psychologist Pierre Janet's work: "The fixed ideas may slumber until some weakening of the nervous system favors their morbid activity. E.g., Col. is victim of a railroad accident, and passes six months in the hospital with a grave abdominal injury . . . [six years later] if the old scar be touched, [he suffers] an hysterical attack . . . consisting in hallucinations of the railroad tragedy."[11] This brief history may serve to contextualize and historicize Freud's references to railway accidents in *Beyond the Pleasure Principle*. When Freud wrote in 1919 about the causes of the traumatic neurosis, he specifically mentioned the two kinds of accidents that the increasingly industrialized world of the previous century had made possible: "A condition has long been known and described which occurs after severe mechanical concussions, railway disasters and other accidents involving a risk to life."[12] Having acknowledged the lengthy history of traumatic neurosis, Freud then proceeds to offer the recent war as the defining moment for diagnosis of psychic shock. On the one hand, Freud indicates a familiarity with the phenomenon of railway trauma; on the other, he seems not to acknowledge the medical studies that had already, for some decades, focused on the absence of gross mechanical force:

> The terrible war which has just ended gave rise to a great number of illnesses of this kind, but it at least put an end to the temptation to attribute the cause of the disorder to organic lesions of the nervous system brought about by mechanical force. . . . In the case of the war neuroses the fact that the same symptoms sometimes came about without the intervention of any gross mechanical force seemed at once enlightening and bewildering. (10)

Freud's study of the dreams of shell-shocked soldiers of the 1914–1918 war provided him with an important insight into the nature of dreams. He noticed that the dreams of the traumatized were markedly different from those of ordinary dreamers in that the dreams woke up the patient "in another fright"; they returned him to the scene of horror, reproducing it repeatedly and literally, whereas ordinary dream work consisted of creating scenarios to express fears and desires. Dreaming allowed ordinary patients to release anxieties and so keep sleeping; traumatic dreams woke the patient and were therefore unable to appease anxiety. This insight in relation to traumatized soldiers allowed Freud to theorize what he had remarked on in a less obvious way in his earlier work on traumatic neurosis. The hallmark of trauma, Freud decided, was the inability to possess memory, to make the event the subject of narrative. The memory seemed to possess the sufferer rather than the other way round. Hence Cathy Caruth's rearticulation of Freud: "[T]o be traumatized is to be possessed by an image or event."[13] It has been suggested that trauma involves the collapse of witnessing, as Dori Laub has put it, in that one can witness the event only at the cost of witnessing oneself. Caruth explains it thus: "Central to the very immediacy of this experience . . . is a gap that carries the force of the event and does so precisely at the expense of simple knowledge and memory. The force of this experience would appear to arise precisely . . . in the collapse of understanding." Trauma is then the experience in which knowledge and cognition are startlingly disjoined. Geoffrey Hartman describes this as the missed encounter, the event "registered rather than experienced," in that "the traumatic event bypasses perception and consciousness, and falls directly into the psyche."[14] The knowledge stored by the traumatized subject is inaccessible to ordinary memory but signals its presence in the form of intrusive return. It is as if the encounter, having been missed, demands recognition through reenactment rather than recall. Both Caruth and Hartman, one may notice, have recourse to spatial images here in delineating the effect of trauma. Caruth's "gap" and Hartman's "bypass" and "fall" resonate with Victorian metaphors of the traffic between conscious and unconscious thinking (and indeed the language of railway accident itself) and underscore the spatial implications of thinking about trauma: as much as trauma has been conceived of as a "disease of time," it is also a disordering of routing and internal space. It is the return that does not recognize itself as return.

Memory, "Unconscious Cerebration," and the Effects of Shock

From the first, Freud's work, unlike that of his Victorian predecessors, emphasized shock's effects on memory. In the review I mentioned earlier, William James writes also of the studies of two "distinguished Viennese neurologists" for whom hysteria "starts always with a shock, and is a 'disease of the memory.'"[15] Although the medical treatises on railway shock and injury move toward a focus on psychic rather than mechanical injury, very little is said about the traumatic shock's effect on memory. Physiologists and psychologists writing about memory are also, however, little interested in the effect of shock, though physical blows to the head prove perennially engaging.[16]

The cornerstone of thinking about memory at this time is that everything that happens to us is recorded—all memory is latent and recuperable. One of the most dramatic and frequently cited instances of this notion is Samuel Taylor Coleridge's account in *Biographia Literaria* of a young woman in Germany who could neither read nor write but, as a result of a fever, began to speak in Latin, Hebrew, and Greek. It was discovered that, as a child, the woman had been looked after by a pastor who knew these languages and used to recite passages from the Latin and Greek fathers and from rabbinical texts.[17] According to Theodule Ribot, well known for his work in France on physiological psychology, the "normal state" of memory is merely a less intense version of the "superintensity" occasioned by an extreme situation, such as the one Coleridge described.[18] The "normal state" of memory is a happy relationship of storage and retrieval, an archive in good working condition. If memory was thought of as a storehouse of previous thoughts, or a kleptomaniac's secret hoard, the assumption that unusual conditions could suddenly bring to light the farther reaches of such stores is consonant and logical. In *Principles of Mental Physiology,* William Benjamin Carpenter, probably the most authoritative Victorian mental physiologist, sets out the prevailing view of the latency or dormancy of all memory:

> It is now very generally accepted by Psychologists as (to say the least) a
> predictable doctrine, that any Idea which has once passed through the Mind

may be thus reproduced, at however long an interval, through the instru-
mentality of suggestive action; the recurrence of any other state of consciousness with which that idea was originally linked by Association, being
adequate to awaken it also from its dormant or "latent" condition, and to
bring it within the "sphere of consciousness."[19]

It is because association or linkage was the key notion in conceptions of memory
that the image of the railway or the train of thought was so apt. Carpenter continues:

> And as our ideas are thus linked in "trains" or "series" which further inos-
> culate with each other like the branch lines of a railway or the ramifica-
> tions of an artery, so, it is considered, an idea which has been "hidden in
> the obscure recesses of the mind" for years—perhaps for a lifetime,—and
> which seems to have completely faded out of the *conscious* Memory...
> may be reproduced as by the touching of a spring, through a nexus of sug-
> gestion, which we can sometimes trace-out continuously, but of which it
> does not seem necessary that all the intermediate steps should fall within
> our cognizance. (429–430)

The railway network metaphor presupposes an essentially spatial and linear map
of the brain. Ideas are linked and retrievable, Carpenter seems to say, even if the
train that takes us from one to another may be such a fast one as to make its travel
seem instantaneous.

The idea of memory as a place, an archive that houses stored and recuperable
past thoughts, was challenged as early as 1862 by Frances Power Cobbe's emphasis on the fallacies of memory. What we remember, she argued, are layered recon-
structions of memories, where each "fresh trace varies a little from the trace
beneath, sometimes magnifying and beautifying it, through the natural bias of the
soul to grandeur and beauty, sometimes distorting it through passion or prejudice;
in all and every case the original mark is ere long essentially changed."[20] Here
the spatial metaphor of the train of ideas dissolves, for to keep it would mean envisioning a surreal world in which each successive visiting of a place changes it.
This train, as it were, can never take you to the same station twice.[21]

The nineteenth-century fascination with altered states of mind, and with the
relation between conscious and "unconscious cerebration," is evident in the complex history of controversy surrounding concepts such as mesmerism, Braidism,
spiritualism, and somnambulism. In part, this fascination with altered states raised
the question of what it means to be human and conscious. What kinds of activities are performed without conscious and voluntary supervision? What force controls mental function in states of mind that are not fully conscious? As John
Abercrombie wrote in the 1830s, there are states of the mind in which ideas and
images follow each other, states over which we have no conscious control: dreams,
somnambulism and double consciousness, insanity and spectral illusion.[22] Mental
functioning in altered states also raised the question of what we know but do not
ordinarily know that we know. The century saw an array of explanations of brain
function and physiology ranging from, on the one hand, small concessions to the
reflex function of the brain to, on the other, theories of the brain as two separate,

rigidly divided hemispheres, an idea, put forward in the 1840s by Arthur Wigan, that offered a striking model for mental dissociation.[23]

Mesmerism was a powerful if controversial influence in England from the 1830s to the 1860s. Proponents of mesmerism claimed a great deal on its behalf: as a manifestation of the mind's power, it was evidence of the power to transmit thoughts from mesmerizer to mesmerized, and in the form of "mental travelling" or clairvoyance, it was able to surmount obstacles of time and space.[24] Like other new technologies—steam and electricity—that were also revolutionizing experiences of time and space, mesmerism was hailed by its adherents in England as progressive and frontier pushing. Alison Winter observes that "this generation, surrounded by astonishing changes wrought by science, set few limits on the powers that might be revealed in electricity, light, magnetism, and gases. . . . The claim that an imponderable fluid could pass from one individual to another, altering the processes of thought, was astonishing, but just as worthy of serious evaluation as other great scientific assertions."[25] Just as the railway was thought of as having "annihilated time and space," so altered states of consciousness played havoc with these less-than-stable coordinates. What Freud came clearly to see, which Victorian mental physiologists did not, was that mental shock or trauma was itself an altered state in which the relations of knowledge, memory, and experience were peculiarly distorted.

Dickensian Railway Trauma

In 1865, Dickens, actress Ellen Ternan, and her mother narrowly escaped death when the train on which they were traveling from Folkestone to London jumped the gap in the line occasioned by some repair work on a viaduct near Staplehurst, Kent. The foreman on the job miscalculated the time of the train's arrival; the flagman was only 550 yards from the works and unable to give adequate warning of the train's approach. The central and rear carriages fell off the bridge, plunging to the river bed below. Only one of the first-class carriages escaped that plunge, coupled fast to the second-class carriage in front: "It had come off the rail and was hanging over the bridge at an angle, so that all three of them were tilted down into a corner."[26] Dickens managed to get them all out of the carriage and then behaved with remarkable self-possession, climbing down into the ravine and ministering for almost three hours to the many who lay injured and dying. With further aplomb, he climbed back into the "swaying carriage and retrieved his manuscript," an account of which brave feat is offered in the memorable postscript to *Our Mutual Friend*. Once back in London, however, he began to develop typical symptoms of what we would call today posttraumatic stress disorder. He was greatly shaken and lost his voice for nearly two weeks: "I most unaccountably brought someone else's out of that terrible scene," he said. He suffered repeatedly from what he called "the shake," and, when he later traveled by train, he was gripped by a persistent illusion that the carriage was down on the left side. Even a year later, he noted that he had sudden vague rushes of terror.[27] At such times, his son and daughter reported, he was unaware of the presence of others and seemed to be in a kind of trance. His son Henry recalled that he got into a state of panic at the slightest jolt; his daughter Mamie opined that his nerves were never really the same again; he would "fall into a paroxysm of fear, tremble all over, clutch the arms of the railway carriage, large beads of perspiration standing on his face and suffer agonies of terror. . . . Sometimes the agony was so great, he had to get out at the nearest station and walk home."[28] An uncanny repetition also characterizes his death, falling as it did five years later on the anniversary of the accident. While these facts have sometimes been acknowledged in criticism of his short story "The Signalman," that acknowledgment is usually by way of a closing gesture to the grim and eerie irony that he died on the anniversary of the accident. I want to argue, however, for a closer link between Dickens's experience of acci-dent trauma and this ghost story, which was published in *All the Year Round* the following year, appearing as part of "Mugby Junction," the special Christmas is-sue of 1866. Through its insistence on uncanny and obsessive return, "The Sig-nalman" apprehends the heart of traumatic experience, not as it was understood in medical literature of the 1860s but, presciently, as it would come to be formulated in the next century. What "The Signalman" signals is the future discourse of trauma.

Some critics have seen in the story a critique of industrialization in Dickens's representation of the alienated labor of the signalman and of the stress his job en-tails.[29] A similar line of interpretation has been to see the signalman himself as a pathological case. In a recent article, Graeme Tytler diagnoses the signalman as suffering from monomania—a clinical condition in which the patient is obsessed by one dominating idea.[30] A man with a one-track mind, the signalman is undeni-ably fixated. But he could equally well be diagnosed as suffering from Aber-crombie's spectral illusion or Wigan's split self, outlined above. Rather than pathologizing the signalman as a "case of partial insanity," or substituting an al-ternate diagnosis stemming from stress in the workplace, or suggesting merely that he displays the symptoms of trauma, I want to look at the narrative itself as shaped by and expressive of the logic of trauma.

The genre of the ghost story and that of trauma narrative have much in com-mon, though, strictly speaking, "trauma narrative" is something of a contradic-tion in terms, for once a trauma has been made the subject of narrative, according to theorists such as Judith Herman, it is no longer traumatic. To be traumatized is arguably to be haunted, to be living a ghost story: to be traumatized is "to be pos-sessed by an image or event."[31] In some measure, then, it is tautological to say that Dickens's story of uncanny possession is a story of trauma, but not all ghost stories are expressive primarily of trauma—*A Christmas Carol,* for example—even though in Dickens they frequently objectify states of mind. "The Signalman" is a story confronting the potential of the railway to dislocate external and internal space; it is about repetition, fixation, hallucination, about powerlessness, interrup-tion, heightened vigilance, and a sense of impending doom, about uncanny re-enactment, terror at the relived intrusion—all legitimate aspects of a tale of horror, all characteristics of trauma. In ghost stories, as in trauma, the sanctity of ordered time is violated as the past intrudes on the present. Dickens's story is intimately concerned with disruptions in chronological and logical time, with belatedness and delayed reactions; it further insists on the gap between knowledge and cognition.

Trauma, too, uncouples the event and the experience of it and in so doing derails a sense of self-possession. Like trauma theory, this story pushes us to ponder the gap between experience and knowledge, signing and meaning, the shocking external occurrence and its internal assimilation and representation. The story, one could argue, is Dickens's way of pondering that fateful and fatal gap in the tracks at Staplehurst, a creative way of articulating his personal experience of railway shock beyond the hermeneutic available in the discourse of the 1860s.

A signalman is hailed one evening by the narrator, who wishes to descend to the box and talk to him. The signalman tells the narrator, whom he initially takes to be a "spectre," of an apparition or ghost who has appeared to him on the line near his signalbox on a number of occasions, once before a terrible collision, another time before the death of a young lady on the train. The signalman imagines that the apparition's reappearance precedes a further tragedy. It turns out to be his own death. On one level, the story enacts a revenge against signalmen for the Staplehurst disaster, though, strictly speaking, it was not the signalman who blundered. The foreman on the job miscalculated the time of the train's arrival; the flagman was too close to give adequate warning of the train's approach. In contrast, Dickens's story focuses obsessively on the signalman's anguish at receiving a warning in time but finding it impossible to heed—because he does not know exactly about what he is being warned. Whereas in the Staplehurst disaster, the signalman was too close to the repair work to warn the train in time, in the ghost story the signalman is too close to the train and does not or cannot heed the warning as the train bears down on him.

Freud postulated two possibilities in relation to traumatic reenactment, both of which seem applicable here: (1) the attempt to master the stimulus retrospectively through repetition and (2) the demonic content of reenactment as evidence of the "death instinct." If "The Signalman" is Dickens's way of trying to master the stimulus retrospectively—his own creative attempt to "clear the way"—the signalman himself may be seen as exemplifying the death drive, not heeding the whistle and literally allowing death to overtake him as the train comes upon him from behind and cuts him down.

At another level, "The Signalman" explores the repetitive cycle of trauma. In Dickens's story the trauma accumulates. Not only is the signalman compelled to witness a terrible train disaster, but he is tantalized, through the specter's visitations, by an impossible clairvoyance. The trauma is compounded as in two instances he has been forewarned but has been unable to avert death and disaster. After the first terrible accident on the line, the signalman thinks he has recovered from witnessing the carnage of the accident: "Six or seven months passed, and I had recovered from the surprise and shock." At that point, the specter appears to him again and the next calamity occurs: "I heard terrible screams and cries. A beautiful young lady had died instantaneously in one of the compartments and was brought in here, and laid down on this floor between us" (*S*, 532). Now the specter has appeared again, signaling to him of some further calamity about to occur on the line. The signalman laments "this cruel haunting of him" (*S*, 533). Haunted not simply by the past but by a past that seems to project itself into the future, the signalman is subjected to relentless repetition and can avail himself of neither hindsight nor foresight.

One of the much discussed characteristics of trauma is that the events of the past continually obtrude on the present in the form of flashbacks and hallucinations. Trauma is signaled by the return that does not recognize itself as return. Like the train disaster that is literally a disruption of linearity, the narrative of "The Signalman" disrupts linear chronology.[32] In part, this sense of disturbed linearity or chronology arises from the clairvoyant specter, whose gestures enact and predict each of three train disasters before they occur. This sense arises also from the fact that the narrator seems to be taking part in something that has already happened.[33] That is, the narrative is itself part of some uncanny repetition. The fact that the narrator uses the words "For God's sake, clear the way," themselves repeated many times in the course of the story, could suggest that he has just repeated his part in the replay of a past he "knows" but does not know he knows. From the first, the narrator seems inexplicably drawn to approach the signalman, all the odder because at the outset the narrator says he is not someone given to starting up conversations. Understandably, the signalman imagines that the narrator is himself a further spectral illusion, especially since the narrator hails him with the exact words that the specter has already used. After a time the signalman seems reassured that the rational, skeptical narrator is not a ghost and confides his story to him. By persistently dismissing as "imagination" what the signalman says he has seen, by construing recurrence as coincidence, by remaining stubbornly unbelieving, the narrator refuses to witness the signalman's hallucination or spectral illusion—refuses to witness the trauma. But it is arguably inscribed upon him nonetheless, and he is now (as narrator) participating in the repetition by telling the story of it. When the narrator arrives at the tracks for the third time, he is struck with a "nameless horror" because he sees the "appearance of a man" in the tunnel and clearly thinks he is seeing a ghost (*S,* 535). The horror that oppresses him passes when he sees that the figure is a real man. Horror gives way to fear that something is wrong. He then learns of the signalman's death. All would appear to be resolved for the rational narrator, except for the fact that the very words the engine driver uses are the ones that the narrator uses mentally. Despite the matter-of-factness of the coda, the narrator too will clearly be haunted by the words "For God's sake, clear the way."

If the specter can be seen as an articulation of the signalman's traumatized consciousness, the narrator shares characteristics of the signalman that suggest he is not just an uninvolved interlocutor, auditor, or reporter. The signalman thinks initially that the narrator is a specter; the narrator has a "monstrous thought" that the signalman is a spirit. Each finds himself in a position that makes him feel compelled to act and assume responsibility for the general safety of those on the line. When the apparition that the signalman has seen returns a third time, the signalman is beside himself to interpret the warning and forestall the disaster. But he cannot. Similarly the narrator feels himself compelled to act: "But what ran most in my thoughts was the consideration how ought I to act, having become the recipient of this disclosure" (*S,* 534). The narrator is worried less about the uninterpretable spectral warnings than about the mental stability of the signalman and his job performance under present stress. The narrator resolves to try and compose him as much as possible and to return the next morning to accompany him on a visit to the "wisest medical practitioner . . . and to take his opinion" (*S,* 535). He is also too late. The specter that the signalman sees appears on three occasions; the narrator descends to the signalman's box three times; the words the narrator uses are the words that the ghost has used and the train driver will use; the gesture that the signalman describes is given words by the narrator. (Significantly, he does not speak these words—"For God's sake, clear the way"—before the engine driver tells him that those are in fact the words he used.) The narrator, the signalman, the specter, and the engine driver are all bound together in a series of overlapping occurrences and verbalizations, in a history that seems to have begun before the narration begins and will continue after it ends. The reader, too, participates in the process by making these associations and puzzling over the relationship of the uncanny elements in the story.

Dickens was engaged throughout his literary career in representations of the railroad, and he used them to various effect, often combining the "humorous and the horrific."[34] As the editor of popular journals, he often featured articles on the railway. Of particular interest is his own essay "Need Railway Travellers Be Smashed?" which appeared in *Household Words* on 29 November 1851, fourteen years before the Staplehurst crash, and which argued vigorously for reforms in railroad safety mechanisms. His own experience of railway disaster was for Dickens a moment of junction and disjunction. In its aftermath he wrote creatively and intuitively about trauma, pushing further than the developing discourse of psychic shock in the 1860s to apprehend the relations between conscious and unconscious memory in unusual circumstances. No stranger to traumatic experience before the railway accident—as his continual, fictive reenactments of abandonment and childhood abuse attest—he was perhaps brought through the accident to a sharper intimation of the nature of trauma than he had ever had before. He lost his voice—a powerful symptom of the inability to speak the unspeakable—to find it later in this story of ghostly clairvoyance and hindsight, a story through which he articulates the characteristics of trauma that have barely begun to be formulated in the psychological and medical discourse of his own time.

Notes

A previous version of this essay has been published as "Trauma, Memory, and Railway Disaster: The Dickensian Connection," *Victorian Studies* 43 (2001): 413–436. I am grateful for permission to reuse the material here. I would like to thank Marjorie Stone, who first turned my attention toward "The Signalman," and Garry Leonard for his reading of an early draft of this paper, which was presented at the Northeast Victorian Society of America conference on Victorian Memory (April 1999) at Yale University.

1. E. S. Dallas, *The Gay Science* (London: Chapman & Hall, 1866), 1:207–208; William Benjamin Carpenter, *Principles of Mental Physiology,* 4th ed. (1876; New York: Appleton, 1890), 429–430; Charles Dickens, *Dombey and Son,* ed. Alan Horsman, World's Classics (Oxford: Oxford University Press, 1982), 449.

2. Henri Ellenberger, *The Discovery of the Unconscious: The History and Evolution of Dynamic Psychiatry* (New York: Basic Books, 1970), 245.

3. Wolfgang Schivelbusch, *The Railway Journey: The Industrialization of Time and Space in the Nineteenth Century* (Berkeley: University of California Press, 1977), 134.

4. Schivelbusch notes that, even after the turn of the century, legal handbooks denied recognition to psychic shock. He cites John Mayne's treatise on damage (1899), which opined that "damages resulting from mere sudden terror or mental shock, could not be considered the natural consequence of the negligence complained of" (*Railway Journey*, 146).

5. See also Ralph Harrington's recent work on railway spine and the traumatic neurosis: "The Railway Accident: Trains, Trauma, and Technological Crisis in Nineteenth-Century Britain," <http//www.york.ac.inst/irs/irshome/papers/rlyacc.htm>, accessed 22 June 1999.

6. Charles Dickens, "The Signalman," in *Christmas Stories,* ed. E. G. Dalziel, The New Oxford Illustrated Dickens (Oxford: Oxford University Press, 1956). Hereafter page references will be given in the text with the abbreviation *S.*

7. See John Eric Erichsen's influential *On Railway and Other Injuries of the Nervous System,* which appeared in 1866 (London: Walton); Thomas Buzzard's series of articles ("On Cases of Injury from Railway Accidents," *Lancet* 1), which appeared in 1867; and John Charles Hall's *Medical Evidence in Railway Accidents,* which appeared in 1868 (London: Longmans).

8. Buzzard, "On Cases of Injury from Railway Accidents," 624. Schivelbusch charts the movement by which accident shock became the focus of systematic study. He draws on testimony from Dickens himself after the Staplehurst crash and that of other nineteenth-century passengers who escaped unscathed but for the fright (*Railway Journey*).

9. Herbert Page, *Railway Injuries, with Special Reference to Those of the Back and Nervous System in Their Medico-Legal and Clinical Aspects* (London: Griffin, 1891), 44.

10. Herbert Page, *Injuries of the Spine and Spinal Cord* (London: Churchill, 1883), 153.

11. William James, review of the essays "L'état mental des hystériques" and "L'amnésie continue," by Pierre Janet, *Psychological Review* 1 (1894): 195.

12. Sigmund Freud, *Beyond the Pleasure Principle,* in *The Standard Edition of the Complete Psychological Works of Sigmund Freud,* ed. James Strachey (London: Hogarth Press, 1955), 18:10.

13. Cathy Caruth, ed., *Trauma: Explorations in Memory* (Baltimore: Johns Hopkins University Press, 1995), 5

14. Dori Laub, "An Event without a Witness: Truth, Testimony, and Survival," in *Testimony: Crises of Witnessing in Literature, Psychoanalysis, and History,* by Shoshana Felman and Dori Laub (New York: Routledge, 1992), 75–92; Caruth, *Trauma,* 7; Geoffrey H. Hartman, "On Traumatic Knowledge and Literary Studies," *New Literary History* 26 (1996): 537.

15. James, review of "L'état mental" and "L'amnésie," 199.

16. See, for example, Carpenter, *Principles,* 443–444.

17. Samuel Taylor Coleridge, *The Collected Words of Samuel Taylor Coleridge,* ed. Kathleen Coburn (London: Routledge and Kegan Paul, 1983), 7:112–113.

18. Theodule Ribot, *Diseases of Memory: An Essay in the Positive Psychology,* trans. William Huntingdon Smith (New York: Appleton and Co., 1887), 176.

19. Carpenter, *Principles, 429.*

20. Frances Cobbe, "The Fallacies of Memory," in *Hours of Work and Play,* anthologized in *Embodied Selves: An Anthology of Psychological Texts, 1830–1890,* ed. Sally Shuttleworth and Jenny Bourne Taylor (Oxford: Oxford University Press, 1998), 151.

21. Acknowledging Cobbe's views on the fallacies of memory, Carpenter too challenges the doctrine of memory's indelibility, suggesting that the doctrine has been too generally applied: it is "questionable whether *everything* that passes through our Minds thus leaves its impression on that material instrument" (*Principles,* 454).

22. John Abercrombie, *Inquiries concerning the Intellectual Powers and the Investigation of Truth* (New York: Collins, 1852), 198–266. For a more detailed discussion of Abercrombie's four states, see Jenny Bourne Taylor, *In the Secret Theatre of Home: Wilkie Collins, Sensation Narrative, and Nineteenth-Century Psychology* (London: Routledge, 1988), 55.

23. For further discussion of fin de siècle fascination with split selves such as Jekyll and Hyde, see Jenny Bourne Taylor, "'Obscure Recesses': Locating the Victorian Unconscious," in *Writing and Victorianism,* ed. J. B. Bullen (London: Longman, 1997), 149.

24. See Alison Winter, *Mesmerized: Powers of Mind in Victorian Britain* (Chicago: University of Chicago Press, 1998); and Taylor, "'Obscure Recesses,'" 150.

25. Winter, *Mesmerized,* 35. As Fred Kaplan has shown, Dickens was involved with mesmerism over a long period and in a number of different ways. He was not only a close friend for many years of Dr. John Elliotson, the great pioneer of mesmerism in England, and witness to a large number of displays of animal magnetism, but also a practicing mesmerist himself (see Fred Kaplan, *Dickens and Mesmerism: The Hidden Springs of Fiction* [Princeton, N.J.: Princeton University Press, 1975]).

26. Peter Ackroyd, *Dickens* (London: Minerva, 1991), 1013.

27. Ibid., 1017; my account draws also on Edgar Johnson's version of this accident and its aftermath in *Charles Dickens: His Tragedy and His Triumph* (London: Gollancz, 1953), 2:1021.

28. Quoted in Kaplan, *Dickens and Mesmerism,* 461.

29. See John Stahl, "The Sources and Significance of the Revenant in Dickens's 'The Signalman,'" *Dickens Studies Newsletter* 11 (1980): 98–101.

30. Graeme Tytler, "Charles Dickens's 'The Signalman': A Case of Partial Insanity," *History of Psychiatry* 8 (1997): 421–432.

31. See Judith Herman, *Trauma and Recovery* (New York: Basic Books, 1992), 175–195; and Caruth, *Trauma,* 5.

32. The identity of the narrator has been variously construed by critics. David Seed, for example ("Mystery in Everyday Things: Charles Dickens' 'Signalman,'" *Criticism* 23 [1981]: 42–57), sees him as one of the Barbox brothers, who narrated the earlier parts of "Mugby Junction"; David Greenman argues that the narrator should be taken as an independent ("Dickens' Ultimate Achievements in the Ghost Story: 'To Be Taken with a Grain of Salt' and 'The Signalman,'" *Dickensian* 85, no. 1 [spring 1989]: 47).

33. Most memorable perhaps is the personification of the engine in Dickens's *Dombey and Son* as a bloodthirsty monster, Death itself. There the railway as a predatory fiend that licks up the tracks and whatever falls in its path not only is identified with the villain Carker, himself predatory and catlike, but also is his nemesis. Extending Robin Atthill's analysis ("Dickens and the Railway," *English* 13 [1961]: 130–135), David Seed suggests that "from the very first the railway was for Dickens associated with violence and mystery" ("Mystery," 47). Further, Seed argues, Dickens used the idea of the railway as an allegory of life. The traveler who appears in the first chapter of "Mugby Junction" is one who has decided to get off the train of his present existence and stay a while in Mugby to see something else of life. Similarly, Seed argues, Dickens used ghosts in his stories as "projections or reflections of mental states" (47), their function being to illustrate states of mind rather than to terrify readers.

34. Atthill, "Dickens and the Railway," 134.

Poetics on the Line

THE EFFECT OF MASS TRANSPORT IN URBAN CULTURE

ANA VADILLO

Mind the Gap

That autumn day in 1893 was brisk and sunny. Windows were open, and what air was possible in London or in Bloomsbury was in the room. Christina [Rossetti] insisted upon sitting up to receive me, although obviously it was a painful effort, and when I protested she lay down again with a sigh of relief. I told her that people wanted to know something about her health. She resigned herself to being interviewed, if I would, without any appearance of resignation. "Ask me whatever you like," she said. It seemed cruel to trouble her, lying there obviously a very sick woman, even though the news was asked not for the general public, but for the public which loved her religious poetry. . . . I spent half an hour with her in quiet conversation, mainly about my own personal affairs. She must always have lived the life of a hermit. She did not even know her London. "Ealing?" she said vaguely. "Where is Ealing? Hammersmith Way?" She knew very little beyond Bloomsbury. People who do not know shudder at the idea of a conventual life. Well, every convent I have ever known was light and bright, exquisitely clean, full of flowers and glimpses of greenery. Christina, I suppose, had originally settled down in Bloomsbury to be near her brothers. But quite close to London were gardens and fields, even at Ealing, not so many stations away from Gower Street by Torrington Square. It must have been at some time or other a mortification of the natural impulses of the poet to be shut up in Torrington Square. Duty to the old might have been done just as well in a little house with a garden, but Christina was probably one of those who find mortification sweet. I asked her that day how she endured the London life. "I have missed the flowers," she said.

Katharine Tynan, *Middle Years*

In 1893 the young Irish poet Katharine Tynan (1859–1931) went to interview Christina Rossetti (1830–1894).[1] But because she was so ill, Tynan decided to talk to her off the record. It was truly one of the most fascinating cultural encounters of the fin de siècle, because the two poets suddenly became conscious of the aesthetic and generational differences between them. And it was their different attitudes and approaches to urban life and urban space that marked the gap. Tynan was fascinated by the city. She loved London, its shops, and its ever moving crowds. In fact, because she was early for her appointment with Rossetti, Tynan passed the time looking at the shops in Tottenham Court Road. She could not resist the temptation, she claimed.[2] By contrast, she found Rossetti rather unfamiliar with, and absent from, the metropolitan experience. In Tynan's own words, "She did not even know her London." Rossetti lived enclosed in her Torrington Square house, willingly disconnected from urban life despite residing in London's city center. Tynan, in contrast, embraced urban life. She was streetwise and both acted and created the city on her walks, on her journeys in the underground, in omnibuses, and in trams.

Fin de siècle women poets were intensely aware both of London's mushrooming growth and of how mass transport was transforming the way in which urban space was understood and lived. By the 1890s remote areas such as Ealing and Hammersmith Way were reached by the underground. Urban distances were now measured not in miles but by railway stations, the new cathedrals of the modern age, as the *Building News* put it.[3] In this sense, mass transport not only facilitated the assimilation and control of London's new geography and growing expansion, but also crucially opened up new spaces in the city. Tynan's interview is a clear illustration of this geographical and, to some degree, psychological change. When Tynan asked Rossetti how she endured her London life, she was surprised by Rossetti's answer: she had missed the flowers that year. Tynan was right to point out the futility of missing flowers—or anything else—in London, because the metropolis could offer to urban dwellers whatever they might need or desire—provided, of course, that they could pay for it. Tynan's response suggests that, for Tynan, knowing the city was an act of aesthetical positioning, and Rossetti's ignorance of London's spatiality was to be understood as a symbolic act by which Rossetti partitioned herself from urban life. If Rossetti "missed the flowers," she could have traveled to greener areas of London. Hammersmith Way, for example, was only ten stations away, and Ealing six.[4] Flowers and gardens, although rare in the polluted city center, could be seen and enjoyed no matter where one lived in London. Urban transport made it possible.

Moreover, the development of structures for mass transportation, especially the underground, had a profound impact upon the constitution of urban social life and forms of cultural identity. Mass-transportation facilities transformed the communication networks in the metropolis and made human connections easier, more fluent and fluid, as Tynan's interview illustrates. Rosalind Williams has argued that "[a]s the new material foundations of industrial and urban life were being laid, so were new social foundations," for technological networks are also human ones.[5] The technological transformation of the metropolis caused a shift in human relations, now deeply dependent upon mass-transportation facilities. Determining how technological networks became human ones is the purpose of the present study, insofar as the appearance of the London underground helped to create a network of fin de siècle women poets affected by the cultural changes that the underground produced. Elizabeth Grosz, in her fascinating study on the presence of bodies in the city, argues thus: "What I am suggesting is a model of the relations between bodies and cities that sees them, not as megalithic total entities, but as assemblages or collections of parts, capable of crossing the thresholds between substances to form linkages, machines, provisional and often temporary sub- or micro-groupings."[6] Grosz is arguing against two very important traditions within urban studies. The first tradition sees the city as a product of the human body: "Humans make cities" (105), and therefore they are active agents in the construction and production of the city. The second tradition understands the city as a mirror, as an isomorphic re-creation of the human body (the city as body politic). Grosz, rather, proposes what she calls an *interface* study, one that would investigate how the body and the city effect and affect each other, producing and transforming each other in their continuous association. This is the method of study of the present work. Although Grosz refers in particular to the "body," my interests are centered on how a particular body of people, a group of women poets in fin de siècle London, created and regulated the city and, vice versa, how the city influenced their work in return. I will first briefly discuss the appearance of mass transport in London and the formation of a group of poets who used urban transport as a device with which to explore urban life and culture at the fin de siècle. I will then discuss in the last section of this chapter Alice Meynell's collection of essays titled *London Impressions*. Here I will suggest that Meynell, as a passenger, captioned the picture of this fin de siècle city by focusing on the ephemerality of urban life. I will also propose that her prose reflected the experience of traveling. Her essays could be compared to a journey in an underground train, an omnibus, or a tram, as her words transport the reader to 1890s London.

Getting Connected: London's Iron Net

> *[I]t is a pain to walk in the midst of all these hurrying and clattering machines; the multitude of humanity, that "bath" into which Baudelaire loved to plunge, is scarcely discernible, it is secondary to the machines; it is only in a machine that you can escape the machines.*
> Arthur Symons, *London: A Book of Aspects*

> *The railways run; their foreshortened sweeps and reaches look like the swinging and swaying of resolute motion. The town would shoulder them, but they evade and slip through, slender and keen, with a stroke of their flying heels.*
> Alice Meynell, *London Impressions*

Until 1863 the only transport system working in London was that of the omnibus, which had been introduced in 1829 by George Shillibeer.[7] Before Shillibeer, only stagecoaches joined London and the surrounding towns. But they were expensive and ran infrequently because most of those who traveled daily to London had their own coaches. Shillibeer, who had imported the idea from Paris, revolutionized urban transport completely. He started by establishing a regular route between Paddington and the Bank of England. The service had no regular stops, and customers picked up the omnibus en route. The cost of a ticket from Paddington to the bank was a shilling, and from Angel, in Islington, to the bank, sixpence. But the new service benefited only the middle classes because the working classes could not afford to travel daily at those prices. As Christopher Hibbert notes, the working classes still walked wherever they had to go, and many would even walk four miles to get to work and four miles back.[8]

The success of the omnibus was, however, formidable, prompting the appearance of numerous omnibus lines in the city. In 1834, there were at least 620 licensed horse omnibuses operating in London, and by 1851, the year of the Great Exhibition, there were at least 150 lines. This expansion coincided with an extraordinary population boom (from one and a half million in the 1830s to five million in the 1880s). The overpopulation worsened the conditions of the metropolis. Since the 1840s, congestion had been one of the biggest problems in London. The swelling traffic, the overcrowded city center, and the worsening conditions of the slums emphasized the necessity for an improved transport system. Charles Pearson, the founder of the Metropolitan Line (the first underground line), argued that London needed a new form of transport that would be affordable for the poorer classes and would allow them to move out of the slums where they lived. He proposed an underground railway line that would join all the mainline railway stations. In 1863, the Metropolitan Line was opened to the public. It was an immediate success, serving more than thirty thousand people on its first day.[9]

The line joined Paddington, in the west, with Farringdon Street, which was the entrance to the East End, via Edgware Road, Baker Street, Portland Road, Gower Street Station, and King's Cross. The Metropolitan Railway's contract with the government ensured that cheap tickets (two-penny return) were issued in the early hours of the day. The middle classes would travel later in the day at a greater cost, and the trains were usually divided into first-, second-, and third-class carriages. The success of the Metropolitan Railway prompted the construction of another underground railway that would join the north and the south banks of the metropolis. The new line, the Inner Circle Line (completed in 1884), surrounded London's city center and joined the north, south, east, and west suburbs of London.

The appearance of the underground shook the foundations of nineteenth-century society, opening new realms of experience by digging up a subterranean world, which until then had only been part of the cultural imagination of the century. As Rosalind Williams argues, subterranean life was now used to question life at surface level.[10] A poem such as A. Mary F. Robinson's "Neurasthenia" shows how this phenomenon entered the imagination of Londoners. As Williams notes, "The undermining of familiar surface patterns with unfamiliar subterranean ones, was vivid evidence that old ways of life were being undermined," and the experience of excavation "gave the form to a revolutionary rupture with past forms of experience, of social order, of human relations."[11] In "Neurasthenia" Robinson describes two levels of life: an external life, visible to everyone, and an underground, subsurface life, hidden from sight but capable of analyzing and, crucially, of challenging life at surface level.

Neurasthenia

I watch the happier people of the house
    Come in and out, and talk, and go their ways;
I sit and gaze at them; I cannot rouse
    My heavy mind to share their busy days.

I watch them glide, like skaters on a stream,
    Across the brilliant surface of the world.
But I am underneath: they do not dream
    How deep below the eddying flood is whirl'd.

They cannot come to me, nor I to them;
    But, if a mightier arm could reach and save,
Should I forget the tide I had to stem?
    Should I, like these, ignore the abysmal wave?

Yes! in the radiant air how could I know
How black it is, how fast it is, below?[12]

Robinson establishes an interesting connection between the unconscious and the underground. This innovative thought reveals the duality and complexities of urban life. For her, the inner world of the unconscious and the underground world (which she associates, significantly, with a "faster" life) share one important role: they are the shadow cast on consciousness and on external life. And they have the power to transfigure and disrupt urban life.

If the underground discovered a new realm of modern life at a subterranean level, at the surface level it drastically transformed London's geography. In the first place it changed the city's urban patterns. As Hibbert argues, "Communities developed around the new outlying railway stations, at first no more than rows of villas, perhaps, but soon small towns with a life—if only a temporary life—of their own."[13] In effect, as Figure 1 shows, communities of women poets developed around the new railway stations.[14] Next to Gower Street Station, for example, lived one of the largest communities of women poets: Christina Rossetti, Amy Levy, A. Mary F. Robinson, Mabel Robinson, the essayist Vernon Lee, Charlotte Mew, and the American poet Louise C. Moulton all moved to the station's neighboring streets. Situated at the entrance of bohemian Bloomsbury, the station was next to University College London (where A. Mary F. Robinson and her sister, the novelist Mabel Robinson, went to college). More important, it was also next to the British Museum, whose reading room, as Judith Walkowitz argues, "became the stomping ground of the 'bohemian set.'"[15] Amy Levy, for instance, used the British Museum as the location for her poem "To Lallie (Outside the British Museum)" and her short story "The Recent Telepathic Occurrence at the British Museum."[16] But at the same time, living near Gower Street Station guaranteed to these poets easy

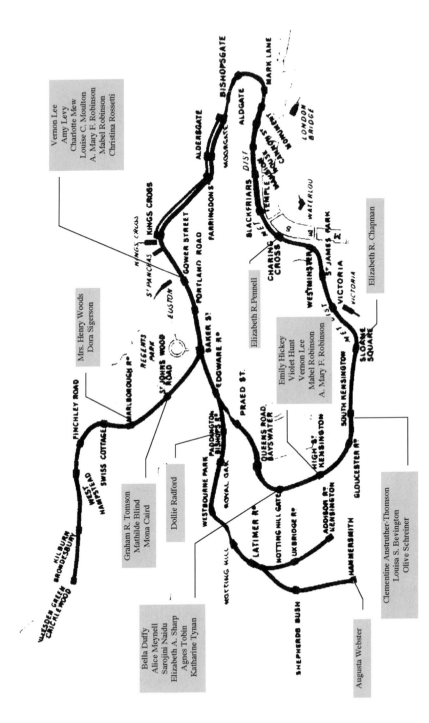

Figure 1. Map of the London Underground and women poets in London, 1880–1900.

access to the rest of the metropolis. Levy, for example, loved going to Kensington Gardens and traveled regularly around London to visit, among others, the poets Ernest and Dollie Radford, "Graham R. Tomson" (Rosamund Marriott Watson), and Olive Schreiner.[17]

Another group of women poets settled in Kensington around Kensington High Street Station and Notting Hill Gate Station. Alice Meynell, her life-long friends Katharine Tynan and Agnes Tobin, the critic Elizabeth A. Sharp (famous for her anthologies of women poets),[18] the translator Bella Duffy, and the Indian poet Sarojini Naidu all lived in Notting Hill Gate. In Kensington High Street lived the poets Violet Hunt and Emily Hickey, the Robinson sisters, and Vernon Lee (the Robinsons and Lee—who was staying with the Robinsons—moved from Bloomsbury to Kensington in 1883). Next to South Kensington Station lived Olive Schreiner, the anarchist poet Louisa S. Bevington, and Clementine Anstruther-Thomson. Kensington was fast becoming one of the most fashionable areas of fin de siècle London. Its excellent communication networks, its new shopping arcades, the famous and fashionable Kensington Gardens, and its museums and libraries—all made it a perfect location for these fashionable and well-known writers.

Another community of poets and writers developed in St. John's Wood. "Graham R. Tomson," Mathilde Blind, and the infamous Mona Caird all lived near St. John's Wood Station. Next to Marlborough Road Station, also in the St. John's Wood area, lived Mrs. Henry Woods and the poet Dora Sigerson, wife of the critic Clement Shorter. Finally, Dollie and Ernest Radford, who had also lived in Bloomsbury and in other areas of London, lived next to Bishop Road Station, on the periphery of St. John's Wood.

Mass-transportation facilities challenged the way in which women poets participated in the new metropolitan experience. Just as the underground connected different areas of London, it also interconnected this network of women poets. They got on the trains and omnibuses to see the city, to travel to the British Museum, or to visit other women poets, as the following letter from Vernon Lee to her mother illustrates:

> Let me see—what has been taken place. Oh—Sunday Mary [Robinson] & I went to Mrs. Clifford. A very friendly woman, & with a pleasant look of Pauline. She wants me to go to Mrs. Orr, who she says is charming. Clodd was there & Pollock, & Miss [Mathilde] Blind, who says she had to leave the theatre at *Tristan,* not (as I unwarily thought) because she was overcome by sleep, but for extreme emotion. Miss [Louisa S.] Bevington came & walked to the station with us.[19]

This passage very tellingly reveals how mass transport facilitated communication among women poets in London. Underground stations became real poetical landmarks within the metropolis, and poets used them to locate themselves. A. Mary F. Robinson, for example, in one of her letters to the poet "Michael Field" (the joint name of Katharine Bradley and Edith Cooper), wrote at the top of her letter not just her postal address but her railway address, so that "Michael Field" would know how to get to her house, which was "5 Minutes off [Kensington] High Street

Figure 2. "Utilitarian London" by William Hyde. *(Alice Meynell,* London Impressions, *with etchings and pictures in photogravure by William Hyde [London: Archibald Constable, 1898], facing p. 10.)*

Station or 5 Minutes East off Addison Road."[20] To get anywhere in the city now required knowledge of its transport system, as Tynan and Robinson realized, and the stations were now geographical as well as poetical markers.

These settlement patterns and the appearance of new suburbs had been foreseen by city planners and underground constructors. Cheap railway fares and fast trains made possible the exodus to newly developed suburbs. Since the Inner Circle Line facilitated the entrance to London's city center from all sides of London, the population could move out from the slums to healthier suburban areas, where accommodation was, at least in theory, cheaper and better. Although the lower classes generally benefited from this exodus, many ended up living in poor accommodations; and suggestions that what the railway had done was to give an excuse to clean up the city—so that the middle classes would not see the appalling conditions in which the lower classes lived—were not infrequent.[21] Gradually, slums started to disappear while the famous colonies of semidetached houses became a typical sight of the London suburb. In "Utilitarian London," one of William Hyde's illustrations for *London Impressions* (see Figure 2), the image that Hyde presents of suburbia is one of rows and rows of semidetached houses linked to the metropolis by railway lines. Another example is Dollie Radford's poem "From the Suburbs," in which she describes the changes in modern life by discussing the appearance of colonies of detached and semidetached houses linked to London's city center by fast trains:

It rushes home, our own express,
So cheerfully, no one would guess
The weight it carries

Of tired husbands, back from town,
For each of whom, in festal gown,
A fond wife tarries.

For each of whom a better half,
At even, serves the fatted calf,
In strange disguises,

At anxious boards of all degree,
Down to the simple "egg at tea,"
Which love devises.

For whom all day, disconsolate,
Deserted villas have to wait,
Detached and Semi-

Barred by their own affairs, which are
As hard to pass through as the far
Famed Alpine Gemini.[22]

Radford's anxieties about suburban life are articulated through the ironical glass
of the fairy tale. A clear-cut gendered division of space tarnishes the image of this
fairyland, and suburbia becomes the "dream prison" of the housewife,[23] as much
as it enslaves men to travel perpetually to and from work. Radford seems to predi-
cate this transformation upon the suggestion that the gendering of urban space is
a consequence of London's fast life. And she seems to point out that this fastness
is a product of underground trains: "I muse on what the spell can be, which causes
this activity," or "if men will always come and go / In these vast numbers, to and
fro, so fast and madly" (84).

But perhaps the most significant change resulting from the introduction and
expansion of mass transport was the emergence of London as a technopolis. At
the fin de siècle, as Arthur Symons so aptly claimed in 1909, London was transformed into a mechanical organism, into a cyborg. Not only did the vehicles for
mass transportation completely take over the streets of the metropolis, but London itself became a machine, as Symons put it.[24] This transformation happened
so fast that by 1900, only thirty-seven years after the first underground line appeared, one could reach anywhere in London by train. The underground expanded
considerably during the last quarter of the nineteenth century and increased its efficiency further when it became connected with local railways that joined the neighboring towns to the metropolis, towns that in turn were absorbed by the metropolis.
Indeed Hyde's "Utilitarian London" illustrates this point. In this photogravure, the
grid of railway roads is presented as the already basic structure of the metropolis
and as a part of its urbanscape. London was populated by trains, omnibuses, and
trams. The inevitable consequence was that the urban dweller struggled to keep
up with the unstoppable pace of the machine, as Amy Levy's poem "A March Day
in London" shows:

From end to end, with aimless feet,
All day long have I paced the street.
My limbs are weary, but in my breast
Stirs the goad of a mad unrest,

I would give anything to stay
The little wheel that turns in my brain;
The little wheel that turns all day,
That turns all night with might and main.[25]

The poem furthermore suggests that this mechanization is producing what Arthur Symons has called "an automobilisation of the mind."[26] Indeed, Levy has used this image ("the wheel that turns in my brain / . . . all day") as a metaphor to explain the struggle of the individual in the metropolis and the increasing mechanization not just of urban life but of urban dwellers. Dollie Radford's poem "Nobody in Town" also expresses this phenomenon. But for Radford, London's mechanization problematizes the presence of individuals in the streets. The urban dweller ceases to be visible in this huge machine:

I stand upon my island home,
    My island home in Regent Street,
And listen to the ceaseless foam
    Of traffic breaking at my feet:
The sky above is clear and sweet,
    The summer day is smiling down,
I muse upon it, and repeat
    That there is nobody in town.[27]

The poem is referring to Dollie Radford's London home in Regent Street NW10, not Regent Street, which has the postcode W1. Here Radford once again suggests the colonization of the metropolis by the machine and the invisibility of urban dwellers. The poet stands alone in a city filled with vehicles that flow unceasingly around the metropolis. The appearance of the underground produced new networks that allowed the individual to travel to previously unattainable urban spaces. But the urban dweller was shadowed by the overwhelming presence of mass-transportation machines. These machines facilitated urban mobility, but they also mediated the urban dweller's experience of modern life. The individual was not anymore in *direct* contact with the world but separated from it by the window-pane of the bus, tube, and tram.

And these machines accelerated the rhythm of the metropolis, creating constant acceleration, that "ceaseless foam of traffic" that Radford so compellingly describes. They have created a "mad unrest," producing "neurasthenia" in the metropolis.[28] In the 1850s a journey between Paddington and the Bank of England by horse-drawn omnibus took from forty to sixty minutes, but by the 1880s the same journey took only eighteen minutes.[29] Space was shrinking. Speed had become the quintessential characteristic of modern life, and the passenger was becoming a new figure in modernity. The *flâneur*'s fascination for walking gave way to the passenger's passion for traveling. As all the above poems suggest, *flâneurs/euses* felt more and more displaced in the mechanized city. The change was substantial: both *flâneurs/euses* and passengers were transient figures, but the movement of passengers was mechanically produced, as they were not agents of their mobility. If *flâneurs/euses* strolled in the city as if they were observing pictures in a museum, passengers sat in the omnibus, train, or tram as if they were sitting in a cinema while the spectacle of the city passed in front of their eyes. Traveling changed the way in which people perceived the city for two main reasons. First, the passenger was separated from experiencing the metropolis first-hand by that which originally made possible its comprehension, the mass-transportation vehicle.[30] Second, the speed of the engines transformed the static appearance of the metropolis. As Alice Meynell put it, "London has a fantastic look, as though there were nothing to do but make haste to be gone."[31] London was now pure speed, pure haste. Inevitably this led to a transformation of perception itself. As Beatriz Colomina argues, "Perception was now tied up to transience," to the fleeting experience of the city.[32] Not surprisingly, the origins of the cinema date to this crucial period. Speed became intrinsically linked to perception. The passenger saw a city more and more organized around the unstoppable movement of the engine.

Spinning through London

In this final section, I will turn my attention to Alice Meynell's collection of essays, *London Impressions*. The book, commissioned by Archibald Constable to accompany photogravures by William Hyde, was a thoughtful, illuminating mixture of photogravures and essays that portrayed and examined different aspects of fin de siècle London, with the narrative being supported by visual references to London's various urban scenes. Meynell included essays describing areas of London such as Chelsea ("Chelsea Reach"); essays that discussed the role of nature in the metropolis ("The Trees," "The Spring"); and essays dedicated to typical aspects of living in London ("The London Sunday," "A Pilgrim"). However, most of the essays (in particular "The Effect of London," "The Climate of Smoke," "Below Bridge," "The Roads," and "The Smouldering City") were dedicated to representing the drastic ontological and material transformations that London had suffered as a consequence of the introduction of new technologies, especially railway roads. For Meynell, this transformation had led to a breach between nature and culture, to the formation of a new economic stage based on reproduction and consumption and not on creation and production, and to a new (visual) aesthetics, centered on the ephemerality of urban life.

In order to sketch out these changes and to capture the speed of modern life, it was necessary to become a passenger, as she indicated in the following note to her husband, written while she was preparing the manuscript for one of the essays of this collection, "The London Sunday": "This morning I took an omnibus drive through London to the farther end of Clapton, thinking to write about the London Sunday."[33] What is interesting about this note is that it clearly reveals Meynell's need of mass-transportation machines, which she used as a tool to visualize and problematize London. In "The Smouldering City," for instance, she recreated a journey from West to East London in order to discuss the social divisions created by technological advancements, arguing that it was "from a high railway" that one might "see the darkened but still soft and charming colour spreading from roof to roof of the cottage-streets of older London" (*LI,* 30). London's industries and the smoke of railways were darkening the skies of the city. But, ironically, only by using railway trains did one became conscious of that transformation. In this sense, urban transport acted as a transient "movie theatre" that she used to comprehend and analyze urban life:

> In spite of the length of London, you may pass from the furthest west to the extreme east, and from the last country field to the first, so quickly as to get a continuous Sunday impression—the day and the people flowing, unfolding, and closing, from suburb to remote suburb, through "town" through the City, through the east, and to the verge of breathless and unfragrant meadows, divided by a league-long tramway line lost in the distances of Epping, whither the smoke, from which a south-west wind has set all London radiantly free, is trailing a broken wing. (*LI,* 2)

Meynell used a cinematic technique to reproduce these journeys. For Meynell, traveling transformed the vast and heterogeneous metropolis into a "continuous impression," into a panorama. Indeed, in *London Impressions,* Meynell masters the position of a passenger-critic attending the opening of a film as she sits on the bus, tram, or underground, pencil and writing pad in hand, to review urban life. Her essays are truly visual records of the spectacle that unfolded in front of her eyes. Moreover, it seems as if the windowpane of the mass-transportation vehicle is the screen onto which the city projects itself. Within the essays, scenes swiftly follow one another as her eyes try to record all that passes by:

> London at night has begun, of late, so to multiply her lights that they make all her scenery. A search-light suddenly draws the eye up to the chimney-pots (sweetly touched, they too, on the westernmost of their squalid sides) and to the unbroken sky; and then at once the eye travels down its shaft, revealing clouded air; and here a puff of steam from some machine at work on the new underground railway takes colour on its curve. Or the search-light makes the programme of a music-hall to shine black and white upon the wall; anon, an advertisement is written in light. (*LI,* 14)

In *London Impressions,* Meynell rendered the metropolis as a complex series of motion pictures, and her eyes recorded the metropolis as if they were the mechanical eyes of the camera.

Notes

I wish to thank the European Social Funds and La Consejería de Educación y Juventud de la Junta de Extremadura for sponsoring this research.
1. Katharine Tynan, *Middle Years* (London: Constable & Co., 1916), 108–109.
2. Ibid., 107.
3. See Donald J. Olsen, *The Growth of Victorian London* (London: Penguin, 1979), 94.
4. Rossetti would have caught the Inner Circle Line in Gower Street, now called Euston Square, to Paddington—four stations away—and then she would have had to transfer

to the Great Western Railway, which would have taken her to Ealing—two more stations away.

5. Rosalind Williams, *Notes on the Underground: An Essay on Technology, Society, and the Imagination* (Cambridge, Mass.: MIT Press, 1990), 53.

6. Elizabeth Grosz, *Space, Time, and Perversion* (London: Routledge, 1996), 108.

7. My brief account of the history of London's urban transport is indebted to T. C. Barker and Michael Robbins, *A History of London Transport,* vol. 1 (London: George Allen & Unwin, 1963); John R. Day, *The Story of London's Underground* (London: London Transport Board, 1963); Richard Trench and Ellis Hillman, *London under London: A Subterranean Guide,* new ed. (London: John Murray, 1993); Alan A. Jackson, *London's Local Railways* (London: David & Charles, 1978); and W. J. Gordon, *The Horse-World of London* (London: Religious Tract Society, Leisure Hour Library, 1893).

8. Christopher Hibbert, *London: The Biography of a City* (1969; reprint, London: Penguin Books, 1977), 184.

9. "Opening of the Metropolitan Railway," *The Illustrated London News,* 17 January 1863, 74.

10. Williams, *Notes on the Underground,* 23.

11. Alan Trachtenberg, quoted in ibid., 53.

12. A. Mary F. Robinson, *The Collected Poems: Lyrical and Narrative* (London: T. Fisher Unwin, 1902), 95.

13. Hibbert, *London,* 185.

14. In this map, I have included novelists, such as Mrs. Henry Woods and Olive Schreiner; journalists, such as Elizabeth Robins Pennell; essayists, such as Vernon Lee; and critics, such as Elizabeth A. Sharp, for their strong connection with this network of women poets. Mrs. Henry Woods was a close friend of the Robinson sisters and Vernon Lee. Amy Levy was a personal friend and follower of Vernon Lee and Olive Schreiner. For a more detailed discussion of this map and women poets, see my Ph.D. thesis, *Women Poets and the Aesthetics of Space and Transport at the Fin de Siècle* (Ph.D diss., University of London, 2000).

15. Judith R. Walkowitz, *City of Dreadful Delight: Narratives of Sexual Danger in Late-Victorian London* (1992; reprint, London: Virago Press, 1998), 69.

16. See Amy Levy, *The Complete Novels and Selected Writings of Amy Levy, 1861–1889,* ed. Melvyn New (Gainesville: University Press of Florida, 1993), 381–383, 431–434.

17. See for example, Amy Levy's 1889 social diary, 7 July 1889. I wish to thank David Bacon and Camellia Plc for allowing me to use the Amy Levy archive in their possession.

18. Elizabeth Sharp, ed., *Women Voices: An Anthology of the Most Characteristic Poems by English, Scotch, and Irish Women* (London: Walter Scott, 1887); and *Women Poets of the Victorian Era* (London: Walter Scott, 1890).

19. Vernon Lee, *Vernon Lee's Letters,* ed. Irene Cooper Willis (privately printed, 1937), 91.

20. Oxford University, Bodleian Library, MS Eng. letts.e.32.f.98.

21. This was the case made by the Rev. William Denton in his *Observations on the Displacement of the Poor by Metropolitan Railways and Other Public Improvements* (London: Bell and Daldy, 1861). See Trench and Hillman, *London under London,* 139.

22. Dollie Radford, *Songs and Other Verse* (London: John Lane, 1895), 82–83.

23. Elizabeth Wilson, *Hallucinations: Life in the Post-Modern City* (London: Radius, 1988), 193.

24. Arthur Symons, *London: A Book of Aspects* (London: privately printed for Edmund

D. Brooks, 1909), reprinted in his *Cities and Sea: Coasts and Islands* (London: W. Collins & Co., 1918), 145.

25. Amy Levy, *A London Plane-Tree and Other Poems* (London: T. Fisher Unwin, 1889), 21.

26. Symons, *London,* 144.

27. Radford, *Songs,* 67–68.

28. For a discussion of neurasthenia as a nervous suffering attributable to the shock of modernity, see Anson Rabinbach, *The Human Motor: Energy, Fatigue, and the Origins of Modernity* (Berkeley: University of California Press, 1990), 154.

29. Barker and Robbins, *History of London Transport,* 22.

30. Michel de Certeau, *The Practice of Everyday Life,* trans. Steven Rendall (Berkeley: University of California Press, 1984), 112.

31. Alice Meynell, *London Impressions,* with etchings and pictures in photogravure by William Hyde (London: Archibald Constable, 1898), 26. Subsequent citations will appear parenthetically in the text with the abbreviation *LI.*

32. Beatriz Colomina, *Privacy and Publicity* (Cambridge, Mass.: MIT Press, 1994), 12.

33. Letter from Alice Meynell to Wilfred Meynell, 1897, in Viola Meynell, *Alice Meynell: A Memoir* (London: Jonathan Cape, 1929), 138.

"Full of Empty"

CREATING THE SOUTHWEST AS "TERRA INCOGNITA"

MARY PAT BRADY

On the Fourth of July in 1849, at a meeting of the United States and Mexican Boundary Commission charged by the Treaty of Guadalupe Hidalgo to survey the newly created United States-Mexico border, Mexican general Pedro Garcia Condé was given what R. P. Effinger describes as a "magnificent reception" by the U.S. delegation:

> [The General] was escorted from the beach by a company of Dragoons who accompanied him to San Diego. He was then met by all the officers of the post and conducted to the head quarters of Colonel Weller and introduced to the head members of the Commission. This ceremony was gone through with on the morning of the 4th of July and was the welcoming in of our glorious celebration—Weeks beforehand we had commenced our preparations for the celebration of independence day and a more magnificent affair I never witnessed. After General Condé had been escorted into town we had a grand military parade—we then formed into a procession lead off by a company of Dragoons then Colonel Weller as orator of the day and Major Emory as President then Ladies and Gentlemen then the Yuma Indians with their squaws—while a company of Infantry brought up the rear. After listening to a very appropriate and eloquent oration from Colonel Weller and the reading of the Declaration of Independence in English and Spanish, we marched to a barbecue which we had exclusively prepared for the Indians—we had an ox and six sheep roasted whole and three barrels of whiskey. The whole affair was a novelty to the people out here and for the first time in their lives many a poor Indian and native Californian got his belly full of beer and whiskey.[1]

As Effinger's notes suggest, far more was at stake in this meeting than the simple plotting of the longitudinal and latitudinal points of the new border. The care taken in preparing what, following Patricia Seed, we might call a "ceremony of possession" indicates that its planners understood that theirs was a task of appropriation—of claiming land in symbolic as well as material terms—and a task inaugurating a process of envisioning the land in a manner that would make it assimilatable to U.S. nationalism as a region *transparently* "American."[2] The "appropriate and eloquent oration" (presumably in English), the bilingual reading of the Declaration of Independence, and the careful timing of the celebration to memorialize a specifically Anglo story of origin all came together to establish a new discursive geography. This geography went beyond the creation of a new border for the United States frontier to rewrite the story of the borderland's history by erasing the traces of an earlier Spanish colonial presence.

Had Commissioner Weller chosen to read from the Monroe Doctrine rather than the Declaration of Independence, the ceremony might have more fittingly marked the completion of one phase of manifest destiny by showcasing Anglo efforts to control the American hemisphere. Oddly, by reading the Declaration of Independence in Spanish as well as English, Weller seemed to enact the promises of the Treaty of Guadalupe Hidalgo to respect the primacy of Spanish in the new U.S. colonies. The records do not suggest, however, that General Garcia Condé, the Mexican boundary commissioner, actually spoke at the event—an indication that U.S. hegemony at this new border was not to be contested or even really masked. Furthermore, almost prophetically, the absence of "Californios" in the parade indicates the extent to which those who had colonized the region, following the Spanish, would not be allowed to play a substantive role in the new story about the region that was about to be told. As if to ensure a broad understanding of the ceremony underway, U.S. soldiers brought up the rear of the parade, underscoring their policing function and signaling their awareness that such ceremonies were not fully welcome. Effinger's proud if patronizing claim that the celebrants were well-fed and well-feted also suggests the ceremonial linkage of commodities with policing (of capitalism under the guise of democracy). Furthermore, the inclusion of Yuma tribespeople in the parade, far from a happy ceremonial multiculturalism, further indicated U.S. power over a community that had fiercely protected its lands for several hundred years. The celebration to begin the work of mapping, of instantiating a new national border, thus predicted the ideological conflicts and demands that would trouble the borderlands over the next 150 years.

By reorienting the land away from the Distrito Federal of Mexico City (which marked it as colonial El Norte) and toward the District of Columbia (which would mark it as the New Mexico Territory), this little-known ceremony of possession marked one beginning of an extensive process of writing the newly conquered land into being as the southwestern United States.[3] The surveyors developed a precise blueprint detailing the struggle necessary to produce, reorganize, successfully occupy, and finally administer a new colonial region. In this sense, the border survey was one among many tools in an already highly evolved imperialist repertoire necessary for the production and management of the newly captured territories. These tools, including ethnography, mapmaking, and the discourses of Anglo superiority and Mexican and Indian otherness, were utilized, first, to raze the region produced by nearly three hundred years of Spanish and Mexican colonial projects as well as more than three thousand years of Indian peoples' development of the area, and, then, to bury the rubble (of bodies and discourses) only to build a new colonial structure and edifice on top of it.

For Commissioner Weller, this Fourth of July ceremony of possession and the subsequent process of measuring and staking land entailed transforming the region from what one nineteenth-century travel writer would describe as a "terra incognita" into something recognizably "American": a region that did not resist Anglo incursions, a region where all "foreignness" had been eradicated, where threats to Anglo hegemony were minimized.[4] In order to achieve this new status, writers repeatedly characterized the land in a manner that would privilege an Anglo sense of superiority in the face of daunting difficulties. William Emory, who would also serve as a boundary commissioner, for example, represents the territory as "extending over a portion of the Continent but little known and diversified with much variety of climate and topography, and infested throughout its whole extent with formidable and hostile bands of Indians."[5] Emory's description exemplifies nineteenth-century Anglo representations of the region: its status as unknown; its variable topography; and its "infestation," as he puts it, of dangerous, seemingly menacing Indians. Yet clearly the region could be understood as "incognita" only through Anglo ignorance of Spanish—since Mexico's northern colonies had been explored and discussed for several hundred years—and through erasure of the nations that had developed and mapped the region for a far longer period, including Pimas, Maricopas, Navajos, and Apaches.[6] On the one hand, this erasure suggests that the Mexican and Indian presence did not pose a threat to Anglo colonialism. On the other, it underscores the refusal to acknowledge Mexican and Indian contributions to developing the mines, roads, and irrigation systems that had settled and organized the territory. The Southwest's status as incognita helpfully masked the range and depth of knowledge compiled and catalogued by others, not Anglos, thereby reinforcing Anglo supremacy and especially the authority of the new narratives compiled by the boundary commissioners and travel writers who followed them. Thus, the doubled narrative of a terra incognita at once justified the imperial project in the well-wrought discourse of adventure, discovery, and conquest, and it foreclosed the extensive and carefully developed knowledge of the region produced in other tongues.[7] In this respect, the border survey functioned in the Southwest as David Lipscomb has argued Cooper's Leatherstocking Tales functioned earlier in the century on the East Coast: remapping occupied territory as virgin soil ready for occupation.

The U.S. boundary survey commissioners' charge then, as practiced by them as well as by a host of writers who would explore the region, was not simply to map the territory or to describe it for eastern armchair travelers and the bankers and investors whose capital they sought to fund mining operations but, rather, "to produce and assemble the physical domain"[8] in such a way that it could be understood as properly "belonging" to the United States. In order for the territory to "belong" to the United States, it had to become more than a "mapped" geographically known area; it had to be empirically and imaginatively produced in such a manner that it could be recognized and woven into the story of U.S. exceptionalism and supremacy. This geomakeover entailed explaining and cataloguing not just the land but the sociospatial matrix of the new colony while personifying that matrix as a collection of haggard "barbarians" awaiting Anglo might and enterprise. These catalogues necessarily described people and land in such a way that their conquest and management (and even, or especially, their death) could be envisioned and strategized not in some fantasy future but immediately.[9] Here, scientific earth description becomes political propaganda. Hence Weller's reading of the Declaration of Independence was a kind of strategic baptism, symbolizing the rebirth of the new territories under U.S. sovereignty; it was also a kind of strategic death warrant, signaling the demise of the long extant sociospatial organizations throughout the new border region.

As much as the boundary commissioners' mapping and the travel writers' narratives made practical the process of Anglo colonization, their ethnographic accounts also made it imaginable. Theirs, perhaps particularly William Emory's and John Russell Bartlett's, fell into the catalogue of nativist modernity or anthropology, which, as José Saldívar argues, reveals the extent to which geographical surveys, anthropology, travel writing, and imperialism constitute one another.[10] Taken together, Bartlett's and Emory's multivolume reports are remarkable documents of literary, cultural, and geographic appropriation. The boundary surveys—Bartlett's was published privately and Emory's as an official government document—were among the first, and maybe among the most effective, to discipline and systematize perceptions of the Southwest. We can see traces of their narratives in everything from subsequent travelogues to dime westerns to the histories both textual and filmic continually proffered about the region.[11] Both sets of volumes offer myriad ways to imagine the Southwest. Both offer ethnographies of the nations encountered during their surveys. Both provide extensive and detailed maps. Both include hundreds of prints—particularly landscapes but also detailed drawings of flora and fauna, of petroglyphs, and of representative "Indians." Both include narrative accounts of the author's travels with descriptions of his "adventures" as well as assessments of the future value and uses of the land under survey. Both include numerous "official" documents such as letters from State and Treasury Department officials, congressional reports, and financial accounts. These discursive registers mutually reinforce each other's authority as objective, scientific accounts. By indexing the Southwest in so many registers, both Bartlett and Emory proclaimed the encyclopedic veracity of their accounts and thus underscored the authority of their productions.

Bartlett's *Personal Narrative of Explorations and Incidents in Texas, New Mexico, California, Sonora, and Chihuahua* (1855), published privately after his tenure as boundary surveyor came to an ignominious close, opens with an exquisite foldout landscape entitled "Fort Yuma on the Colorado River." Thus, before readers even encounter a written narrative, Bartlett offers them an image that centralizes U.S. power and authority while also signaling the vastness of the new terrain, the smallness and scarcity of its permanent inhabitants, the "uncivilized" nature of the Yumas (indicated by their apparent lack of permanent dwellings), and the extraordinary and strange mountainous formations and local vegetation. The rendering of the landscape emphasizes the difficulty of the terrain upon which Bartlett's survey was conducted and the enormity of the region available for very specifically Anglo conquest. Put differently, from the very first image onward, Bartlett produced the Southwest, in Helena Viramontes's words, as "full of empty."[12] But we should note that the territory was empty only for Anglos.

Not surprisingly, Bartlett conducted his survey with an eye toward future U.S. possession of the untapped resources of Sonora and probably Chihuahua as well:

> The time is not far distant, either, when crowds as large as those now press-ing on to California and Australia will be "prospecting" among the moun-tains of Texas, New Mexico, Chihuahua, and Sonora, attracted by similarly rich mineral deposits, and probably with the like splendid success. This will not be the result of accidental discovery, as was the gold in the mill-race near the Sacramento; for the existence of such treasures is already known, as well as the localities where they are to be found. My journeys through Sonora, Chihuahua, and other Mexican States, are given with much detail on the topics mentioned.[13]

His inclusion of Sonora and Chihuahua alongside Texas and New Mexico (the newly purchased and already existing colonies) indicates his sense of their likely eventual possession by the United States. Thus he aligns his mapping/ethnographic project directly with future colonization and the imperial work of collecting addi-tional land. This text serves as a resource in this effort. The land surveyed was vast, according to Bartlett, but not vast enough. More and broader expanses of land awaited further colonization. Emory would also write his survey with an eye toward keeping the frontier open. Commenting at the end of his survey that the United States and Mexico now had a militarily feasible border, he also notes the likely possibility that the border will be changed once again: "In other respects, the boundary is a good one; and if the United States is determined to resist what appears to me the inevitable expansive force of her institutions and people, and set limits to her territory before reaching the Isthmus of Darien, no line traversing the continent could probably be found which is better suited to the purpose."[14] Both Bartlett and Emory thus conceptualize the border they are creating not as a terminus but as a staging ground for U.S. expansion.

Bartlett's text entails three different narratives of the landscape: as fruitless and forbidding, as desirable and potentially lucrative, and as a wasteland of ear-lier failed colonial efforts. Looking out over one portion of the yet-to-be border, he writes: "Is this the land which we have purchased, and are to survey and keep at such a cost? As far as the eye can reach stretches one unbroken waste, barren, wild, and worthless."[15] Yet this recoil is followed by more hopeful assessments: "The soil is exceedingly rich, and is capable of producing abundant crops of maise and wheat (the only cereals cultivated), fruits of various sorts, and, with pains, ev-ery kind of vegetables" (266). Hucksterism supplants wonder, which supplants de-spair. Throughout the *Narrative* the land is alternately emptied and filled up again.

The drawings repeat this alternating rhythm. In nearly all, the sky occupies a third of the area, shrinking the scale of people and buildings. In most panoramas, rocky outcroppings and vegetation dwarf any people present. People are nearly always represented singly or in small groups. Even when a large town is the subject of a landscape, the view offered hardly shows the permanent structures. For example, the drawing of Tucson, a Mexican settlement dating back to the end of the seventeenth century, emphasizes the enormity of saguaro cacti. The settlement's buildings are barely discernable from the distant foliage. If these drawings collectively suggest the "nature" of the region, they also withhold from the imagination any concept of Mexican and Indian cultures—reifying them as stereotypically nomadic and backward, as culture-less. Indeed, portraits of small groups of Indians, drawings of ruins, shards of pottery, and hieroglyphics further reinforce the concept of Indians as premodern. The frequent allusions to the "Indianness" of Mexicanos similarly places them out of time through the drawings as well as through narrative inference. By tacitly ignoring Mexicano settlements and Indian communities, these drawings collectively emphasize the placelessness of the people the survey commission encounters. Such a maneuver, as Ana Alonso perceptively writes, "condensed diverse significations. Placelessness was a privileged sign of the barbarians 'quality,' of their animality and incivility, of their wild potency and indomitableness, of their location outside the social: the unbounded spaces of nature."[16] Seemingly outside of modernity, without apparently fixed urban centers, the occupants captured in Bartlett's drawings are placeless beings without history.

While Bartlett travels through New Mexican territory, he does note signs of human populations—yet these populations are remainders, nearly ghosts that make visible a sense of loss. Such loss lends picturesque qualities to the region. It works in other ways as well: by listing abandoned towns, ranches, and mines, Bartlett suggests that the region is indeed inhabitable but that there are no longer any settled, civilized communities to prevent the arrival of enterprising Anglos. The emphasis on the loss rather than on the historical presence of Mexicanos or their ongoing habitation of the region had another insidious result as well. The denial of the ongoing presence of the Mexicanos contributed to the characterization of Mexicanos as "degraded, filthy, destitute,"[17] too inept to build a thriving, sustainable culture. Through omission and selective reporting, Bartlett disarticulates a considerable Mexicano culture and history. A different characterization might have highlighted the explorations that had already been carried out by the Mexicanos, the complex relationships they had developed with the Pima, Maricopa, and Tohono O'odham peoples, and their sophisticated mining and smelting techniques. But these stories are left untold.

The cumulative effect of Bartlett's two volumes is to assert a kind of typical atypicality for the region: ruined and abandoned buildings, bizarre outcroppings of rock, treacherous though bountiful terrain, stunning Mexican failure, and fierce Apache aggression. He asserts this typicality through the repetition of similar scene after similar scene, each staged as part of the progress of his journey. Thus his "progress," his movement, is framed against the repeated typicality of the landscape, which finally does not change and is locked outside of modernity. Action is lodged with him and, by inference, the viewer, while the timeless landscape awaits modernizing capital and the progress Bartlett embodies.

Bartlett stylized himself as gentleman-explorer, traveling in a carriage and accompanied by Emory, who eschewed the trappings of gentility and presented himself instead as a scientist and military man, a technologically astute and dispassionate collector of data. Yet Emory's work is remarkably similar to Bartlett's. Emory also produced a two-volume border survey that included ethnographic narratives, drawings, mining information, and horticultural observations. Emory's *Report on the United States and Mexican Boundary Commission* follows his earlier, 1846 survey of the region conducted under the auspices of the U.S. Cavalry and designed as a reconnaissance of the Southwest to determine whether it should be incorporated into U.S. territory.[18] Emory introduces his report with a critique of Bartlett's tenure as commissioner and then follows with another critique of the famous explorer Baron Alexander von Humboldt. His criticisms of his predecessors bolster his own position as an objective observer while locating his narrative within the tradition of what José Rabasa terms "the mode of a master discourse—writing as a conquering and capitalist enterprise."[19]

Despite his critique of Bartlett, Emory repeats his predecessor's formula: he describes barren and fruitful land, fantastic natural formations, "savage" foes, and abandoned settlements. He, too, emphasizes the barren quality of the region:

> Imagination cannot picture a more dreary, sterile country, and we named it the "Mal Pais." The burnt lime-like appearance of the soil is ever before you; the very stones look a sickly vegetation, more unpleasant to the sight than the barren earth itself; scarce an animal to be seen—not even the wolf or the hare to attract the attention, and save the lizard and the horned frog, naught to give life and animation to this region. The eye may watch in vain for the flight of a bird; to add to all this the knowledge that there is not one drop of water to be depended upon from Sonyta to the Colorado or Gila. All traces of the road are sometimes erased by the high winds sweeping the unstable soil before them, but death has strewn a continuous line of bleached bones and withered carcasses of horses and cattle, as monuments to mark the way.[20]

Like Bartlett, Emory also narrates his survey in the competing lexicons of absence and potential: "[The Ranch of San Bernardino] when in its most flourishing condition boasted as many as one hundred thousand head of cattle and horses. They have been killed or run off by the Indians, and the spacious buildings of adobe which accommodated the employees of this vast grazing farm are now washed nearly level with the earth" (94). One of his assistants similarly reports, "Tubac is a deserted village. The wild Apache lords it over this region, and the timid husbandman dare not return to his home" (118). Emory thus urges his readers to see the Apaches as a threat to U.S. colonization—but only if the government fails to send troops.

Not surprisingly, Emory's drawings follow Bartlett's in suggesting the boundlessness of the terrain. Emory's racial narratives, themselves caught up in images of boundlessness, underscore the point. Early on he complains about the dangers of "mixed breeds," worrying about the incapacity of the "mongrel race of Indians" and their "obfuscated" intellects. Here we see Emory's need to maintain a bounded identity against the potential boundlessness of *mestizaje*. Hence he describes in some detail the "pure" blooded Indians, depicting them and cataloguing them in the same manner as the local fauna. Mexican *mestizaje,* on the other hand, presents a peculiar threat to Emory's investment in purity and boundaries. Not surprisingly then, Mexican skill and culture cannot be praised or really even mentioned, despite the commissioner's dependence on Mexican guides, laborers, and hospitality.

Emory's cultural assessments function as interludes between spectacularly detailed drawings of the desert terrain, extraordinary depictions of fossils, charts detailing geological data, contour maps, hundreds of tables of longitudinal and latitudinal measurements, and drawings of various "Indians" in "normal" garb. His descriptions of Mexican and Indian peoples rely on connecting them with objects in the landscape around them: geological fossils and skeletons and snakes. The people he describes, however complex their cultures and languages, are taken out of the context of history. They exist not as the producers or possessors of cultures but, like the snakes that populate his landscapes of relentless, rocky, and steep mountains, as parahuman signs of a regional otherness.

The official and semiofficial reports by Emory and Bartlett were not alone in their production of the Southwest as uncivilized territory—nor in their razing of its Sonoran history. A chorus of other writers would join them over the next forty years to create a seemingly unified portrait of the Southwest in general and Arizona in particular as largely devoid of Mexicanos and in deep need of Anglo intervention to develop its resources and fight the wars that would end Apache (and Mexican) control over the region. Speaking before the New York Geographical Society in 1857, Sylvester Mowry, for example, echoed Bartlett and Emory, noting that the Mexican population had been "extirpated" by Apaches who "roamed uninterrupted and unmolested."[21] Twenty years later, Hiram Hodge in his *Arizona as It Is; or, The Coming Country* (1877) described his 1874 sojourn across Arizona in a survey of territorial assets that emphasizes their potential for investors and the need for new colonists.[22] So profoundly does Hodge apparently feel this potential that he concludes his narrative with a celebratory prediction of Arizona's future:

> With the completion of railroads, and the development of mines, all other industries will prosper. Her rich agricultural lands will be settled and tilled, her millions of acres of grazing lands will be covered by numerous flocks of sheep and herds of cattle . . . brisk and prosperous towns and cities will spring into existence, and in less than a decade of years, a prosperous, wide-awake, and energetic American population will have centered in her borders, and she will be knocking at the doors of Congress for admission into the Union, where she will become a bright star in the galaxy of free and independent States of the great American Union. (252–253)

Arizona's potential as a colony is narrated in teleological terms: it will provide a productive environment for both capital and Anglos; its productivity will be rewarded by admission into the Union. For Hodge, Indians are more annoying than threatening to this project (see 161), and Mexicans are largely nonexistent. Although he claims that "[g]eneral good feeling exists between the white and Mexican population" (155), he is careful to note that only one "Spanish Mexican land grant" is said to be "legal and valid" and that with this "freedom from land grants" Arizona can expect "future prosperity and freedom from litigation and strife, which has been so prolific a source of trouble in California" (47). By acknowledging the "strife" and declaring Arizona free of it, Hodge implicitly refers, even in his denial, to a kind of Gramscian "war of positions" underway between Anglos and the newly colonized Mexicans in Arizona for possession of the land.[23]

Raphael Pumpelly, in *Across America and Asia: Notes of a Five Years' Journey around the World, and of Residence in Arizona, Japan, and China* (1870), recalls his 1860 experiences in Arizona as a mining engineer.[24] Pumpelly narrates his experiences in the form of an adventure tale in which he is beset by the inhabitants despite his "honest" desire merely to mine for silver. Thus, in Pumpelly's text Arizona is defined in terms of its natural resources and through a "social disorganization" that results from "the effect of the absence of the usual restraints upon society" (v). He complains about Mexican "treachery" (32) and accuses Mexicans of murdering whites "at almost every mine in the country" (22) and forming "the most degraded class in a land where social morality was, in every respect, at its lowest ebb" (29). Pumpelly is adamant in his hatred of Mexicanos and his desire to exclude them from the emerging territory. Noting that the territory "was almost entirely depopulated excepting the Indian tribes," he argues against recruiting Mexican labor even though "it seems doubtful whether Americans can be profitably used for hard work in the climate of Arizona" (32). Pumpelly is careful to note, "By Americans I refer throughout to the white natives of the United States" (32). Even as he addresses himself to "white natives," his narrative makes clear that Anglos have depended upon Mexican mining techniques and expertise, as well as Mexican laborers and guides, for taking possession of this land. If the oblique references suggest a prior Mexican presence and traces of Mexican history in the region, his overtly hostile depictions relegate that presence not to the margins of his text but to a structural unconscious. That is to say, where "Mexicans" are the subject of the text, they are subjected to vitriolic, stereotypical descriptions. Yet their productive labor is everywhere in evidence. In the distinctions he draws between Anglos, Mexicans, and Indians, Pumpelly's adventure narrative produces Arizona as a "marvelous" land of mystery and intrigue."[25] Arizona, however, must not remain only an exotic place like China or Japan, the scenes of his further travels. The resolutions of the conflicts he cites will allow Arizona to be fully incorporated into the United States—as "a bright star in the galaxy" of "the great American Union."

Journalist J. Ross Browne, who toured Arizona during the Civil War and wrote a series of travelogues for *Harper's Monthly,* used the trope of devastation to describe the region, but, even more than other writers, he enhanced that trope with rehearsals of vicious stereotypes of Mexicanos. In Browne's vision, if the Gadsden Purchase region had been devastated by war with the Apaches, the Mexicans who remained were barely distinguishable from the devastated landscape. Indeed Browne's text makes explicit what Bartlett and Emory only imply; the representation of the landscape necessarily describes those that inhabit it. The Arizona landscape is formulated as so fantastic, so extraordinary, and so uninhabitable that those who do inhabit it are necessarily at least as extraordinary, weird, and perverse as Bartlett's "strange formations."

In offering his travelogue Browne promises the reader "extraordinary advantages in the way of burning deserts, dried rivers, rattlesnakes, scorpions, Greasers, and Apaches; besides unlimited fascinations in the line of robbery, starvation, and the chances of sudden death by accident."[26] Folded into the list of natural objects are "Greasers" and "Apaches." Their position in the sentence as syntactical hinges emphasizes their centrality, however, because what follows in the sentence is a series of human catastrophes. In this manner Browne suggests that Mexicans and Apaches serve as a kind of bridge between a terrifying environment and the catastrophic possibilities of life apart from "civilization."

What the reader is to take away from all these accounts is a sense of failed promise. As Browne writes:

> Scarcely three years ago . . . the works were in active operation; vast piles of silver were cast up daily from the bowels of the earth; wagons were receiving and discharging freights; the puff and whistle of the steam-engine resounded over the hills; herds of cattle, horses, mules, and other stock ranged over the valleys. At the time of our visit it was silent and desolate—a picture of utter abandonment. (265–266)

The pairing of past production and failed reproduction is repeated so frequently by commentators that it functions as a kind of advertisement for renewed Anglo colonization to realize the unfulfilled potential. Browne writes, "Situated as it is at the junction of the two main roads from Sonora, the Santa Cruz and Magdalena, it might be made a very valuable piece of property in the hands of some enterprising American" (154). The folding of past failure with current crisis into fantastic potential offers a rhetorical opportunity for the representation of both Mexicanos and Apaches, whose very failures suggest a prospect for American enterprise.

If everywhere the commentators blame the desolation and abandonment directly on the Apache wars, they simultaneously make a second, implied charge. Echoing the familiar formula, Browne describes the land as a fecund paradise: "The ground is rich and the climate unsurpassed, and with the rudest cultivation abundant crops of wheat, maize, pomegranates, and oranges might be produced" (168). But elsewhere he points out that, due to the "Bedouins of the desert" known for their "depredations upon stock, robbing the ranches, killing the ranchers, and harassing emigrant parties" (21), "no industry could prosper under their malign influence. The whole State of Sonora was devastated, and the inhabitants in a starving condition" (21).

Built into such descriptions of devastation is the charge that the Mexicans lost these wars. This charge is no simple one. Made during the buildup toward the Civil War, in the aftermath of a victorious war of aggression against Mexico, and, later, during the height of the Civil War, when the United States was defining itself not as agrarian but as an industrialized warfare society bent not on failure but on victory, the assessments by Bartlett and Browne and others do more than render Indians animal-like and Mexicans incapable. The critique of Mexico as a failed military power offers a measuring stick against which the United States can reassert Anglo superiority—especially in Browne's case—at the very moment that superiority is endangered.

The contests over the production of Arizona as a region and the constitutive need to "disappear" Mexicanos from that formulation in order to structure the space as peculiarly available to Anglo colonizers depended in large measure on utilizing the mediation of "the Apaches" as the agents who cleared the soil, in literal terms, of Mexicanos, so that Anglos could reap the fruits of three centuries of explorations, mining development, and ranching labor and expertise. The Apaches were used as icon and agent, as something like a precursor to the Cold War client state that would fight a war for U.S. interests and with U.S. supplies but not with explicit U.S. acknowledgment. By this I mean that Anglo colonists regularly, and with official U.S. support, armed and supplied Apaches with the tacit understanding that they would focus their attacks on Mexicanos. As one 1859 editor, complaining of the tactic, noted, by offering Apaches "safe retreat" and a "market for their booty" Arizonans were committing "nothing more nor less than legalized piracy."[27] Yet this policy (public enough that Cochise, who led Apaches in a series of battles in Sonora and across the Southwest, armed his expeditions in full view of U.S. government officials)[28] disappears from the narratives for East Coast readers that I have been discussing. Apaches were initially encouraged to attack Mexicanos, especially in Sonora, to clear the way; the Apaches were never intended to be incorporated into Anglo concepts of citizen-subject. Thus if Mexicanos were quickly incorporated into the narratives of Arizona's natural resources, Apaches were described as waiting to be eliminated.

These narratives of Apache destruction and Mexicano unproductivity hide the extent to which Anglo colonialists and the U.S. government supported, encouraged, and even funded Apache battles against the Sonorenses. The narratives also make it clear that Mexico in general and Sonora in particular would not be formidable forces should the United States decide to seize the mineral-rich territory of Sonora. Thus the writers work their way out of their contradictions. If the land is barren and dangerous, why colonize it? Why expend capital on its development? But if the land can be declared to have once been productive, its current barrenness blamed on a longstanding war, an animalized and predatory nation, and the failings of those who first colonized it, then Anglos can be made to appear heroic and the region ripe for yet another round of development.

Thus, the blueprint that developed over the course of roughly thirty years was one whose impact would continue to influence the narrative of the Southwest for the next one hundred. Mexicano history continued to be omitted from the cultural geography of the Southwest even as Mexicanos and Indian peoples contested characterizations of them as uncivilized. The tools of nineteenth-century scientific geography—the map, the survey, and the boundary commission itself—were effectively deployed to mythologize the territory as virgin soil ready to be possessed, as wilderness waiting for order and organization. And while Mexicanos in Tucson, Sonora, and New Mexico immediately took up their pens to contest these narratives, it would be another hundred years before the Southwest would begin to be broadly remapped and the imperial strategies for producing the Southwest widely deconstructed.[29]

Notes

1. Letter dated 1 August 1849, from Robert Patterson Effinger, MS643, California Historical Society Manuscript Collection, San Francisco, California.
2. Patricia Seed, *Ceremonies of Possession in Europe's Conquest of the New World, 1492–1640* (Cambridge: Cambridge University Press, 1995).
3. My discussion here is indebted to Gearóid O Tuathail, *Critical Geopolitics* (Minneapolis: University of Minnesota Press, 1996).
4. J. Ross Browne, *Adventures in the Apache Country: A Tour through Arizona and Sonora, 1864* (Tucson: University of Arizona Press, 1974), 16
5. William H. Emory, *Report on the United States and Mexican Boundary Survey, Made under the Direction of the Secretary of the Interior* (Washington, D.C.: A.O.P. Nicholson, 1857), 1:1
6. Even travel writer J. Ross Browne acknowledged this erasure when he commented that "few people in the United States knew anything about it, save the curious book-worms who had penetrated into the old Spanish records" (*Adventures,* 16).
7. Iris H. W. Engstrand, for example, notes in *Arizona Hispánica* (Madrid: Editorial Mapfre, 1992) that beginning with Cabeza de Vaca's travels through southern Arizona in 1536 and continuing to Escalante's mission in 1776, at least a dozen official expeditions were conducted across Arizona (83). Countless unofficial expeditions were conducted by explorers and adventurers seeking new land to colonize and new mines to develop (55–120). Ellen Trover, in *Chronology and Documentary Handbook of the State of Arizona* (Dobbs Ferry, N.Y.: n.p., 1972), identifies the first Spanish exploration as early as 1526 with Don José de Basconales's exploration through Zuñi territory (1). For an additional account of narratives of exploration, see Oakah Jones, "The Spanish Written Word: Changing Images and Neglected Legacy of the Southwest," in *Essays on the Changing Images of the Southwest,* ed. Richard Francaviglia and David Narrett (College Station: Texas A&M University Press, 1994), 40–71.
8. David Barker, quoted in O Tuathail, *Critical Geopolitics,* 3.
9. My discussion here is indebted to the insights into the colonial process offered by Ana Alonso in *Thread of Blood* (Tucson: University of Arizona Press, 1995).
10. John Russell Bartlett, *Personal Narrative of Explorations and Incidents in Texas, New Mexico, California, Sonora, and Chihuahua* (New York: D. Appleton and Co., 1856); and José Saldívar, *Border Matters: Remapping American Cultural Studies* (Berkeley: University of California Press, 1997). Saldívar's comment, which comes in a thoughtful revaluation of John Gregory Bourke's ethnographic-military reconnaissance of South Texas, may be seen in part within the context of the renewed work on tourism and the construction of the Southwest that has been influenced by Edward Said's *Orientalism* as well as by the work of Michel Foucault and Dean McCannel. Among the most helpful

analysts of the production of the Southwest is Barbara Babcock; see, for example, the special issue on Southwest tourism that she coedited for the *Journal of the Southwest* 32, no. 4 (winter 1990). See also Sylvia Rodríguez, "The Tourist Gaze, Gentrification, and the Commodification of Subjectivity in Taos," in *Essays on the Changing Images of the Southwest,* ed. Richard Francaviglia and David Narrett (College Station: Texas A&M University Press, 1994); Martin Padgett, "Travel, Exoticism, and Writing the Region: Charles Fletcher Lummis and the Creation of the Southwest," *Journal of the Southwest* 37, no. 3 (autumn 1995): 421–449. For other very helpful analyses of the production of the U.S.-Mexico border, see also Dawn Hall, ed., *Drawing the Borderline: Artist-Explorers of the U.S.–Mexico Boundary Survey* (Albuquerque, N.M.: Albuquerque Museum, 1996); and especially, Oscar Martínez, *Troublesome Border* (Tucson: University of Arizona Press, 1988). For an early, popular discussion, see Edward Wallace, *The Great Reconnaissance: Soldiers, Artists and Scientists on the Frontier, 1848–1861* (Boston: Little, Brown and Co., 1955). Also helpful are Burl Noggle, "Anglo Observers of the Southwest Borderlands, 1825–1890: The Rise of a Concept," *Arizona and the West* 1, no. 2 (summer 1959): 105–131; and Raymund Paredes, "The Mexican Image in American Travel Literature, 1831–1869," *New Mexico Historical Review* 52, no. 1 (January 1977): 5–29

11. For a wonderful discussion of popular culture and the 1848 war, see Shelley Streeby, "American Sensations: Empire, Amnesia, and the U.S.-Mexican War," *American Literary History* 13, no. 1 (March 2001): 1–40.
12. Helena Viramontes, *Under the Feet of Jesus* (New York: Dutton, 1995), 20.
13. Bartlett, *Personal Narrative,* iv–v.
14. W. H. Emory, *Notes of a Military Reconnaissance* (New York: H. Long and Brother, 1848), 39.
15. Bartlett, *Personal Narrative,* 247.
16. Alonso, *Thread of Blood,* 20.
17. Bartlett, *Personal Narrative,* 277.
18. Emory, *Notes of a Military Reconnaissance.*
19. José Rabasa, *Inventing America: Spanish Historiography and the Formation of Eurocentrism* (Norman: University of Oklahoma Press, 1993).
20. Emory, *Report on the United States and Mexican Boundary Survey,* 115.
21. Sylvester Mowry, *Memoir of the Proposed Territory of Arizona* (Washington, D.C.: Henry Polkinghorn Printer, 1857).
22. Hiram Hodge, *Arizona as It Is; or, The Coming Country* (New York: Hurd and Houghton, 1877).
23. Hubert Bancroft would more accurately indicate the numerous land grants that could be contested, and indeed the Spanish-language newspapers published between 1870–1890 indicated that battles over land grants preoccupied the entire Arizona territory. Hubert Howe Bancroft, *History of Arizona and New Mexico: 1530–1888* (1889; Albuquerque, N.M.: Horn and Wallace, 1962). Similarly, Sylvester Mowry, in his *Memoir,* predicts that "disorder & anarchy will reign supreme" over Arizona "if Congress fails to settle disputes over land titles" (10).
24. Raphael Pumpelly, *Across America and Asia: Notes of a Five Year's Journey around the World, and of Residence in Arizona, Japan, and China* (New York: Leypoldt and Holt, 1870).
25. Samuel Woodworth Cozzens, *The Marvellous Country; or, Three Years in Arizona and New Mexico, the Apaches' Home, Comprising Description of this Wonderful Country, Its Immense Mineral Wealth, Its Magnificent Mountain Scenery, the Ruins of Ancient*

Towns and Cities Found Therein, with a Complete History of the Apache Tribe, and a Description of the Author's Guide, Cochise, the Great Apache War Chief. The Whole Interspersed with Strange Events and Adventures (1874; reprint, Minneapolis, Minn.: Ross and Haines, 1967), 45.

26. Browne, *Adventures,* 11.
27. Quoted in Joseph Park, "The Apaches in Mexican-American Relations, 1848–1861: A Footnote to the Gadsden Treaty," *Arizona and the West* 3, no. 2 (summer 1961): 142.
28. Ibid., 129.
29. For helpful discussions of Arizona, in addition to Park see also Miguel Tinker Salas, *In the Shadow of the Eagles: Sonora and the Transformation of the Border during the Porfiriato* (Berkeley: University of California Press, 1997). For a discussion of the Mexicano responses to the Anglo narrative of Arizona, see "Razing Arizona," in my *Extinct Lands, Temporal Geographies: Chicana Literature and the Urgency of Space* (Durham, N.C.: Duke University Press, in press). See also Gabriel Meléndez, *So All Is Not Lost: The Poetics of Print in Nuevomexicano Communities, 1834–1958* (Albuquerque: University of New Mexico Press, 1997).

Empire's Second Take

PROJECTING AMERICA IN *STANLEY AND LIVINGSTONE*

JON HEGGLUND

This twinning of power and legitimacy, one force obtaining in the world of direct domination, the other in the cultural sphere, is a characteristic of classic imperial hegemony. Where it differs in the American century is the quantum leap in the reach of cultural authority, thanks in large measure to the unprecedented growth in the apparatus for the diffusion and control of information. . . . Whereas a century ago European culture was associated with a white man's presence, indeed with his directly domineering (and hence resistible) physical presence, we now have in addition an international media presence that insinuates itself, frequently at a level below conscious awareness, over a fantastically wide range.

Edward Said, *Culture and Imperialism*

$\mathcal{L}$ooking backward from the end of the twentieth century, Edward Said comments on the historical shift from a nineteenth-century European acquisitive imperialism to the global domination of the "American century."[1] Inquiring into different modes of imperial power, Said suggests that, while European imperialisms of the "classic" age relied upon a "directly domineering" presence, United States imperialism is characterized by an expansion and dissemination of the international mass media. This epistemic shift in the history of geopolitics also marks a decisive turn in the history of spatial representation. On the one hand, the direct domination of European—particularly British—imperialism was ideally expressed by one spatial form: the map. On the other hand, the rise of the "cultural authority" accorded to U.S. imperialism relies on the cultural forms of the mass media. In particular—and especially before the advent of television—the culture of U.S. global imperialism has been embodied in the medium of narrative film. What Said identifies as a historical shift from British territorialism to U.S. cultural imperialism can be traced back to two simultaneous developments in the history of space, both occurring at around the turn of the nineteenth century: the loss of faith in the map as a reliable indicator of geographical truth and the rise of cinema as a persuasive, dynamic, almost magical "space and time machine."[2] This essay examines the decline of British imperialism and the rise of U.S. imperialism as a representational rivalry between cartography and cinema.

At the turn of the nineteenth century, the global map gave ample evidence of British territorial supremacy, as Great Britain laid claim to the Indian subcontinent, Canada, large portions of Africa and Australasia, and territories in the Middle East, South America, and the Caribbean. The map of the world had been more or less completed by then—the "blank spaces" of Marlow's youth had been filled—and the British maintained possession of more colonial territory than any other imperial power. The cartographic visibility of British dominance testifies to the characteristic space of British imperialism: the map both measured and represented the British Empire. Indeed, British imperial cartography *produced* imperial space. Drawing on Enlightenment ideals of rationality and instrumental knowledge, the science of cartography converted the terra incognita of the colonies into an archive that presented a coherent, and thus governable, projection of world space. Seen through the grid of the map, the unknown world was suddenly legible. The opacities created by differences of race, culture, climate, and topography were dissolved through the mathematical transparency of longitude and latitude. Maps, by transforming the heterogeneity of place into the abstraction of space, created a grid onto which other imperial sciences—anthropology, archaeology, and biology, for example—could be plotted and inscribed. The map did not merely represent the empire; the map was the empire. As the geographical historian Matthew Edney writes of overseas British conquest, "The empire exists because it can be mapped."[3]

While the science of cartography defined the episteme of territorial imperialism, the map's power to represent the world could not keep up with the dynamism of modern space. The space of cartography is static and fixed; as soon as a map is drawn, its moment is fixed, and it becomes instantaneously obsolete. By the turn of the twentieth century, space was being conceived as a fragmentary, shifting, dynamic entity that was ill suited to the totalizing, fixed, static projections of cartography. Stephen Kern characterizes an epistemic shift from "homogenous" to "heterogeneous" space; that is, space as substantial and differentiated rather than empty and homogenous. Kern lists just a few of these disciplinary reimaginings of space: "Biologists explored the space perceptions of different animals, and sociologists, the spatial organization of different cultures. Artists dismantled the uniform perspectival space that had governed painting since the Renaissance and reconstructed objects seen from several perspectives. Novelists used multiple perspectives with the versatility of the new cinema."[4] Ironically, the years that Kern identifies as decisive in the transformation of space—1880 to 1918—coincide with the heyday of British imperial dominance. Yet, just as the British map testified to the territorial dominance of the British Empire, the map was already being rendered obsolete by a more dynamic mode of representation.

As Kern suggests, film was ideally suited to the modern transformations of spatial experience. As spatial and temporal projections presented for the gaze of an individual subject, films were much more amenable than maps to the new heterogeneity of modern space, especially space's new interrelationship with time. In its ability to project a mimetic image, represent motion, shift perspective, and create diegetic space through editing, film assumed an almost magical quality in its representation of space and time. Cinema was a "space and time machine" that brought all parts of the globe into the locality of the movie house—transporting viewers to France, the Philippines, Cuba, and South Africa, among other places— with the simultaneity and immediacy of direct experience.[5] Through its mimetic representation of various localities throughout the world, film created a new register of global space: seemingly all parts of the living, moving, breathing world could be made accessible in a dark room containing projections of light and shadow on a two-dimensional screen. Although this transportation of the cinema viewer was not physical, it was no less real. As the cultural geographer Jeff Hopkins has claimed, "[T]he time and space portrayed on the screen are indeed imaginary, but the temporal and spatial experience is genuine." Film viewing, Hopkins concludes, is "an experience that is first and foremost geographical."[6]

This essay posits a relationship between two developments in the history of space: first, the shift from map to film as the most culturally resonant representation of geographical space and, second, the transition from the British form of imperialism based on territorial acquisition to a United States form of imperialism based on the manipulation of image and spectacle.[7] Although these movements occurred around the turn of the twentieth century—Edison's kinetoscope and the Lumières' cinematograph appeared in 1891 and 1895, respectively—they become culturally legible *after* a decisive shift in the balance of power from Great Britain to the United States. They are especially visible in one Hollywood narrative film of 1939, Twentieth Century–Fox's *Stanley and Livingstone,* produced by Darryl Zanuck, directed by Henry King, and starring Spencer Tracy and Cedric Hardwicke as Henry Stanley and David Livingstone, respectively. In telling the story of the British geographer and missionary "lost" in Africa and the American reporter who undertakes the expedition to find him, *Stanley and Livingstone* dramatizes the competition between the British geographical establishment and the American mass media in their search for Livingstone. Through its comparative representations of African space through both maps and landscapes, the film aligns British imperialism with the spatial form of cartography while associating U.S. imperialism with the visual powers of cinema. The film thus transforms the legendary story into a study of contrasts between the arrogance, elitism, and inefficiency of the British imperial project and the pragmatism and commercial interest of United States involvement in the exploration of the African interior. These contrasting portrayals of British and U.S. imperialism resonate with Anglo-American diplomacy on the eve of World War II: U.S. imperialism recalls the glories of British imperialism only to dismiss it as an anachronistic mode of spatial representation. In its criticism of the British mode of imperial rule, the film justifies a more mobile, influential, "disinterested" global presence, paving the way for an empire that could prosper without imperialism.

Historically, the story of Stanley and Livingstone has been portrayed as a tale of lost and found.[8] In 1866, the geographer and missionary David Livingstone left England to make a final expedition to find the source of the Nile. He did not return, nor did he send word of his progress. In 1870, he was reported to have been in Ujiji, Tanzania, on the shores of Lake Tanganyika, incapacitated with illness. The popular presses of Great Britain and the United States portrayed Livingstone as a helpless figure "lost" in the wilds of Africa, even though he had been doing cartographic and missionary work in Africa for decades. The Welsh-born, American-raised Henry Stanley, then a field reporter for various newspapers in the American West, was hired by James Gordon Bennett Jr., the editor of the *New York Herald,* to "find" Livingstone. Although having had no experience as an explorer, Stanley managed to lead a group of nearly two hundred men into the African jungle with a fairly specific idea of where Livingstone would be. The expedition reached Livingstone on November 10, 1871, more than nine months after leaving Zanzibar, on the east coast of Africa. The legendary meeting of the two men bespeaks the heroism of two white men miraculously finding each other on the "Dark Continent." Even Stanley's infamous greeting—"Dr. Livingstone, I presume"—provoked amusement precisely because of its apparent superfluity and incongruity, as if Stanley and Livingstone had met in Fleet Street rather than in the African interior. After sending word that Livingstone had been met, Stanley returned to a divided reception: some thought him a brave hero; some thought him a conniving reporter who fabricated part or all of the events of his "discovery." Livingstone, in poor health, remained in Africa by choice, and he died while exploring the Lualaba River basin in 1873.

The popular reception of the Stanley and Livingstone story as a tale of the glory of two white men finding each other in the vast African jungle helped to shape the Victorian imaginative geography of Africa as the quintessentially savage dark space in opposition to the civilized white world of the West. Yet within this story of Western mastery over African space, a tale of imperial rivalry was also unfolding. Stanley was enlisted not only as an American journalist but also as an amateur explorer, assigned by Bennett to report on Livingstone's geographical discoveries in the African interior. This secondary mission trespassed upon the professional territory of the Royal Geographical Society (RGS), Britain's association of geographers and cartographers that included the African explorers Richard Burton, John Speke, and Livingstone himself. When Stanley was invited to present Livingstone's geographical findings to the RGS, he was met with a skepticism that bordered on outright hostility. Not only was Stanley considered an amateur geographer, he was also vilified for his American brashness and indifference to the "gentlemanly" protocols of the RGS. Geographical method and national temperament were aligned: the British method of "scientific" exploration and mapmaking had been usurped by a brash American who had indiscriminately hacked his way through the jungle in search of a journalistic scoop.

Henry Morton Stanley's own ambiguous Anglo-American identity as well as his self-transformation from reporter to geographer suggest that United States commercialism and British territorialism were complementary rather than contradictory interests in Africa. Stanley, who was born John Rowlands to an impoverished Welsh family, spent much of his childhood in a workhouse before escaping on a sailing ship to New Orleans at the age of sixteen. A relentless self-fashioner, Rowlands changed his name to Henry Stanley in commemoration of an early American benefactor, and he later experimented with invented middle names such as Morlake and Morland before eventually deciding on Morton. With few fond memories of his childhood in Wales, Stanley began to identify himself as an American, even adopting an "American twang" in his travels throughout the United States as a young journalist. Stanley's autodidacticism—he taught himself science, history, and literature—further establishes his persona as a "self-made man" so consonant with U.S. popular mythology. Although Stanley always sought acceptance and respect from the British imperial establishment, especially the Royal Geographical Society, his entrée into imperial exploration was for purely mercenary reasons: just as he was hired by Bennett, he would later undertake further expeditions up the Congo in the pay of King Leopold of Belgium. Stanley, although doing more than any other person to solidify European territorial domination of Africa, was himself a mercenary imperialist not bound to the service of any one nation. Stanley's multiple national affiliations suggest a man who cannot be claimed as an authentic national hero by any imperial power. Yet it is precisely in his commercial motivations that Stanley perfectly embodies an American imperialism that *denies* national self-interest, claiming enterprise rather than conquest as the sole reason for its international presence.

Although having taken place more than a half-century prior to its cinematic adaptation, the story of Stanley and Livingstone had a particular Anglo-American resonance in the years approaching World War II. In the 1930s, Hollywood studios began to produce numerous films adapted from British imperial stories. Major studio films such as *Charge of the Light Brigade, Clive of India, Wee Willie Winkie, The Sun Never Sets,* and *Gunga Din* all depicted various episodes from the nineteenth-century history of the British Empire.[9] Perhaps surprisingly, Hollywood was just as enthusiastic about the British Empire as was the English film industry: as one reviewer from the London *Times* wrote in 1937, "The Union Jack has in the last few years been vigorously and with no little effect waved by Hollywood."[10] Hollywood's approval of imperial rule was not lost on American critics; of the 1935 film *Lives of a Bengal Lancer,* starring Gary Cooper, one reviewer for the *New York Times* commented, "[I]t is so sympathetic in its discussion of England's colonial management that it ought to prove a great blessing to Downing St."[11] The American success of these pro-empire films suggests that American audiences were happy to sympathize with an ally whose national ideologies were opposed to the official democratic anti-imperialism of the United States. Although these films seem at odds with the anti-imperial political stance of the United States with regard to the British Empire, they resonate with an American culture learning to negotiate the apparent contradiction between an anti-imperial national mythology and aspirations of global commercial and cultural dominance. During the 1930s, American self-representations claimed a difference both from the totalitarian fascism of the Axis powers and from the traditional class hierarchies and acquisitive imperialism of Great Britain. Yet the United States had long maintained its own informal empire and was beginning to emerge from its post–World War I isolationism as a world power on a scale rivaling that of the British Empire in its heyday. Understanding this singular historical context helps to explain why Hollywood might take a special interest in the past exploits of the British Empire. The ideological disjunction communicated by these films could be sutured only as long as any U.S. connection to British imperialism was expurgated; that is, as long as narratives of Britain's imperial past were packaged as the innocent adventures of a distinctly "other" national culture and historical period. Because of this apparent contradiction between United States democracy and British imperialism, the imaginative attraction of imperial stories to American audiences depended upon a temporal disjunction projected by these films: the British Empire existed in the past; the United States in the present and future.[12]

Stanley and Livingstone was one of the more successful films inspired by Hollywood's imperial craze of the 1930s; yet the film avoids the uncritical Anglophilia of many prewar adventure films by establishing a rivalry between English and American methods and motives of exploration. The film thus walks a fine diplomatic line: it projects an underlying image of Anglo-American entente while also drawing obvious contrasts between British and American overseas interests in Africa. The cinematic partnership between the two nations resonates with its contemporaneous geopolitical situation because it paints the informal imperialism of the United States with the complimentary hues of an enterprising commercialism catering to the needs of American consumerism. The American presence in Africa thus appears as a function of an expanding domestic consumer market rather than as the primary means of national identity and expression. The entrepreneurial, commercial, democratic United States "happens" to be in Africa simply because that is where the best human interest story is. By contrast, Britain is the stodgy imperial presence whose very national identity is invested in the occupation and administration of its overseas colonies. Although this contrast is projected back onto the 1870s, a time when the British Empire was in its ascendance, an audience familiar with the state of the empire in 1939 would recognize the inflexible quality of imperial administration as one of the main reasons for its contemporaneous state of decline. The layering of the twentieth-century political present onto the nineteenth-century past separates *Stanley and Livingstone* from the other imperial films produced by Hollywood in the 1930s: the film explicitly confronts the historical differences between American and British styles of rule rather than casts the British Empire into the ahistorical realm of romance and adventure.

From its first sequence, *Stanley and Livingstone* appeals to the cultural sensibilities of an American audience by placing the well-known story of the two explorers in the most American of film genres: the western.[13] Beginning in the Wyoming Territory in 1870, the film opens with a long shot of the western landscape as Stanley rides into a military camp, proclaiming to exasperated American generals that he has just returned from an interview with the hostile Kiowa chief, Santanta.[14] From this scene, the film establishes an analogous relationship between the American West and the African interior: both are wild, lawless territories in which the roguish Stanley uses his resourcefulness to find his journalistic story. The parallel between the Wild West and the "Dark Continent" is more firmly established later in the film, when Stanley, narrating the progress of his expedition, remarks that Africa's "open mountainous country is little different from our own great West." Africa's geography, or more properly, its lack of geography, thus becomes comprehensible only by its superimposition onto a map of the United States. When the newspaper mogul Bennett persuades Stanley to undertake the search for Livingstone, Bennett points to a map of Africa and remarks, "The Dark Continent . . . a vast jungle in which you could lose half of America." Stanley also appeals to specific American geographical referents when he explains the difficulty in searching for Livingstone: "It's as if we had left New York on foot and hoped to find Livingstone somewhere between Chicago and New Orleans." Even as we learn of the British Empire's control of vast stretches of Africa, the film annexes the continent to the United States by imagining it through the lens of an American geography.

This evocative, imaginative geography of an "American" Africa is contrasted to the narrowly cartographic interests of the British. In *Stanley and Livingstone,* maps are more than set pieces: they support institutions that are instrumental in defining both individual and national identities. Ironically, the plentitude of knowledge produced by the British cartographical establishment signifies the weakness rather than the strength of the empire that it serves. Caught up in the science of mapmaking for its own sake, British geographers have forgotten their practical purposes. This point is emphasized in the film's climactic sequence, in which Stanley returns to present Livingstone's preliminary maps of the Lualaba River basin to the Royal Geographical Society. When members of the society criticize the crudeness of Livingstone's maps (suggesting that they are in fact Stanley's forgeries), Stanley defends Livingstone by explaining that the missionary had indicated that his map was one "possible" version of the area's geography. One of the cartographers dismisses Stanley's protest, exclaiming that "the word 'possible' does not exist on the map." Wedded to scientific methods of exploration and cartography, the geographers cannot see the jungle for the trees: the maps, though crudely drawn, indicate possible sources for several significant waterways, including the Congo and the Nile, that would in fact greatly assist British penetration of the African interior. In his impassioned speech to the assembly, Stanley rails against the impractical excesses of the RGS, claiming that none of the corpulent geographers has ever "looked into anything deeper than a plum pudding." Stanley's speech points out the visually obvious: while practical men like Stanley have been hacking through the African jungle with a specific purpose, the academic geographers of the RGS have literally gotten fat off of the growth of geography as a field of academic rather than practical knowledge. Divorced from its real-world practice, the institutionalized science of imperial cartography illustrates the overgrown, ineffectual nature of Great Britain's territorial mode of imperial rule.

Not only does the film criticize the British imperial enterprise of mapmaking as an impractical, ineffectual activity; it takes issue with the objective, scientific status accorded to cartographical representation in general. While the film employs the frequently used device of showing maps as both diegetic and nondiegetic markers of the spatial progress of the narrative, the maps in *Stanley and Livingstone* assume a more central importance. As mentioned above, the film's plot in fact turns on the RGS's adjudication of the maps that Livingstone has drawn and asked Stanley to present. Beyond the explicit importance of the map as a plot device, the map is also implicitly compared to film itself as a medium of representing both history and geography. The film's own mise-en-scène subsumes the British imperial cartographic project into a geography of affect, through which images of Africa's pristine landscape replace the map as the primary representation of geographical space. This geography of affect is achieved by the use of location shooting to project "authentic" wide-angle landscapes of an empty African continent, thus replacing the scientific credibility of the map with the aesthetic immediacy of the filmed image. Both narratively and formally, then, *Stanley and Livingstone* offers viewers an important lesson in the differences between British and U.S. imperialism: the former treats geographical space as a matter of truth and falsehood to be determined by the cartographer's pen, whereas the latter presents the geography of empire as a cinematic image deployed for the aesthetic judgment of the viewer. For an audience in 1939, the anachronism of cartography as a medium of imperial power is made manifest by the visual immediacy of the cinematic projection of imperial landscape. Once presented as part of a larger on-screen spectacle, maps can no longer represent the objectivity on which Britain had established its territorial control of Africa.

Yet while the film criticizes British imperial cartography, it also places an investment in the map's status as factual knowledge. This double movement offers the most telling illustration of the contradictory American attitudes toward the colonial methods of the British Empire. The film initially uses maps as objective representations of world geography (in scenes, for example, that plot Stanley's journey to and return from Africa). This much accords with standard Hollywood practice; films borrow the authenticating device of the map from novelistic adventure tales such as Robert Louis Stevenson's *Treasure Island* and travel narratives such as Stanley's own writings. The filmic display of the map, however, recalls its literary antecedents only to draw attention to its more spectacular visualization of narrative geographies. The dynamic display of the map in the film serves to distinguish the medium of film from "the novel's reliance upon words or static drawings."[15] While maps on film draw attention to the formal differences between visual and verbal narrative, the cinematic map is still an arbiter of "real" space, a signifier of some actual geographic location in the world of the film. Seldom, if ever, is the practice of cartography itself questioned. The display of the map offers evidence of what J. B. Harley refers to as a primary rule of cartography: "that the objects in the world to be mapped are real and objective, and that they enjoy an existence independent of the cartographer."[16] This use of the map relies on the assumption that cinema, like cartography, unproblematically represents the world. Conventional cinema draws on the epistemological authority of the map to assist the viewer in believing in the referential power of the images projected onto the screen.

While references to cartography ground the fiction of film in the facts of geographical reference, the cinematic display of the map cannot be dissociated from the political and historical context of imperial cartography. The objectivity of the map has its own particular history, which is inextricably bound up with the project of European colonial conquest. The cinematic use of the map is thus not an innocent endeavor wherein maps lend a neutral facticity to a fictional narrative; it is no accident that virtually all the maps displayed in films of this period depict the terrae incognitae of the colonies rather than well-known areas of the metropole or home country. Maps on film thus memorialize a specifically nineteenth-century moment of Western exploration of the "Dark Continent" or the "mysterious East." In light of the craze for both British and American films of empire in the thirties, we can read this memorialization of cartography as a compensatory nostalgia in an age in which the world's geography was not only a more or less known quantity but also subject to appropriation and reinscription by nascent anti-imperial national movements. The commemoration of the map in the film makes formal as well as historical sense. As Ella Shohat points out, "By associating itself with the visual medium of maps, cinema represents itself as a twentieth-century continuation of the cartographic science."[17] In light of the declining cultural authority of pro-imperial institutions such as the Royal Geographical Society, however, Shohat's claim demands a proviso: that such a continuation is indicative of institutional and geopolitical shifts arising from the decline of British imperialism and the rise of a global market for cinematic production. In other words, the Anglocentric cultural authority of institutions like the RGS gives way to the more global designs of the new arbiters of colonial images: Hollywood's major film studios.

Even as *Stanley and Livingstone* uses the epistemological authority of the map to establish its relation to geographical space, the film suggests that the cultural importance of the map belongs to a bygone age of British imperialism, which must now give way to the visual immediacy of the Hollywood image-machine. Opposed to the British "armchair geographers" ensconced safely within the libraries and studies of the metropole, Stanley is portrayed marching purposefully through wide-angle African landscapes in several safari sequences. As early as the opening titles, the film establishes the geographical authenticity of these scenes, which were in fact filmed on location in Kenya, Uganda, and Tanzania—then British colonies. Such great expense in location shooting—the film's budget was more than $1.3 million[18]—was packaged as part of the film's attraction: the opening credits make a point of thanking "the officials of His Majesty's government in East Africa" for cooperating in the filming of the safari sequences. These safari scenes generally consist of long shots of the African landscape as Stanley's expedition marches through the frame; at times the camera lingers on the spectacle of animal life in the Serengeti. Although these sequences are not narratively well integrated into the film, they help to establish the difference between American and British cultural sensibilities. The safari scenes lend an immediacy to Stanley's explorations that enlists the audience's support in his battle against the cartophilia of the Royal Geographical Society; when we see how the imperial geographical establishment scoffs at Livingstone's crudely drawn maps, we cannot help but side with Stanley because we have seen him marching through what we know to be the *actual* African landscape. The epistemological power of the map is weakened by its association with the narrow-minded, effete geographers. Yet, rather than explicitly criticize British imperial domination, the film suggests a replacement of one spatial representation of empire with another. In place of the distanced, mediated representation of the map, the film projects an empire of "authentic" cinematic landscapes devoid of any native population or culture. When Stanley remarks that the African savannahs are "swarming with life," he refers only to the animal population of the "hunter's paradise." Tracy's voice-over during the safari sequences makes the associations between landscape, display, and American consumerism explicit: "What old P. T. Barnum wouldn't give for a few of these specimens. This is the greatest show on earth." The film offers a telling self-commentary: Africa is a "show" presented for the pleasure of a U.S. consumer audience. By turning the presumed authenticity of location shooting into the glitzy surface of spectacle, the film not only continues but also supersedes cartographic science; the guarantee of authenticity is established no longer by institutions of "science," however, but by the spectacular evidence of the viewer's eyes.

The competition between the authority of the map and the affect of the image leads to a conclusion that unites these visions of empire both symbolically and visually in the figure of Stanley. In the film's final sequence, we see a montage of Stanley marching across the African terrain superimposed on a map of the continent. Stanley's imperial legacy leaps out at the viewer: geographical places he names during his later expedition down the Congo—Stanleyville, Stanley Pool, Stanley Falls—shoot to the foreground of the frame. As the film draws attention to the inscription of his name onto the African map, the geography of Stanley's expedition becomes the triumphant story of Stanley the hero. Stanley's tune has even changed: where he was previously identified with the ur-American "Oh! Susanna," he now marches into the African sunset to Livingstone's theme, "Onward Christian Soldiers." The film's conflict between the institutionalized imperialism of British mapmaking and the visual landscape of the Hollywood film is resolved in a final merger of the cartographic and cinematic image, as the map's iconic representation of Stanley Falls dissolves into an actual shot of the waterfall. The image swallows up the map, leaving a close-up image of a rushing waterfall where the map of Africa had previously been. The map still signifies the facticity of Stanley's expedition, but cinematic technique possesses an even greater power: to transform the map into a mere artifact or cinematic prop. Far from dismissing the British cartographic project outright, then, *Stanley and Livingstone* concludes by revitalizing the "armchair geography" of British imperialism with the visual immediacy so readily available in the idiom of the Hollywood studio movie.

Stanley and Livingstone's particular appropriation of British imperial history serves U.S. ends in two ways. First, by retrospectively presenting the United States' benign imperialism in contrast with Britain's acquisitive territorial colonialism, the film suggests that future U.S. interventions around the globe can likewise be cast according to the innocent ideals of democracy and progress rather than the violent practices of conquest and domination. Second, by privileging the immediacy of the image at the expense of specialized geographical and historical knowledge, the film encourages viewers to see colonial space as a mise-en-scène rather than as a contested territory of political and cultural struggle. The use of the map in films of empire encourages this vision by always recalling a past geography in order to elide the more complicated and less triumphant geographies of decolonization. In imperial cartography's moment of cultural decline, it is thus preserved by the more global appeal of cinematography. By subsuming the map into the film, *Stanley and Livingstone* works through the ideological problem posed at its outset. Just as the film recasts the exploration of Africa as a tale of Anglo-American imperial rivalry, so does it also address its late-imperial context by enfolding the British imperial science of cartography into the emergent technology of film, incorporating the signifiers of the British domination of space only to announce their final reduction into the two dimensions of the screen.

By using the temporal displacement necessary in the historical drama, the film is better able to speak about the Anglo-American tensions of the pre–World War II present. *Stanley and Livingstone* enables Americans to embrace Great Britain as a worthy ally against the totalitarianism of the Axis powers, first, by explicitly acknowledging the ideological differences between each nation and, then, by suggesting that U.S. geopolitical power arises from the very qualities that distinguish it from the British Empire: enterprise, pragmatism, and individualism. These "American" qualities—clearly in opposition to British traits—form the ideological foundation of the postwar mythology that has allowed the United States to maintain a global imperial presence based on strategic influence and covert intervention rather than on territorial acquisition. As long as the global interventions of the United States are screened as the pragmatic helping hand of a beneficent power, as they are in *Stanley and Livingstone,* the Hollywood production of American mythology can stand in for a more complex—and less innocent—imperial history.

Notes

1. Edward Said, *Culture and Imperialism* (New York: Vintage, 1993), 291. For a more extensive treatment of the relationship between British imperialism and U.S. imperialism, see Anne Orde, *The Eclipse of Great Britain: The United States and British Imperial Decline, 1895–1956* (New York: St. Martins, 1996).
2. This phrase is used by Ian Christie in *The Last Machine: Early Cinema and the Birth of the Modern World* (London: British Film Institute, 1994).
3. Matthew Edney, *Mapping an Empire: The Geographical Construction of British India, 1765–1843* (Chicago: University of Chicago Press, 1997), 2. See also Matthew Edney, "Reconsidering Enlightenment Geography and Map Making: Reconnaissance, Mapping, Archive," in *Geography and Enlightenment,* ed. David N. Livingstone and Charles W. J. Withers (Chicago: University of Chicago Press, 1999), 165–198.
4. Stephen Kern, *The Culture of Time and Space, 1880–1918* (Cambridge, Mass.: Harvard University Press, 1983), 132.
5. Some of the first films shown internationally were the Lumière Brothers' 1895 films of everyday life in France, including scenes of workers leaving a factory and (famously) a train arriving at a station. Within five years, Thomas Edison's film company mixed actual with re-created footage to depict both the Anglo-Boer War in South Africa and the Spanish-American War in the Philippines and Cuba. For histories of the relationship

between cinema and the Boer War, see Colin Harding and Simon Popple, eds., *In the Kingdom of Shadows: A Companion to Early Cinema* (London: Cygnus Arts, 1996). For an account of Edison's role in producing films of the Spanish-American War, see Amy Kaplan, "The Birth of an Empire," *PMLA* 114 (October 1999): 1068–1076.

6. Jeff Hopkins, "A Mapping of Cinematic Places: Icons, Ideology, and the Power of (Mis)Representation," in *Place, Power, Situation, and Spectacle: A Geography of Film,* ed. Stuart C. Aitken and Leo E. Zonn (Lanham, Md.: Rowman & Littlefield, 1994), 57.

7. For a complete account of the economic and institutional history of the globalization of American cinema, see Kerry Segrave, *American Films Abroad: Hollywood's Domination of the World's Movie Screens from the 1890s to the Present* (Jefferson, N.C.: McFarland & Co., 1997).

8. For the historical account of the Stanley and Livingstone story and biographical information about Stanley, I have relied upon the following: Richard Hall, *Stanley: An Adventurer Explored* (Boston: Houghton Mifflin, 1975); and John Bierman, *Dark Safari: The Life behind the Legend of Henry Morton Stanley* (New York: Knopf, 1990). On Stanley's importance within late-nineteenth-century geographical cultures, see Felix Driver, *Geography Militant: Cultures of Exploration and Empire* (Oxford: Blackwell, 2001), esp. chap. 6.

9. *Wee Willie Winkie* and *Gunga Din* are adaptations of a Kipling short story and poem, respectively. *Wee Willie Winkie* (starring Shirley Temple!) is a domestic melodrama about a British family who has relocated to India. *Gunga Din* tells the story of a native who turns traitor against his tribe to assist the British in their war against Indian insurgents. *Charge of the Light Brigade,* loosely based on Tennyson's poem, depicts events of the Crimean War (1860s). *Clive of India* is a biographical picture about Robert Clive, who was instrumental in securing British rule in the subcontinent. *The Sun Never Sets* is a fictional tale of two brothers trying to prevent native insurgency in an African colony.

10. Quoted in Jeffrey Richards, "Boy's Own Empire: Feature Films and Imperialism in the 1930s," in *Imperialism and Popular Culture,* ed. John M. MacKenzie (Manchester, U.K.: Manchester University Press, 1986), 154.

11. Ibid.

12. The heyday of Hollywood's Anglophilia was brief, however. As with the British film industry, the production of these pictures was halted with the coming of the war, albeit for different reasons in Hollywood. With the onset of World War II, Hollywood's imperial craze had pronounced geopolitical consequences; the Office of War Information, for example, called off planned reissues of the immensely popular *Gunga Din,* which had been banned in India and Southeast Asia for its religious and racial representations. Realizing India's important role in maintaining opposition to Japanese aggression in Asia, the U.S. government was forced to recognize the dangers of its imperial nostalgia in the previous decade (Richards, "Boy's Own Empire," 156–157). In 1942, *Life* magazine voiced a similar opposition to twentieth-century British imperial policy, an opposition that clarifies the sudden halt to the release of Hollywood's imperial films: "One thing we are sure we are not fighting for," one contributor writes, "is to hold the British Empire together" (quoted in Bernard Porter, *The Lion's Share: A Short History of British Imperialism, 1850–1955,* 3d ed. [London: Longman, 1996], 313–314).

13. Given Stanley's multiple national affiliations, the writers of the film's screenplay, Sam Hellman and Hal Long, had some leeway in constructing their version of Stanley. Significantly, the choice to situate the film on American cultural terrain was not the in-

tention of Hellman and Long, who were originally commissioned by Zanuck to write an outline of the script. In their initial version, Hellman and Long begin the story in 1854 with a parliamentary meeting establishing a plan for England's imperial expansion and worldwide domination. The meeting is interrupted by one David Livingstone, who reminds the assembly that the outward expansion of England will come at the expense of the rapidly growing underclass population within Britain's borders. The story then shifts to St. Asaph's workhouse in Wales, where the young Henry Stanley resides, abused by James Francis, "the brutal one-armed schoolmaster." Although the Dickensian tenor of this narrative clearly identifies its subject matter as English rather than American, the eventual screenplay rewrote Stanley's story into a much more recognizably American context—revealing his British origins only in a climactic scene in front of the Royal Geographical Society. See Sam Hellman and Hal Long, *Stanley and Livingstone* (Beverly Hills, Calif.: Twentieth Century–Fox Film Corp., 1937).

14. Henry King, dir., *Stanley and Livingstone* (Beverly Hills, Calif.: Twentieth Century–Fox, 1939).
15. Ella Shohat, "Imaging Terra Incognita: The Disciplinary Gaze of Empire," *Public Culture* 3, no. 2 (spring 1991): 46.
16. J. B. Harley, "Deconstructing the Map," *Cartographica* 26, no. 2 (summer 1989): 4.
17. Shohat, "Imaging," 46.
18. Aubrey Solomon, *Twentieth Century–Fox: A Corporate and Financial History* (Metuchen, N.J.: Scarecrow Press, 1988), 114.

Notes on Contributors

MARY PAT BRADY is an assistant professor of English and Latino/a studies at Cornell University. Her study of Chicana literature, *Extinct Lands, Temporal Geographies,* is forthcoming from Duke University Press.

MARTIN BRÜCKNER is an assistant professor of English and American literature at the University of Delaware. He has published essays on maps, geography, nationalism, and the early American novel. He is currently completing a book on geographical literacy and identity formation in early America.

DIANE DILLON is completing two books about the World's Columbian Exposition of 1893. One focuses broadly on the visual and economic culture of the fair, and the other examines race, technology, and the production of amusement on the Midway Plaisance. She has taught art history and American studies at George Mason, Northwestern, and Rice Universities.

MONA DOMOSH is a professor of geography at Dartmouth College. She is the author of *Invented Cities: The Creation of Landscape in 19th-Century New York City and Boston* (Yale University Press, 1996); the coauthor, with Joni Seager, of *Putting Women in Place: Feminist Geographers Make Sense of the World* (Guilford Publications, 2001); and the coauthor, with Terry Jordan-Bychcov, of *The Human Mosaic: A Thematic Introduction to Cultural Geography* (Longman, 1999).

JULIE FROMER, at the University of Wisconsin-Madison, focuses on Victorian literature and culture. Her research interests include Victorian novels, material culture, and the study of consumption and commodities. She is currently finishing her dissertation, titled "A Necessary Luxury: Tea in Victorian Fiction and Culture." In addition to her interest in tea, she has published on the role of tobacco in Victorian culture.

JON HEGGLUND is an assistant professor of English at Washington University. While completing his Ph.D. from the University of California at Santa Barbara, he was a visiting scholar in the Department of Geography at Royal Holloway, University of London. He is currently working on two book-length studies: one on British modernist narrative and geographical knowledge, the other on images of "Englishness" in twentieth-century American popular culture.

BETTY JOSEPH is an associate professor of English at Rice University. Her recent projects include a book on gender and the colonial archive, forthcoming from the University of Chicago Press, and articles on gender and globalization in historical and contemporary contexts.

JULES LAW is an associate professor of English and comparative literature at Northwestern University, where he teaches Victorian literature and contemporary literary and cultural theory. His book *The Rhetoric of Empiricism* (Cornell, 1993) examines the relationship between philosophy and aesthetic criticism. He has published numerous essays on theoretical and political topics in *SIGNS, Critical Inquiry, New Literary History,* and other journals, and he is currently at work on a study of fluids in the Victorian novel.

PHILIPPA LEVINE is a professor of history at the University of Southern California. Her current research focuses on race and sex, with a particular emphasis on the regulation of prostitution in the British Empire.

DAVID C. LIPSCOMB teaches writing at Georgetown University, where he is an adjunct assistant professor, and at Ketchum, a communications firm. He earned his Ph.D. from Columbia University in 1998, and he lives in Washington, D.C.

JOSEPH LITVAK is a professor of English at Tufts University. He is the author of two books, *Caught in the Act: Theatricality in the Nineteenth-Century English Novel* (University of California Press, 1992) and *Strange Gourmets: Sophistication, Theory, and the Novel* (Duke University Press, 1997), as well as numerous articles on Victorian literature and culture and queer theory.

USSAMA MAKDISI is an associate professor of Arab and Middle Eastern history at Rice University. He is the author of *Culture of Sectarianism: Community, History, and Violence in Nineteenth-Century Ottoman Lebanon* (University of California Press, 2000).

JILL L. MATUS, a professor of English at the University of Toronto, is the author of *Unstable Bodies: Victorian Representations of Sexuality and Maternity* (1995) and *Toni Morrison* (1998), as well as several articles on literature and culture in Victorian Britain. Her current project is a book on trauma, the unconscious, and Victorian literature.

HELENA MICHIE is a professor of English at Rice University. She is the author of *The Flesh Made Word* and *Sororophobia,* both about Victorian cultural representations of the body, and the coauthor of *Confinements,* on contemporary reproductive technologies. She is finishing a book about Victorian honeymoons.

ROBERT L. PATTEN, Lynette S. Autrey Professor in Humanities at Rice University, edits *Studies in English Literature 1500–1900 (SEL).* He has written about nineteenth-century fiction, book illustration, and publishing history, and he is currently coediting *The Palgrave Guide to Dickens Studies* and writing a book about the formation of the "industrial-strength author."

RONALD R. THOMAS is a professor of English at Trinity College in Hartford, Connecticut. He is the author of numerous articles on nineteenth-century literature and culture, the novel, photography, and film, and he has published two books: *Detective Fiction and the Rise of Forensic Science* and *Dreams of Authority: Freud and the Fiction of the Unconscious.* Thomas has also taught at the University of Chicago and Harvard University and served as department chair, vice president, and acting president at Trinity College.

ANA VADILLO teaches at Birkbeck College, University of London. She has published a number of articles on fin-de-siècle women poets and is currently writing a monograph entitled *Passengers of Modernity: Women Poets and the Aesthetics of Urban Mass-Transport.*